MESENCHYMAL STEM CELLS - ISOLATION, CHARACTERIZATION AND APPLICATIONS

Edited by **Phuc Van Pham**

Mesenchymal Stem Cells - Isolation, Characterization and Applications
http://dx.doi.org/10.5772/65212
Edited by Phuc Van Pham

Contributors

Thanaphum Osathanon, Waleerat Sukarawan, Thilakavathy Karuppiah, Norshariza Nordin, Gurbind Singh, Rohayu Izanwati Mohd Rawi, Peyman Ghoraishizadeh, Rajesh Ramasamy, Sherif Amr, Harish C Chandramoorthy, Vishnu Balaji Radhakrishnan, Narasimman Gurusamy, Shishu Huang, Radwa Mehanna, Aydin Gulses, Mustafa Ayna, Ana Colette Maurício, Rui Damásio Alvites, Artur Varejão, Rita Caseiro, Lizeth Fuentes-Mera, Alberto Camacho, Nidia K. Moncada-Saucedo, Victor Peña-Martinez, Sabrina Valente, Gianandrea Pasquinelli, Hechmi Toumi, Marija Mazor, Tomoki Aoyama, Padraig Strappe

Published by InTech
Janeza Trdine 9, 51000 Rijeka, Croatia

Publishing Process Manager Romina Skomersic
Technical Editor SPi Global
Cover InTech Design team

First published November, 2017
Printed in Croatia
Legal deposit, Croatia: National and University Library in Zagreb

Additional hard copies can be obtained from orders@intechopen.com

Mesenchymal Stem Cells - Isolation, Characterization and Applications , Edited by Phuc Van Pham
p. cm.
Print ISBN 978-953-51-3615-6
Online ISBN 978-953-51-3616-3

Meet the editor

Phuc Van Pham received his PhD degree in Human Physiology from Vietnam National University, Ho Chi Minh City, Vietnam. He is currently a professor of Biology at the University of Science (Vietnam National University, Ho Chi Minh City, Vietnam), the director of the Stem Cell Institute, and the vice-director of Key Laboratory of Cancer Research. He is a long-standing lecturer and translational scientist at the University and is a member of several societies and journal editorial boards focused on stem cells. Dr. Pham and his colleagues have established one of the first multidisciplinary stem cell centers in Vietnam, and he has successfully launched an array of technologies in stem cell isolations. His research interests include stem cell isolation, stem cell therapy, mesenchymal stem cells, cancer stem cells, immunotherapy, and regenerative medicine, and he has published extensively in these areas. After many years of experience as an embryologist, cell biologist, and molecular biologist, collaborating with leading researchers in Singapore, Japan, and the United States, Dr. Pham is a student again, keen to reach beyond the traditional boundaries of biology.

Contents

Preface IX

Section 1 Isolation and Characterization 1

Chapter 1 **Physical versus Immunological Purification of Mesenchymal Stem Cells 3**
Radwa Ali Mehanna

Chapter 2 **Phenotypic and Functional Mapping of Mesenchymal Stem Cells Harvested from Different Portions of the Human Arterial Tree 19**
Valente Sabrina and Pasquinelli Gianandrea

Chapter 3 **Characteristics of Full-Term Amniotic Fluid-Derived Mesenchymal Stem Cells in Different Culture Media 39**
Karuppiah Thilakavathy, Norshariza Nordin, Rajesh Ramasamy, Peyman Ghoraishizadeh, Izanwati Mohd Rawi Rohayu and Gurbind Singh

Chapter 4 **Stem Cells from Human Exfoliated Deciduous Teeth: Biology and Therapeutic Potential 55**
Waleerat Sukarawan and Thanaphum Osathanon

Chapter 5 **Olfactory Mucosa Mesenchymal Stem Cells and Biomaterials: A New Combination to Regenerative Therapies after Peripheral Nerve Injury 77**
Rui Damásio Alvites, Ana Rita Caseiro Santos, Artur Severo Proença Varejão and Ana Colette Pereira de Castro Osório Maurício

Section 2 Perspectives and Clinical Applications 103

Chapter 6 **Mesenchymal Stem Cells: A Future Option for Intervening Disease Management 105**
Harish C. Chandramoorthy, Vishnu Balaji Radhakrishnan and Narasimman Gurusamy

Chapter 7 **The Proangiogenic Potential of Mesenchymal Stem Cells and Their Therapeutic Applications 125**
Nadeeka Bandara, Shiang Lim, Haiying Chen, Shuangfeng Chen, Le-Xin Wang and Padraig Strappe

Chapter 8 **Current Applications of Mesenchymal Stem Cells for Cartilage Tissue Engineering 149**
Lizeth Fuentes-Mera, Alberto Camacho, Nidia K. Moncada-Saucedo and Víctor Peña-Martínez

Chapter 9 **Mesenchymal Stem Cells for Optimizing Bone Volume at the Dental Implant Recipient Site 185**
Mustafa Ayna, Aydin Gülses, Jörg Wiltfang and Yahya Açil

Chapter 10 **Stem Cell-Based Therapies for Osteoarthritis: From Pre-Clinical to Clinical Applications 205**
Hechmi Toumi, Eric Lespessailles and Marija Mazor

Chapter 11 **Prerequisites for Mesenchymal Stem Cell Transplantation in Spinal Cord Injury 227**
Sherif M. Amr

Chapter 12 **Mesenchymal Stem Cell in the Intervertebral Disc 267**
Dan Long, Shaoyu Liang, Haiyan Liu, Xiuping Wu, Zhongyang Li, Haozhong Wang, Shishu Huang and Jiancheng Zeng

Chapter 13 **Transportation of Mesenchymal Stem Cells for Clinical Applications 279**
Tomoki Aoyama

Preface

The first study about the existence of the mesenchymal stem cells was carried out by a Russian-born morphologist—Alexander A. Maximow. He found a unique type of precursor cell within mesenchyme that could develop into some types of blood cells. In the next discovery, Ernest A. McCulloch and James E. Till firstly revealed the clonal nature of marrow cells in the 1960s. However, the first ex vivo assay for examining the clonogenic potential of multipotent stem cells from the bone marrow was later reported in the 1970s by Friedenstein and colleagues. In 2006, Dominici and colleagues suggested the minimal criteria for defining mesenchymal stem cells. Besides hematopoietic stem cells, mesenchymal stem cells were clinically used from 1995. Since then, over 500 clinical trials using mesenchymal stem cells have been performed. Mesenchymal stem cells have become the most common type of stem cells in both research and application nowadays.

Mesenchymal Stem Cells: Isolation, Characterization, and Applications thoroughly presents the isolation, characterization, and some applications of mesenchymal stem cells in the clinic. The book has two parts: "Isolation and Characterization" and "Clinical Perspectives and Applications." In Part I, the subsequent chapters introduce some techniques in isolation, characterization, and purification of mesenchymal stem cells in different tissues. In Part II, some applications of mesenchymal stem cells in the popular diseases, which include cartilage regeneration, spinal cord injury, and osteoarthritis, are discussed.

Many people have contributed to making our involvement in this project possible. We are extremely thankful to all the contributors of this book. Many people have had a hand in the preparation of this book. We thank our readers, who have made our hours putting together this volume worth it. We are indebted to the staff of InTech Publisher that published this book.

Prof. Phuc Van Pham, PhD
Stem Cell Institute
University of Science, VNUHCM
Ho Chi Minh City, Vietnam

Section 1

Isolation and Characterization

Chapter 1

Physical versus Immunological Purification of Mesenchymal Stem Cells

Radwa Ali Mehanna

Additional information is available at the end of the chapter

http://dx.doi.org/10.5772/intechopen.69295

Abstract

A prerequisite before dealing with any cell type is to identify it and isolate it from the heterogeneous cell population that it belongs to. Mesenchymal stem cells (MSC) can be found in nearly all tissues and are mostly located in perivascular niches.

MSC isolated from Bone marrow, adipose tissue, peripheral blood and different organs had shown promising potential for proliferation and differentiation into different cell types. They exhibit plastic-adherence under standard culture conditions, and this physical method of isolation is widely used as it is the most economic method and yet reveals relatively purified populations of cells after 3 or 4 passages. The complete purification still needs a specific call to different MSCs subsets. This could be achieved by immunological sorting, which depends on identifying cell marker(s) of such cells. Selecting these cells using antibodies against their specific markers then sorting the cells either by Magnetic or florescence based techniques named Magnetic Activated Cell Sorting (MACS) or Florescence Activated Cell Sorting (FACS) respectively is the principle of such purification techniques.

The aim of this chapter is to thoroughly define MSCs and compare between the different available methods for their purification

Keywords: MSC surface markers, MSC isolation, purification techniques

1. Definition of MSC

In 1970, Friedenstein [1] discovered in the bone marrow a rare stromal cell population forming around 0.0001 to 0.01% of nucleated cells. These cells are having the ability to proliferate in culture, and now commonly called mesenchymal stem or stromal cells (MSCs).

No unique cell surface marker clearly distinguishes MSCs, which makes a uniform definition difficult. The International Society for Cell Therapy proposed criteriathat comprise (1) adherence to plastic in standard culture conditions; (2) expression of the surface molecules CD73 (ecto-5′-nucleotidase marker), CD90 (Thy1 marker), and CD105 (endoglin marker) in the absence of CD34 (hematopoietic stem cell marker), CD45 (leukocyte marker), HLA-DR (human leukocyte antigen class II), CD14 or CD11b (monocyte and macrophage markers), CD79 or CD19 (B cell marker), and (3) a capacity for differentiation to osteoblasts, adipocytes, and chondroblasts *in vitro* [2].

These criteria were established to standardize human MSC isolation but may not apply uniformly to other species. The expression of these markers may decline over sub-passaging yet with the preservation of its proliferative, self-renewal and multilinage differentiation capability. Although the latter criteria are more consistent in defining MSCs, the above mentioned definition is discussed thoroughly as follows:

1.1. Adherence to plastic in standard culture conditions

MSCs grow as adherent monolayers, and unless they have transformed and become anchorage independent, after tissue disaggregation or subculture they will need to attach and spread out on the substrate before they will start to proliferate, thus giving rise to the criterion of adherence to plastic in standard culture conditions.

Cell adhesion is a complex event that refers to binding of cells to a surface. This surface may be another cell, the surrounding extracellular matrix (ECM) or a substrate. Mammalian cells coexist *in vivo* in intimate contact with each other and the surrounding ECM. Adhesion between these surfaces is directed at the molecular level by two different types of interactions. One is the "cell-cell adhesion" which is regulated by membrane expression of specialized integral membrane proteins called "cell adhesion molecules" (CAMs) that are generally clustered together at specialized points of cell contact with the cytoplasm of neighboring cells and thus can regulate signal transduction. A large number of CAMs exist and fall into four major families: the cadherins, immunoglobulin (Ig) superfamily, integrins, and selectins [3–5].

While the other is the "cell- matrix adhesion" through which the cells adhere indirectly by binding of a membrane adhesion receptor to specified components of ECM. The ECM is an organized network of proteins and polysaccharides secreted by cells that play a key regulatory role in determining the development, organization, and biological behavior of cells. In mammalian systems, three types of molecules are abundant in the ECM of all tissues: collagens, multi-adhesive matrix proteins, and proteoglycans. While collagen fibers and proteoglycans provide mechanical support, they are primarily the adhesive matrix proteins that bind to cell-surface adhesion receptors and other ECM components.

By way of these two types of interactions, cells can communicate bidirectionally with each other and respond to changes in the extracellular environment [6].

The process of adhesion regulates cell shape and biomechanics and is required for a variety of other cellular processes including proliferation, differentiation, migration, and invasion [7].

Originally, it was found that MSCs would attach to, and spread on, glass that had a slight net negative charge. They would also attach to some plastics, such as polystyrene, if the plastic was appropriately treated with strong acid, a plasma discharge, or high-energy ionizing radiation.

As cell adhesion is mediated by specific cell surface receptors for molecules in the extracellular matrix, so it seems likely that spreading may be preceded by the cells' secretion of extracellular matrix proteins and proteoglycans. The matrix adheres to the charged substrate (glass or treated plastic), and the cells then bind to the matrix via specific receptors. Thus, glass or conditioned plastic in which previous cells were grown upon can often provide a better surface for attachment, and substrates pretreated with matrix constituents, such as fibronectin or collagen, or derivatives such as gelatin, will help more demanding cells' attachment and proliferation.

1.2. Expression of the surface molecules (cell markers)

As previously mentioned, the definition of MSCs included the expression of certain cell markers together with the other criteria of their adherence and differentiation capacity.The selection of such criteria was to obtain easier comparisons between different studies and to adapt standards for the characterization of MSC. Nevertheless, these markers represent differentiation potential of MSC. Furthermore, these criteria apply to human MSCs, but do not necessarily extend to other species [8], also following culture, these markers may be lost or new markers may arise. So, some results fail to meet these criteria, making the comparison difficult. Thus, it was more convincing to agree on referring to human MSCs as stem cells when they prove self-renewal capability and showing their capacity for multilinage differentiation [9].

The expression of surface molecules (**Table 1**) and thus the phenotyping of human MSCs have been illustrated by many researchers based on the characterization of cultured cells.

MSCs have immunomodulatory properties as they express moderate levels of human leukocyte antigen (HLA), major histocompatibility complex class I, lack major histocompatibility complex class II expression, and do not express costimulatory molecules B7 and CD40 ligand [11–13]. The allogeneic transplantation of MSCs is well tolerated due to this unique immunophenotype together with the powerful immunosuppressive activity via cell-cell contact with target immune cells and secretion of soluble factors, such as nitric oxide, indoleamine 2,3-dioxygnease, and hemeoxygenase-1 [14–17]. MSCs produce an immunomodulatory effect by interacting with both innate and adaptive immune cells.

The innate immune cells (neutrophils, dendritic cells, natural killer cells, eosinophils, mast cells, and macrophages) are responsible for a nonspecific defense to infection, and MSCs have been shown to suppress most of these inflammatory cells. The adaptive immune system, composed of T and B lymphocytes, is capable of generating specific immune responses to pathogens with the production of memory cells. MSCs have been shown to suppress T cell proliferation in a mixed lymphocyte culture [18, 19].

Common name	CD locus	Detection
Adhesion molecules*		
ALCAM	CD166	Positive
ICAM-1	CD54	Positive
ICAM-2	CD102	Positive
ICAM-3	CD50	Positive
E-selectin	CD62E	Negative
L-selectin	CD62L	Positive
P-selectin	CD62P	Negative
LFA-3	CD58	Positive
Cadherin 5	CD144	Negative
PECAM-1	CD31	Negative
NCAM	CD56	Positive
HCAM	CD44	Positive
VCAM	CD106	Positive
Hyaluronate receptor	CD44	Positive
Growth factors and cytokine receptors*		
IL-1R (α and β)	CD121a, b	Positive
IL-2R	CD25	Negative
IL-3R	CD123	Positive
IL-4R	CD124	Positive
IL-6R	CD126	Positive
IL-7R	CD127	Positive
Interferon γ R	CDw119	Positive
TNF-α-1R	CD120a	Positive
TNF-α-2R	CD120b	Positive
FGFR		Positive
PDGFR	CD140a	Positive
Transferrin receptor	CD71	Positive
Integrins*		
VLA-α1	CD49a	Positive

http://dx.doi.org/10.5772/intechopen.69295

Common name	CD locus	Detection
VLA-α2	CD49b	Positive
VLA-α3	CD49c	Positive
VLA-α4	CD49d	Negative
VLA-α5	CD49e	Positive
VLA-α6	CD49f	Positive
VLA-β chain	CD29	Positive
β4 integrin	CD104	Positive
LFA-1 α chain	CD11a	Negative
LFA-1 β chain	CD18	Negative
Vitronectin R α chain	CD51	Negative
Vitronectin R β chain	CD61	Positive
CR4 α chain	CD11c	Negative
Mac1	CD11b	Negative
Additional markers*		
T6	CD1a	Negative
CD3 complex	CD3	Negative
T4, T8	CD4, CD8	Negative
Tetraspan	CD9	Positive
LPS receptor	CD14	Negative
LewisX	CD15	Negative
—	CD34	Negative
Leukocyte common antigen	CD45	Negative
5′terminal nucleotidase	CD73	Positive
B7-1	CD80	Negative
HB-15	CD83	Negative
B7-2	CD86	Negative
Thy-1	CD90	Positive
Endoglin	CD105	Positive
MUC18	CD146	Positive
BST-1	CD157	Positive

*Data are from Pittenger et al. [9] and Azizi et al. [10], or are previously unreported communication.

Table 1. Phenotyping of MSCs.

2. Sources, isolation, and types of MSCs

Firstly, MSCs were found to be isolated from BM [9], adipose tissue [20], synovial tissue [21], lung tissue [22], umbilical cord blood [23], and peripheral blood [24] are heterogeneous, with variable growth potential, but all have similar surface markers and mesodermal differentiation potential [25]. Later, MSCs have also been isolated from nearly every tissue type of adult mice, suggesting the existence of such cells in almost postnatal organs [26].

The bone marrow (BM) is the major source of hematopoietic stem cells (HSCs), the precursors of red blood cells, platelets, monocytes, and granulocytes. MSCs found in the BM act as a support to the microenvironment termed the "hematopoietic niche" through which HSCs are housed. This microenvironment is necessary for development and differentiation of HSCs [27, 28]. Physiologically, MSCs do not migrate easily in the peripheral blood, and available protocols are not very successful in inducing the translocation of this cell pool from the BM to the periphery. Therefore, isolation and culture expansion of MSCs is usually necessary for therapeutic purposes.

3. Methods for isolation and purification of MSCs

3.1. Plastic adherence of MSCs

Plastic adherence of MSCs, as discussed before, is now the most adapted method of their isolation from more heterogonous cell population sample as bone marrow or mononuclear cell layer known as the buffy coat. The advantage of this isolation technique lies in its feasibility. The only limitation is the inability of selecting, thus culturing a named subpopulation of MSCs, and also it needs several passaging to purify more and more MSCs from non-MSCs in the cell culture. This procedure resulted in a heterogeneous population, which contains both single stem cell-like cells as well as progenitor cells with different linage commitment.

3.2. Magnetic-activated cell sorting

Knowing that cells could be selected by their markers, different mechanisms by which these cells can be sorted without affecting their viability, morphology, or function are developed. One of these mechanisms is the use of magnetic power for attracting these cells when labeled with antibodies conjugated to magnetic beads.

Magnetic beads are microscopic, synthetic beads provided with a core of magnetite or other magnetic material, and coated with a thin polymer-shell, are subjected to chemical modification, facilitating covalent protein attachment.

The magnetic particles used for labeling of the cells, are divided into micro and nanobeads. Microbeads range from 0.5 to 5 nm in diameter, while nanobeads range from 100 to 500 nm. Such beads are provided commercially, for example, as Dynal (microbeads 1–3 nm;

http://dx.doi.org/10.5772/intechopen.69295

Invitrogen, Carlsbad, USA), MACS (nanobeads 20–100 nm; Miltenyi Biotec, Bergisch Gladbach, Germany), IMAG nanobeads 100–500 nm; BD Biosciences, San Jose, USA), EasySep (nanobeads about 150 nm; Stem Cell Technologies, Vancouver, Canada), or MagCellect beads (nanobeads about 150 nm; R&D Systems (Techne), McKinley Place NE, USA) [29–32].

In a classic practice, magnetic beads are added to the media in which the cells are cultured. They are then incubated for variable duration according to the given protocol. The magnetic beads then attach to cells mostly via antibodies but sometimes through other substances. The selection of the right biomarker for a given cell population guarantees the labeling of only desired cells. When these labeled cells are placed with the entire mixed-cell population into a biomagnetic separation system, the targeted cells are attracted by magnetic force to the tube wall or paramagnetic column, separating them from other cells in the culture.

3.2.1. *Methods of cell labeling*

Labeling of cells can be either direct or indirect. Direct labeling is when cells are labeled with antibodies that are readily conjugated to the magnetic beads. It is the fastest way of magnetic labeling as only one incubation step is required. Direct magnetic labeling requires a minimal number of washing steps and therefore minimizes cell loss.

While, indirect labeling is done in two-step procedure. Firstly, cells are labeled with a primary antibody directed against a cell surface marker. Secondly, the cells are magnetically labeled with magnetic beads, which either bind to the primary antibody or to a molecule that is conjugated to the primary antibody.

The primary antibody can either be unconjugated, biotinylated, or fluorochrome-conjugated. These antibodies will be further labeled with the magnetic beads that will be antiimmunoglobulin, antibiotin, or antifluorochrome beads, respectively.

3.2.2. *Positive versus negative selection for cell separation*

The selection can be positive by labeling the cells targeted for analysis or culture and thus the unlabeled cells are discarded. Alternatively, negative selection labels unwanted cells that are left in biomagnetic separation system and the unlabeled cells are extracted without them; it's also called cell depletion method. Comparison between the two methods is shown in **Table 2**.

In the context of magnetic cell separation technologies, two main methods are provided: the tube-based method and column-based separation method (**Figure 1**).

3.2.3. *Methods of separation technology*

3.2.3.1. *Tubular cell separation method*

Tubular cell separation is fully implemented in a single vessel. Magnetic beads are added to a cell-sample, which is incubated. Targeted cells are pulled into the tube wall toward the magnet when its power is applied, effectively separating cells with attached beads.

	Positive selection	Negative selection, cell depletion
Pros	Only one antibody is required that binds to the targeted cell marker (easy, cheap, fast)	No bound antibodies to the cells of interest
	High purity of sorted cells	Purification of cell population with unknown specific marker
		Combination with subsequent positive selection is possible
Cons	Potential interference with biological function of antibody-bound marker	Relatively impure
	Antigen expression must be unique to the cells of interest	Many antibodies necessary

Table 2. Comparison between the positive and the negative selection for magnetic cell sorting.

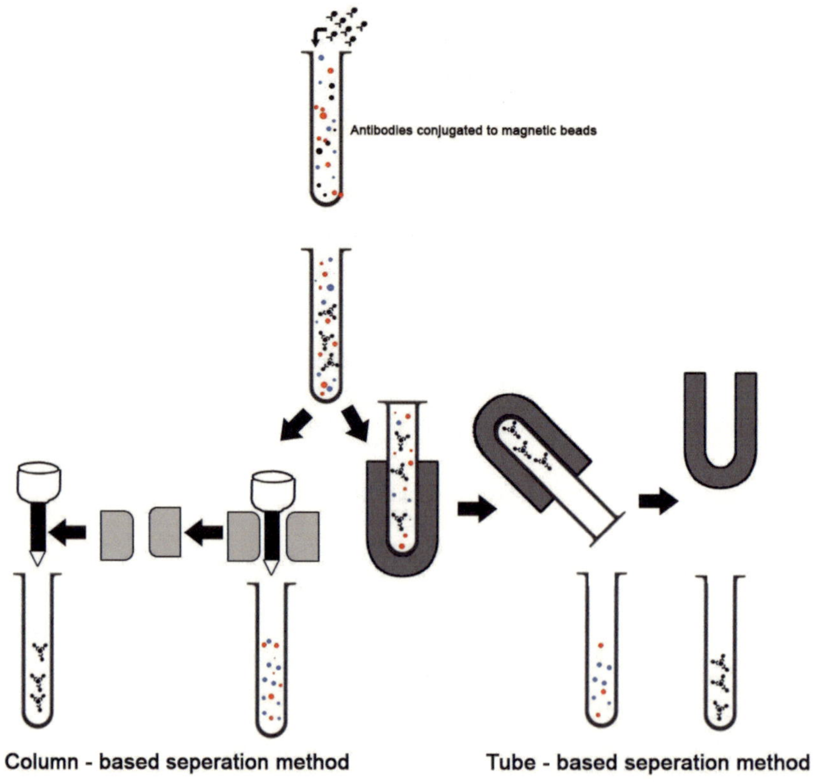

Figure 1. Tube-based magnetic separation method and column-based magnetic separation method.

3.2.3.2. Column-based separation method

Column is a vessel that contains an optimized matrix to generate a strong magnetic field when placed in a permanent magnet. Magnetic beads are added to a cell-sample, which is incubated.

Targeted cells are pulled into the surfaces of the magnetic spheres forming the matrix when the column is placed in the magnetic field.

The advantages and disadvantages of the two methods are shown in **Table 3**.

3.3. Fluorescence-activated cell sorting (FACS)

A significant improvement has been made since the initial commercialization of flow cytometry (FC) and fluorescence-activated cell sorting (FACS) in 1968. However, numerous points of weakness still exist, starting with the high cost and ending with the acceptance of the technology by many laboratories.

Flow cytometry is a widely used method for characterizing and defining different cell types in a heterogeneous cell population. It analyzes the expression of cell surface and intracellular molecules as well as the size and the shape of the cell. It also assesses the purity of isolated subpopulations.

In conventional laser flow cytometry, cells after passing through the flow cell will be treated as a waste. In fluorescence-activated cell sorting (FACS), the characteristics of the cells determined in the flow cell is the tool by which these cells will be further sorted into different paths in the equipment. Thus based on fluorescent labeling, FACS will separate a population of cells into subpopulations.

Sorting involves more complex mechanisms in the flow cytometer than a nonsorting analysis. Cells stained using fluorophore-conjugated antibodies can be separated from one another depending on which fluorophore they have been stained with.

Fluorescent dyes, or fluorochromes, are dyes that absorb light energy of a certain wavelength and reemit it at a longer wavelength. The main types of these dyes are; small dyes (e.g., fluoroscein isothiocyanate/FITC and alexa dyes), protein dyes (phycoerythrin [PE] allophycocyanin

	Tubular cell separation method	Column-based cell separation method
Advantages	Eliminates undue cell stress that can be generated by column-based separation methods or from exposure to iron spheres forming the column matrix	Minimal cell labeling with nanosized beads is sufficient to isolate cells effectively due to the high surface area and the generated strong magnetic field
	Diminishing the risk of experimental procedures negatively impacting cell function and phenotype	Gain the benefits of minimal labeling; no nonspecific labeling and no cell activation
Disadvantages	Low gradient of magnetic force that is only applied to the tube wall	Exposure of the cells to undue stress due to the exposure to iron particles
	Massive labeling required that may lead to nonspecific labeling and/or cell activation	The high cost and waste of disposable columns that must be periodically changed after a limited number of cell separation runs

Table 3. Comparison between the tubular and column-based cell separation methods.

[APC] GFP), tandem dyes, where a protein dye collects laser light, transfers it to a small dye, and the tandem emits at the wavelength of the smaller dye (e.g., perCP, APC-Cy7), quantum dots, and polymer dyes (brilliant violet). All have advantages and disadvantages, but the protein and small molecule dyes have been the mostly used in flow cytometry.

The choice of fluorochromes to use in an experiment is based on the lasers and filters available on your flow cytometer or FACS, the relative richness of the targets—brighter fluorochromes should be used on less abundant molecules—and if any of the targets are intracellular. Intracellular targets need brighter dyes than that used for the cell surface. PE is typically the brightest, followed by APC, so they should be conjugated to antibodies to intracellular or low abundance targets.

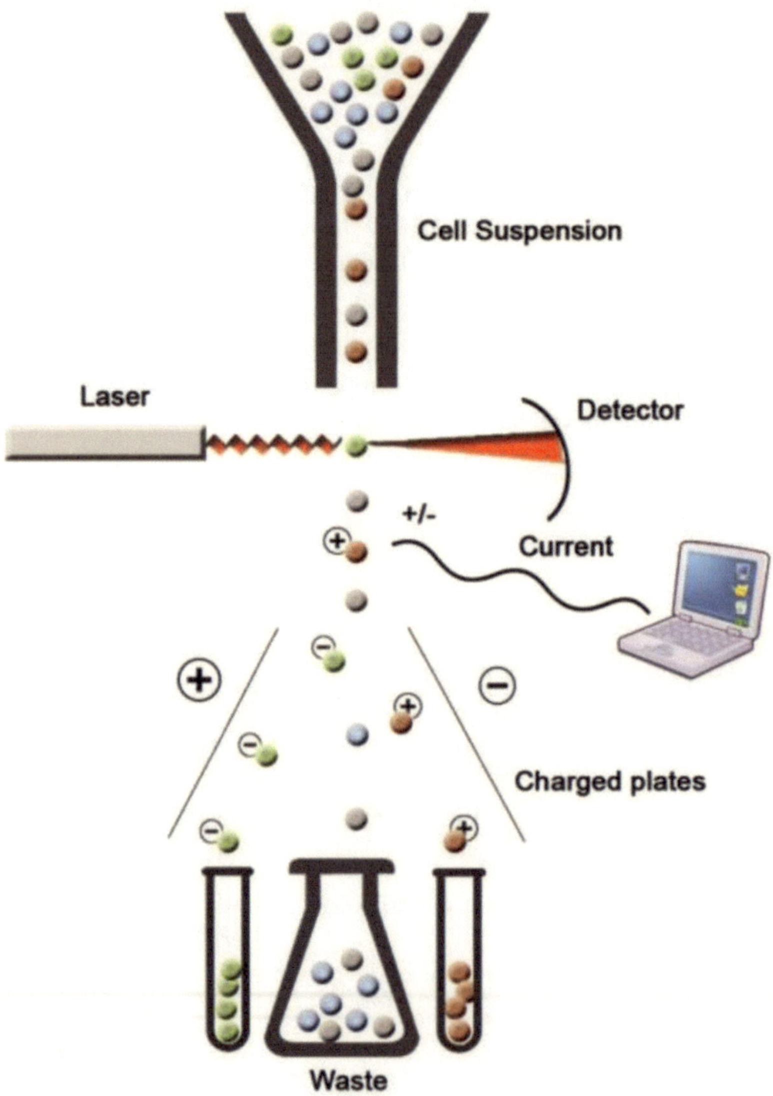

Figure 2. Principle of fluorescence-activated cell sorting (FACS).

Cells stained using fluorophore-conjugated antibodies are then taken into a column of pressurized sheath fluid, and as they emerge from the nozzle, they pass through one or more laser beams. At this point, the moment of analysis, the cytometer collects information about the fluorescence characteristics of the cell. After passing through the stream for the break-off distance, the stream is charged when the cell breaks off into a drop (moment of charging). Charged drops then pass through two high-voltage deflection plates and are deflected into collection vessels or aspirated to waste (**Figure 2**).

3.3.1. Points of weakness for FACS

Technical weakness could be like, the difficulty in detecting low abundance molecules in intracellular compartments, the great variability in cell permeabilizing chemistries, confounding effects from cell autofluorescence, overlap of emission spectra between used fluorochromes, and sometimes the unavailability of reagents for targeting molecules of interest.

Specifically for cell sorters, cell survival after pressure stress during droplet formation and collection, dilution of the sorted cells prior to reanalysis or culture, and the long duration it takes to obtain sufficient number of viable cells are considered to be some of the major problems. Lastly, data analysis is complicated, especially when dealing with low abundance targets.

4. Comparison between MACs and FACS cell sorting techniques

Although both methods are efficient, knowing their relative strengths and weaknesses can help make an informed choice on the technique used.

Each technique has "what it's best for" that gives it a privilege over the other. FACS is best in the following conditions: (1) when you want your sorted cell population to have a higher purity and recovery; (2) when sorting is based on an intracellular molecule (to which magnetic beads would not have access); FACS can sort cells labeled with fluorescent probes for intracellular targets; (3) when an information is needed about cell surface molecules, such as membrane protein receptors especially if these are of low density. It also can sort cells according to presence, absence, and density of the receptors.

On the other hand, sorting cells using magnetic beads is suitable for the separation of cells according to one separation criterion or characteristic, rather than several. It is also best as a method that classifies and sorts simultaneously and not sequentially as FACS separation. Magnetic beads separation is often used as a preparatory step prior to FACS.

MACs is a must use method when cells exhibit a high level of intrinsic cell fluorescence (auto fluorescence), which would disrupt the ability of a FACS instrument to detect signals.

Some aspects of comparison between the two techniques are shown in **Table 4**.

Fluorescence-activated cell sorting (FACS) can be combined with **magnetic-activated cell sorting** (MACS) if fluorescent magnetic microspheres are bound to the cells of interest. This added specificity can be useful in complicated sorts. Magnetic-activated cell sorting relies on

	MACs	FACS
Technology complexity	Low	High
Purity	Intermediate (90–98%)	High (98%)
Specificity	High	High
Negative selection	Possible (low purity)	Possible
Positive selection	Possible	Possible
Multimarker selection	Very limited	Possible
Risk for bacterial contamination	Low	Intermediate
Sorting for distinct expression levels	Not possible	Possible
Sorting of cells with intracellular fluorescence (e.g., eGFP)	Not possible	Possible
Simultaneous sorting of different populations	Very limited and not simultaneous	Possible

Table 4. Comparison between the MACs and FACS techniques for cell sorting.

the introduction of an external magnetic field to control the movement of magnetic-particle-bound cells in a cell lysate. Typically, the magnetic field traps the cells of interest on the sides or bottom of a tube while the unwanted solution and contaminants are washed away. While MACS alone is less expensive than FACS, it is unable to provide information about individual cells and cannot isolate one cell at a time. Therefore, **it is beneficial to use a combination of MACS and FACS**. Magnetic-activated cell sorting is used to obtain the purest sample possible before sending it through the FACS machine. This means that the sample must be incubated with magnetic particles and fluorophores. To save time, one can purchase fluorescent magnetic particles. These particles allow MACS and FACS to be performed sequentially with only one incubation period.

5. Sorting specific MSC subsets before culture

As per the definition of MSCs mentioned in the beginning of this chapter, all the criteria mentioned perfectly define MSC only in culture; however, how to identify these cells *in vivo* is still unrecognized. This, mainly due to their minimal existence among other cell populations *in vivo*, forms only 0.001–0.01% of cells in the BM as described by Pittenger et al. [9]. Also, undifferentiated cells with no specific phenotype make them rather more complicated. Many investigators directed their efforts to find markers for the identification of these cells, which help their purification through specific selection, rather than the adherence based purification method.

CD271 (LNGFR) has been described as one of the most specific markers for the purification of human BM-MSCs [33, 34]. CD271, also known as low-affinity nerve growth factor receptor (LNGFR), nerve growth factor receptor (NGFR), or p75NTR (neurotrophin receptor), belongs to the tumor necrosis factor superfamily [35], yet it would not be considered as a universal marker to identify MSC before culture, as it is not adequate in the isolation of MSC from some tissues such as umbilical cord or umbilical cord blood.

6. Conclusion

MSCs express several cell markers that differ according to the cells source, also, these markers could be lost or changed with further culturing of these cells. Thus, the immunologic isolation of all MSC subsets may be a difficult thing to do. Accordingly, the adherence to plastic in standard culture conditions is still the gold standard method for MSC isolation and purification. Yet their characterization before use is a must, either by using these cell markers or more expedient by proving their multilinage differentiation capability.

Abbreviations

MSCs	Mesenchymal stem cells
MACS	Magnetic-activated cell sorting
FACS	Fluorescence-activated cell sorting
CD	Cluster of differentiation
ECM	Extracellular matrix

Author details

Radwa Ali Mehanna

Address all correspondence to: radwa.mehanna@alexmed.edu.eg

Centre of Excellence for Research in Regenerative Medicine and its Applications CERRMA, Faculty of Medicine, Alexandria University, Alexandria, Egypt

References

[1] Friedenstein AJ, Chailakhjan RK, Lalykina KS. The development of fibroblast colonies in monolayer cultures of guinea-pig BM and spleen cells. Cell and Tissue Kinetics. 1970;**3**:393-403

[2] Dominici M, Le Blanc K, Mueller I, Slaper-Cortenbach I, Marini F, Krause D, Deans R, Keating A, Prockop D, Horwitz E. Minimal criteria for defining multipotent mesenchymal stromal cells: The International Society for Cellular Therapy position statement. Cytotherapy. 2006;**8**:315-317

[3] Clark EA, Brugge JS. Integrins and signal transduction pathways: The road taken. Science. 1995;**268**:233-239

[4] Pardi R. Signal transduction by adhesion receptors. Nature Education. 2010;**3**(9):38

[5] Aplin AE, Howe AK, Juliano RL. Cell adhesion molecules, signal transduction and cell growth. Current Opinion in Cell Biology. 1999;**11**(6):737-744

[6] Miranti CK, Brugge JS. Sensing the environment: A historical perspective on integrin signal transduction. Nature Cell Biology. 2002;**4**(4):E83-E90

[7] Kim B-S. Park I-K, Hoshiba T, Jiang H-L, Choi Y-J, Akaike T, Cho C-S. Design of artificial extracellular matrices for tissue engineering. Progress in Polymer Science. 2011; **36**:238-268

[8] Peister A, Mellad JA, Larson BL, Hall BM, Gibson LF, Prockop DJ. Adult stem cells from bone marrow (MSCs) isolated from different strains of inbred mice vary in surface epitopes, rates of proliferation, and differentiation potential. Blood. 2004;**103**:1662-1668. [PubMed: 14592819]

[9] Pittenger MF, Mackay AM, Beck SC, Jaiswal RK, Douglas R, Mosca JD, Moorman MA, Simonetti DW, Craig S, Marshak DR. Multilineage potential of adult human mesenchymal stem cells. Science. 1999;**284**:143-147. [PubMed: 10102814]

[10] Azizi SA, Stokes D, Augelli BJ, DiGirolamo C, Prockop DJ. Engraftment and migration of human bone marrow stromal cells implanted in the brains of albino rats—similarities to astrocyte grafts. Proceedings of the National Academy of Sciences of the United States of America. 1995;**95**:3908

[11] Devine SM, Hoffman R. Role of mesenchymal stem cells in hematopoietic stem cell transplantation. Current Opinion in Hematology. 2000;**7**:358-363

[12] Majumdar MK, Keane-Moore M, Buyaner D, Hardy WB, Moorman MA, McIntosh KR, Mosca JD. Characterization and functionality of cell surface molecules on human mesenchymal stem cells. Journal of Biomedical Science. 2003;**10**:228-241

[13] Tse WT, Pendleton JD, Beyer WM, Egalka MC, Guinan EC. Suppression of allogeneic T-cell proliferation by human marrow stromal cells: Implications in transplantation. Transplantation. 2003;**75**:389-397

[14] Krampera M, Cosmi L, Angeli R, Pasini A, Liotta F, Andreini A, Santarlasci V, Mazzinghi B, Pizzolo G, Vinante F, Romagnani P, Maggi E, Romagnani S, Annunziato F. Role for interferon-gamma in the immunomodulatory activity of human bone marrow mesenchymal stem cells. Stem Cells. 2006;**24**:386-398

[15] Ren G, Zhang L, Zhao X, Xu G, Zhang Y, Roberts AI, Zhao RC, Shi Y. Mesenchymal stem cell-mediated immunosuppression occurs via concerted action of chemokines and nitric oxide. Cell Stem Cell. 2008;**2**:141-150

[16] Chabannes D, Hill M, Merieau E, Rossignol J, Brion R, Soulillou JP, Anegon I, Cuturi MC. A role for heme oxygenase-1 in the immunosuppressive effect of adult rat and human mesenchymal stem cells. Blood. 2007;**110**:3691-3694

[17] Sato K, Ozaki K, Oh I, Meguro A, Hatanaka K, Nagai T, Muroi K, Ozawa K. Nitric oxide plays a critical role in suppression of T-cell proliferation by mesenchymal stem cells. Blood. 2007;**109**:228-234

[18] Di Nicola M, Carlo-Stella C, Magni M, Milanesi M, Longoni PD, Matteucci P, Grisanti S, Gianni AM. Human bone marrow stromal cells suppress T-lymphocyte proliferation induced by cellular or nonspecific mitogenic stimuli. Blood. 2002;**99**:3838-3843

[19] Aggarwal S, Pittenger MF. Human mesenchymal stem cells modulate allogeneic immune cell responses. Blood. 2005;**105**:1815-1822

[20] Zuk PA, Zhu M, Mizuno H, Huang J, Futrell JW, Katz AJ, Benhaim P, Lorenz HP, Hedrick MH. Multilineage cells from human adipose tissue: Implications for cell-based therapies. Journal of Tissue Engineering. 2001;**7**:211-228

[21] De Bari C, Dell'Accio F, Tylzanowski P, Luyten FP. Multipotent mesenchymal stem cells from adult human synovial membrane. Arthritis & Rheumatology. 2001;**44**:1928-1942

[22] Sabatini F, Petecchia L, Tavian M, Jodon de Villeroche V, Rossi GA, Brouty-Boye D. Human bronchial fibroblasts exhibit a mesenchymal stem cell phenotype and multilineage differentiating potentialities. Laboratory Investigation. 2005;**85**:962-971

[23] Erices A, Conget P, Minguell JJ. Mesenchymal progenitor cells in human umbilical cord blood. British Journal of Haematology. 2000;**109**:235-242

[24] Zvaifler NJ, Marinova-Mutafchieva L, Adams G, Edwards CJ, Moss J, Burger JA, Maini RN. Mesenchymal precursor cells in the blood of normal individuals. Arthritis Research. 2000;**2**:477-488.

[25] Baksh D, Yao R, Tuan RS. Comparison of proliferative and multilineage differentiation potential of human mesenchymal stem cells derived from umbilical cord and bone marrow. Stem Cells. 2007;**25**:1384-1392

[26] da Silva ML, Chagastelles PC, Nardi NB. Mesenchymal stem cells reside in virtually all post-natal organs and tissues. Journal of Cell Science. 2006;**119**:2204-2213

[27] Prockop DJ. Marrow stromal cells as stem cells for nonhematopoietic tissues. Science. 1997;**276**:71-74

[28] Pittenger MF, Martin BJ. Mesenchymal stem cells and their potential as cardiac therapeutics. Circulation Research. 2004;**95**:9-20

[29] Neurauter AA, Bonyhadi M, Lien E, Nokleby L, Ruud E, Camacho S, Aarvak T. Cell isolation and expansion using Dynabeads. Advances in Biochemical Engineering / Biotechnology. 2007;**106**:41-73

[30] Abts H, Emmerich M, Miltenyi S, Radbruch A, Tesch H. CD20 positive human B lymphocytes separated with the magnetic cell sorter (MACS) can be induced to proliferation and antibody secretion in vitro. Journal of Immunological Methods. 1989;**125**:19-28

[31] Peters CE, Woodside SM, Eaves AC. Isolation of subsets of immune cells. Methods in Molecular Biology. 2005;**302**:95-116

[32] Zborowski M. Commercial magnetic cell separation instruments and reagents. In: Zborowski M, Chalmers JJ, editors. Magnetic Cell Separation. Vol. 32, Laboratory Techniques in Biochemistry and Molecular Biology. Amsterdam, Oxford: Elsevier B.V.; 2007. pp. 265-292

[33] Bühring HJ, Battula VL, Treml S, Schewe B, Kanz L, Vogel W. Novel markers for the prospective isolation of human MSC. Annals of the New York Academy of Sciences. 2007; **1106**:262-271 [PMID: 17395729]

[34] Quirici N, Soligo D, Bossolasco P, Servida F, Lumini C, Deliliers GL. Isolation of bone marrow mesenchymal stem cells by anti-nerve growth factor receptor antibodies. Experimental Hematology. 2002;**30**:783-791 [PMID: 12135677]

[35] Thomson TM, Rettig WJ, Chesa PG, Green SH, Mena AC, Old LJ. Expression of human nerve growth factor receptor on cells derived from all three germ layers. Experimental Cell Research. 1988;**174**:533-539 [PMID: 2828087]

Chapter 2

Phenotypic and Functional Mapping of Mesenchymal Stem Cells Harvested from Different Portions of the Human Arterial Tree

Valente Sabrina and Pasquinelli Gianandrea

Additional information is available at the end of the chapter

http://dx.doi.org/10.5772/intechopen.68427

Abstract

The human arterial wall contains progenitors and mesenchymal stem cells (MSCs) acting as a postnatal reservoir of stem cells during lifetime. They are nestled in distinct arterial zones close to blood support, that is, the intima and the media-adventitia vasa vasorum plexus, representing vascular stem cell niches. In previous studies, MSCs were successfully isolated from fresh and cadaveric human large- and middle-sized arteries; these cells have a mesenchymal phenotype, self-renewal ability, and tri-lineage plasticity with high endothelial and smooth muscle cell differentiation potential. Here we present an overview of human MSCs derived from the vascular wall (hVW-MSCs) of different anatomical sites focusing on their phenotypic expression, multilineage potency, and stemness properties based on the localization in the arterial tree. We describe the isolation protocols as well as immunophenotyping, functional, and ultrastructure methods used to investigate these cell properties. hVW-MSCs from distinct portions of the vascular tree exhibit distinct phenotypic expression, multilineage potency, and stemness properties. This observation may contribute to explain the regional differences seen in vascular disease; moreover the different attitudes that hVW-MSCs exhibit in vascular differentiation should be taken in consideration whenever cell therapy, regenerative medicine, and tissue engineering strategies are attempted to replace tissues and organs.

Keywords: human arteries, vascular wall, mesenchymal stem cells, endothelial progenitors, smooth muscle progenitors, stem cell niche

1. Introduction

1.1. Mesenchymal stem cells: phenotype, mesodermal differentiation, and immunomodulation properties

The scientific community has been investigating since decades the stemness properties of the mesenchymal stromal/stem cells (MSCs). Due to their multiple properties, MSCs are the favorable candidates for cell- and gene-based therapy, regenerative medicine, and tissue engineering applications. They are a rare and multipotent cell population characterized by self-renewal and multilineage differentiation such as bone, cartilage, and adipose tissue as well as myocytes and bone marrow stromal cells [1, 2]. In 1976, Friedenstein discovered MSCs in the bone marrow [3] as adherent cells able to form colonies starting from single cells and to differentiate in osteoblasts. Several studies demonstrated that it is possible to successfully isolate a similar populations in multiple adult tissues other than the bone marrow including the amniotic fluid [4], cartilage [5], peripheral blood [6, 7], adipose tissue [8, 9], dental pulp [10, 11], fetal membranes [12], umbilical cord [13, 14], and human large- and medium-sized blood vessels such as the aorta and femoral artery [15, 16], pulmonary artery [17], internal mammary artery [18], and saphenous vein [19, 20]. According to the minimal criteria proposed by the International Society for Cellular Therapy [21], human MSCs are in vitro defined by the following properties: spindle-shape fibroblast-like morphology, the capacity to grow in adhesion on plastic surfaces, and to expand under appropriate experimental conditions. Phenotypically, MSCs express an array of surface markers usually detected by flow cytometry and exhibit differentiation capacity toward the tri-potential mesodermal adipogenic, osteogenic, and chondrogenic lineages. Due to the absence of specific markers useful to discriminate MSCs from other cytotypes, many attempts have been made to develop a mesenchymal profile in order to improve the purification and identification of MSCs. MSCs express numerous mesenchymal antigens such as CD73, CD90, CD105, CD44, and CD106 and are negative for the most common hematopoietic lineage markers like CD34, CD45, CD14, CD19, and HLA-DR.

MSCs isolated from different tissues show minimal changes in phenotype and growth; moreover they have been reported to be heterogeneous for their multilineage differentiation potential [22]; tissue-specific MSCs are also more prone to differentiate into one specific type of lineage. A similar behavior was seen in clones derived from MSCs in relation to the state of early commitment [23].

In addition to multilineage mesodermal differentiation, several studies reported the high immunosuppressive property of MSCs both in vitro and in vivo [24]. Although initially described in BM-derived cells [25], the immunomodulatory functions were also described in different human sources [26–28]. The MSC therapeutic effect is exerted not only by their low immunogenicity, migratory capacity, and direct reparative differentiation into cells of the residing tissue but also by the secretion of several bioactive molecules capable to inhibit the inflammatory milieu [29, 30].

1.2. Arterial wall structure

Three concentric layers compose the arterial wall: the intima, the media, and the adventitia. A single and continuous layer of endothelial cells leaned on basal membrane, and a thin subendothelial matrix characterizes the tunica intima, the most internal layer, in contact with the flowing blood. The tunica media is sandwiched between the intima and adventitia, from which it is separated by the internal and external elastic lamina, respectively. It represents the major component of the vessel wall and contains smooth muscle cells embedded in a matrix rich in elastic fibers, collagens, and proteoglycans. The adventitia is placed externally to the external elastic lamina; it is a loose connective tissue containing fibroblasts, adipocytes, small vascular structures (vasa vasorum), and nerve fibers; the adventitia is critical for numerous functions, that is, dampening the systolic force, nurturing the outer portion of the media, modulating the contractile response, and regulating vascular wall homeostasis.

Based on the architecture, diameter, and function, the arteries are divided into elastic and muscular arteries. Elastic arteries are characterized by large diameters, richness in elastic tissue, and low contractile ability. Pulmonary trunks, aortic arch, and their principal branches, that is, pulmonary, common carotid, subclavian, and common iliac arteries, belong to this category. The medium-sized arteries, called muscular arteries, are characterized by a low blood flow; they have a thin intima, a well-developed internal elastic lamina, and a media that is composed by concentric layers of smooth muscle cells. The peripheral arteries and those of the internal organs such as femoral arteries, external carotid artery, bronchial arteries, and mesenteric arteries are medium-sized arteries.

1.3. Mesenchymal stem cells resident in the human artery wall

Recent findings indicate that the adventitia of large- and medium-sized adult human arteries contains resident MSCs with multilineage differentiation capacity acting as a postnatal reservoir of stem cells.

In the human pulmonary artery, human vascular adventitial fibroblasts (hVAFs) were isolated from adventitia showing a strong ability to differentiate in mesenchymal cells. Immunophenotypically, these multipotent cells express vimentin, type-1 collagen, CD29, CD44, and CD105 markers and are negative for the most common monocyte markers. Under appropriate differentiation medium, the hVAFs were committed to adipocytes and osteocytes as well as myogenic cells positive to calponin and alpha smooth muscle cells (αSMA) in response to transforming growth factor-beta 1 (TGF-β1) [17].

Our group has demonstrated for the first time the presence of MSCs in large- and medium-sized vessels, including the thoracic aorta, aortic arch, and femoral artery from healthy and heart-beating donors [15, 16]. The vascular wall resident mesenchymal stem cells (VW-MSCs) were isolated from the adventitia with mechanical and enzymatic digestion and selected using plastic adherence-based cultures. These cultured-isolated cells expressed stemness markers (Notch-1 and Oct-4) and mesenchymal antigens (CD44, CD90, CD105, CD73, CD29, CD166, and STRO-1). As the bone marrow-derived MSCs, these multipotent cells displayed mesengenic potential to differentiate into the cartilage, adipose tissue, and, albeit to a lesser extent, also bone; consistent with their vascular localization, VW-MSCs were able to originate endothelial and smooth muscle cells [16].

In an interesting morphogenetic study performed on adult fresh human internal thoracic artery fragments, the authors [31] identified a CD44+ multipotent stem cell population (VW-MPSCs) residing in the arterial adventitia; these cells exhibited typical MSC properties, including cell surface antigens (CD44+, CD90+, CD73+, CD34−, and CD45−) without expression of CD146 and platelet-derived growth factor receptor beta (PDGFR-β) pericyte markers and multilineage plasticity into adipocytes, chondrocytes, and osteocytes, when cultured under appropriate differentiation media. Moreover, VW-MPSCs were able to generate smooth muscle cells particularly after TGF-β1 stimulation and pericytes. In vivo experiments performed on SCID mice, coculture of VW-MPSCs, and human umbilical vein endothelial cells (HUVECs) in a (three-dimensional) 3D Matrigel model resulted in the formation of a spontaneous vascular network where pericytes or smooth muscle cells derived from implanted VW-MPSCs cells were incorporated into new capillary-like structures.

The search for inexhaustible sources without ethical restrictions allowed to identify and isolate a population of VW-MSC residents in the human epiaortic wall collected from cadaveric donors; these progenitors were able to support prolonged ischemic injury and to survive in the explanted vascular tissues after 4 days of donor death and 5 years of cryopreservation in liquid nitrogen without losing their stemness characteristics. These multipotent human cadaveric mesenchymal stem cells (hC-MSCs) showed rapid expansion, clonogenic capability, immunomodulatory function, and ability to originate vascular and mesodermal tissues [28]. The possibility of obtaining stem cells from cadavers also represents a demonstration of the ability of these cells to survive adverse conditions, including long-time cryopreservation.

As a further demonstration of this capability, VW-MSCs obtained from abdominal aneurysms and exposed to extremely adverse culture conditions, for example, media acidification, hypoxia, starving, drying, and hypothermia, remained alive while keeping their morphology and stemness features [32].

1.4. Other stem cells resident in the human vascular wall

Other studies have reported the existence of stem cell and stem cell-like populations residing in the vascular wall.

Pericytes or mural cells represent a distinct cell embracing endothelial cells which share the basal membrane [33, 34]; although considered a contractile cell, when seen with electron microscopy, they contain only small quantities of assembled contractile filaments, raising uncertainties as to their actual vascular role. Using functional studies, pericytes have been found crucial for the control of endothelial cell growth and differentiation, capillary tone, caliber, and permeability; they are essential for supporting the capillary stability establishing mutual contacts with endothelial cells and producing proteins of the basal lamina [35].

Pericytes have a heterogeneous morphology, phenotype, and embryological origin (mesodermal and neuroectodermal). These peri-endothelial cells were found in intimal and adventitial niches sharing a common phenotype and multilineage plasticity with MSCs [36, 37]. In situ, they are identified through the expression of CD146, neural glial antigen (NG2), and PDGFR-β; they also express MSC markers (CD44, CD73, CD90, CD105, CD29, and alkaline phosphatase); pericytes do not express hematopoietic and endothelial cell antigens (CD31, CD34, CD45, CD14, and von Willebrand factor (vWF)).

In vitro pericytes acquire several MSC-like properties including a spindle-shape morphology; high proliferation; clonogenicity; ability to differentiate in several mesodermal lineages including the bone, cartilage, adipose tissue, smooth muscle cells, and skeletal muscle [38, 39]; immunomodulation functions [40]; as well as paracrine activity, promotion of the angiogenesis, and tissue regeneration [41, 42]. This finding supports the recent evidences that pericytes may represent the MSC in situ counterpart.

Recently, the presence of a novel stromal cell type called telocyte was documented in several organs and tissues [43]. The main features used to distinguish them from other stromal and interstitial cells are the presence of thin and long telopodes [44] and the co-expression of CD34 and CD117/c-kit, vimentin, PDGFR-α, or PDGFR-β markers [45].

Based on electron microscopy techniques, telocytes appear to be located in the stem cell niche [46, 47] where they probably serve as nursery for stem and progenitor cells influencing their survival and destiny. Here, telocytes form an intricate 3D network by contacting the resident stem cells, vessels, nerve endings, and neighboring stromal and immune reactive cells; this suggests that they have a potential role in tissue repair and regeneration [43, 48] as well as in tissue homeostasis, development, and immunosurveillance [43].

1.5. Vasculogenic niches

Progenitor cells are nestled in a three-dimensional (3D) hypoxic microenvironment localized in a specific anatomic district keeping them in their native undifferentiated and quiescent state, regulating their self-renewal, differentiation, and destiny. In the better-characterized niches, that is, the bone marrow stem cell niche, progenitors are found close to blood-bedewed areas.

In adult human vascular wall, a "vasculogenic zone" localized in between the media and adventitia was identified. In this hypothetical "vascular niche," endothelial progenitor cells (EPCs) and MSCs have been described along with hematopoietic progenitor cells (HPCs) as well as precursors of smooth muscle cells, fibroblasts, and pericytes [18]. According to this view, the vasculogenic zone contains a complete hierarchy of resident stem cells.

Despite EPCs have been intensely studied for years, there are conflicting results on their true identity; they were initially discovered in the peripheral blood [49] as circulating angioblasts involved in new blood vessel generation in response to various stimuli; it is still unclear whether these cells can also reside permanently in the vessel wall where they are expected to contribute to vascular homeostasis.

Studies performed on human aortic endothelial cells [50], coronary [51], and internal thoracic arteries [15, 18] have demonstrated that postnatal EPCs are localized in between the endothelium and in the innermost portion of adventitia; these observations corroborate the existence of EPCs resident in the human vascular wall. Peripheral blood EPCs express CD45, CD31, CD34, CD133, KDR (vascular endothelial growth factor receptor-2 (VEGFR-2)), Tie-2, the ligand for lectin ulex europaeus agglutinin-1 (UEA-1), and the low density lipoprotein (LDL) receptor [52]; different methods of EPC isolation have been proposed including the colony formation as spindle adherent cell [53]. Moreover, EPCs are hierarchy organized showing different clonogenicity, variable proliferative potential, ability to differentiate into functional, differentiated, and mature endothelium and to form capillary-like structures under appropriate induction [50, 54].

Within the human media, the presence of postnatal smooth muscle cell progenitors have been hypothesized but not sufficiently demonstrated yet. Most of the knowledge about resident smooth muscle cell progenitors comes from animal models even if some studies hypothesize their presence and role in the human vascular wall. After enzymatic digestion of the human carotid arteries, a multipotent vascular stem cell (MVSC) with in vitro self-renewal, clonogenicity, and plasticity to differentiate into mesodermal and neural lineage was discovered in the tunica media. Additionally, these vascular progenitor cells showed a propensity to give rise to smooth muscle cells after stimulation with basic fibroblast growth factor (bFGF), PDGF-BB, and TGF-β1. Furthermore, in a vascular disease model such as endothelial denudation, the MVSC contributed to the formation of neointima producing new synthetic smooth muscle cells, and deposition of extracellular matrix [55].

The human adventitia of large vessels also contains a branched plexus of vasa vasorum or "vessels of vessels" in close proximity with the vasculogenic niche. These capillary vessels ensure the oxygenation and the nourishment of the deeper layers of vascular wall as well as the removal of waste products. Recently hot spot areas of intensely positive nestin and WT1 vasa vasorum were described by our group [56]; nestin, an intermediate filament of neural stem cells that is under WT1 control, marks endothelial cells that are functional to the vascular niche, possibly regulating mononuclear cell traffic as demonstrated in an ApoE knockout murine model of atherosclerosis [57].

2. Immunophenotype and plasticity of hVW-MSCs derived from human arterial segments

In this chapter, we present an overview of human VW-MSC derived from the vascular wall of different anatomical sites focusing on their phenotypic expression, multilineage potency, and stemness properties based on the localization in the arterial tree. For this purpose, several human variously sized arteries as subclavian, brachiocephalic, common carotid, aortic arch, thoracic, renal, and femoral collected from multiorgan or multitissue donors were employed to recover hVW-MSCs.

2.1. Isolation procedure

After decontamination for 72 hours in an antibiotic mixture, each arterial segment was washed in physiological solution, cut lengthways and into small pieces, and enzymatically digested with 0.3 mg/ml liberase type II (Roche, Milan, Italy) in serum-free Dulbecco's Modified Eagle's Medium (DMEM) (Lonza, Basel, Switzerland) overnight at 37°C using a rotor apparatus. The following day, the remaining digested tissue was filtered using cell strainers (100-70-40 μm) (Becton Dickinson, Franklin Lakes, NJ) pelleted, counted, seeded at $1 \times 10^5/cm^2$ on T75 flasks plates with DMEM plus 20% fetal bovine serum (FBS), and incubated at 37°C in a humidified atmosphere with 5% CO_2. After the removal of nonadherent cells using phosphate-buffered saline (PBS), the cells were cultured near confluence changing culture medium every 2 days. For expansion, the cells were detached with trypsin-EDTA (Sigma, Milan, Italy), replated in a new culture flasks in DMEM supplemented with 10% FBS, and expanded in vitro for immunophenotype and differentiation analysis.

2.2. Phenotype characterization

To determinate their mesenchymal and stemness identity, early passages (P2 or P3) of each hVW-MSCs cell population was analyzed using flow cytometry accompanied by immunofluorescence detection. For surface antigen, the cells were washed in PBS and stained using the following extensive conjugated monoclonal antibody (moAb) panel: anti-CD90-phycoerythrin-cyanine 5 (PC5), anti-CD105-phycoerythrin (PE), anti-CD73-PE, anti-CD44-fluorescein isothiocyanate (FITC), anti-CD146-PE, anti-CD34-FITC, anti-CD31-PE, anti-CD14-FITC, anti-CD45-allophycocyanin (APC) (all from Beckman Coulter), anti-vWF (DakoCytomation), anti-NG2 (R&D System), anti-PDGFR-β (R&D System), anti-STRO-1 (R&D System), anti-Notch-1 (Santa Cruz Biotechnology), and anti-Oct-4 (Santa Cruz Biotechnology). The anti-mouse IgG-APC (Beckman Coulter) and anti-rabbit IgG-FITC (DakoCytomation) were used as a secondary antibody for the detection of unconjugated primary moAbs. For nuclear or cytoplasmic antigens, the cells were permeabilized with IntraPep Kit (Beckman Coulter). Negative controls were stained with secondary antibodies only. Samples were analyzed using a Navios FC equipped with two lasers for data acquisition (Beckman Coulter) and Kaluza FC Analysis software (Beckman Coulter) for data analysis.

In addition, further antigens were analyzed using a single immunofluorescence staining. In parallel experiments to flow cytometry, 6×10^5 hVW-MSCs were seeded on glass overnight that allowed to adhere, fix, and permeabilize in 2% paraformaldehyde plus 1% Tryton X-100 in PBS for 4 minutes at room temperature (rt). After washing in PBS, the sample was blocked with 1% bovine serum albumin (BSA) in PBS for 30 minutes at rt in humid chamber to reduce nonspecific staining and incubated with antihuman αSMA (1:9000, Sigma); Nestin (1:400, Millipore); fibroblast surface protein (FSP) (1:100, Abcam); and ki-67 (1:100, Novocastra) antibodies. After prolonged washing, the cells were stained with Alexa Fluor-488 (1:250, Life Technology; Carlsbad, CA, USA) secondary antibody in the dark and counterstained with ProLong antifade reagent with DAPI (4,6-diamidino-2-phenylindole, Molecular Probes). All incubations were performed for 1 hour at 37°C in humid chamber; antibodies were diluted in 1% PBS/BSA. Negative controls were stained with secondary antibodies only. Specimens were observed and the pictures captured with Leica DMI6000 B inverted fluorescence microscope (Leica Microsystems; Wetzlar, Germany). No signal was detected in the negative controls.

2.3. In vitro multilineage differentiation

The mesengenic potential of hVW-MSCs was proved inducing the differentiation into osteogenic, adipogenic, chondrogenic as well as angiogenic lineage considering their vascular origin.

For adipogenic differentiation, 6 × 10^4 hVW-MSCs/well were plated in a 24-well culture plate using the Mesenchymal Stem Cell Adipogenesis Kit (Chemicon International, Temecula, CA, USA) in accordance to the manufacturer's instructions. Induction medium was replaced every 2–3 days alternating with maintenance medium (DMEM 10% FBS and 10 μg/mL insulin). After three complete cycles of induction/maintenance medium (about 3 weeks), the presence of cytoplasmic lipid droplets was assessed by Oil Red O staining and confirmed by transmission electron microscopy (TEM) analysis. Control cells were cultured in DMEM basal medium plus 10% FBS.

For osteogenic differentiation, 6 × 10^4 hVW-MSCs per well were seeded in a 24-well culture plate using the osteogenic induction medium Mesenchymal Stem Cell Osteogenesis Kit (Chemicon International) plus 10% FBS and cultured for 3 weeks changing the medium every 2–3 days according to manufacturer's recommendations. Control cells were cultured in basal medium (DMEM with 10% FBS). The identification of calcium salt extracellular deposition was evaluated using Alizarin Red staining and confirmed by TEM analysis.

For chondrogenic differentiation, a 3D model was employed; the hVW-MCS cells were pelleted at the concentration of 2.5 × 10^5 in 15 ml polypropylene conical tubes containing 500 μl of differentiation basal medium chondrogenic (Poietics, Lonza) supplemented with hMSC Chondrogenic Single Quotes (Poietics, Lonza) and 10 ng/ml transforming growth factor-beta 3 (SIGMA, Lonza). For control cells, the same medium without TGF-β3 was used. The medium was refreshed every 2 days for 3 weeks of culture. Each pellet was formalin-fixed, paraffin embedded, and stained with Alcian blue to identify the deposition of extracellular matrix rich in sulfated proteoglycans and confirmed by TEM analysis.

For angiogenic differentiation, 6 × 10^5 hVW-MSCs were cultured in T25 flasks for 7 days in DMEM plus 2% FBS with 50 ng/ml vascular endothelial growth factor (VEGF; Sigma) as well as in DMEM plus 10% FBS for 25 × 10^4 control cells. To demonstrate whether VEGF could prompt MSCs to differentiate into the endothelium, a tube formation assay (Matrigel assay) was used for evaluating the ability to form capillary-like structures. At the end of induction, 50 μl of Matrigel (BD Bioscience) solution was dispensed into a 96-well plate and incubated for 30 minutes at 37°C to allow the solidification of the Matrigel solution. Meanwhile, the cells were detached from plastic supports and counted in order to have a final cellular suspension containing 5 × 10^3 in 50 μl of DMEM. The cellular suspension was seeded onto Matrigel and incubated at 37°C 5% CO_2 taking images after 2, 4, 6, and 24 hours with a camera connected to CKX41 Olympus inverted microscope. HUVECs were used as a positive control.

For each differentiation assay, the control and induced hVW-MSCs were fixed with 2.5% buffered glutaraldehyde directly in culture plate for 20 minutes at rt, scarpered, collected in an microtubes, pelleted, fixed again for 24 hours at 4°C, and processed for TEM analysis.

2.4. Results and Discussion

Human VW-MSCs derived from the vascular wall of different anatomical sites such as subclavian, brachiocephalic, common carotid, aortic arch, thoracic, renal, and femoral arteries showed the distinctive features of mesenchymal stem cells such as the fibroblast-like spindle-shaped morphology, growth in adhesion on plastic culture flasks, and high proliferative capacity. Further stemness skills such as the capability to form spheroids when grown in suspension were found in all vascular segments, while the clonogenic activity was reserved to the brachiocephalic artery and thoracic aorta only (**Figure 1**).

Focusing on hVW-MSC phenotype, flow cytometry and immunofluorescence analysis revealed that the 90% of hVW-MSCs derived from each vascular segment express the typical mesenchymal markers (CD44, CD90, CD105, CD73, CD146, and STRO-1) even if the same antigens were reduced to 80% in renal and femoral arteries, and no expression of CD146 and STRO-1 was seen in these same segments.

After cell isolation, a contamination with mature endothelial cells (CD31+ and vWF+) was found when hVW-MSCs were derived from the aorta and its branches; they were losing during the culture passages and completely absent in isolates from distal and peripheral arteries. In each segment, vascular and hematopoietic antigens (CD31, CD14, and CD45) were expressed by a minority of the isolated cells; CD31 was seen in less than 11% of hVW-MSC and CD14 and CD45 in less than 3%; in contrast, the vWF endothelial marker expression gradually increased from 26% in aortic branches to 60% in the thoracic aorta, while it was

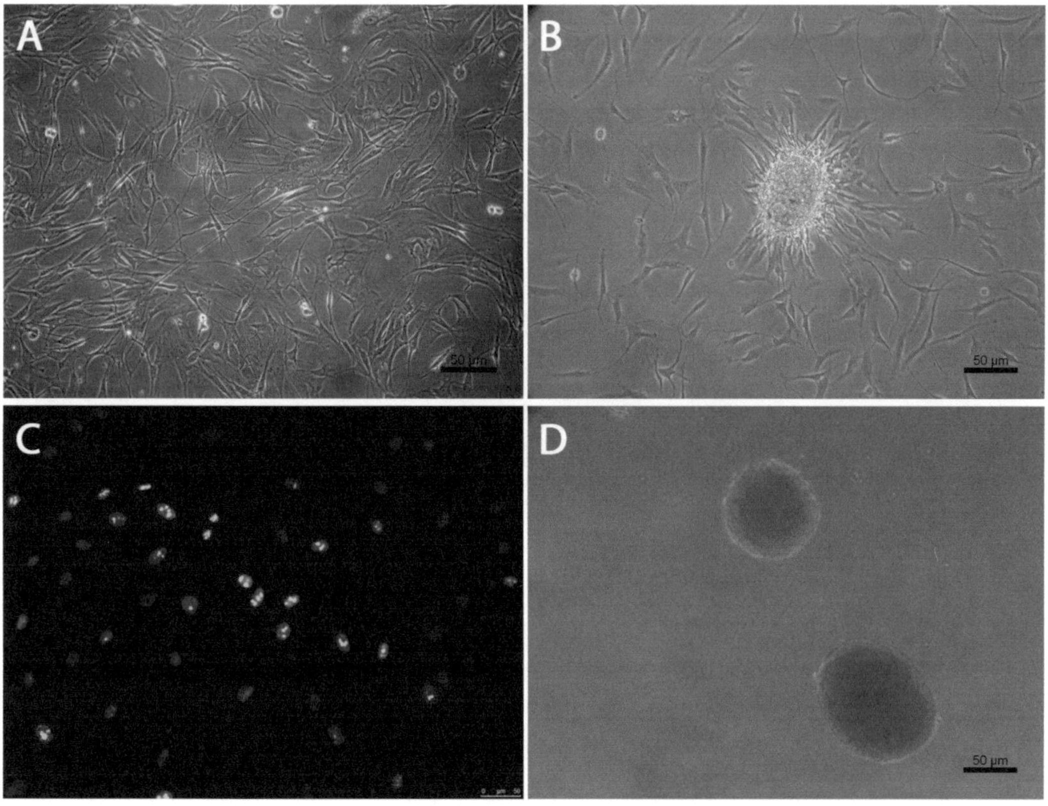

Figure 1. Representative images of hVW-MSC in adhesion to plastic support (A) and their stemness features including clonogenicity (B), high proliferation (C) and ability to form spheroids (D). Scale bars: 50 μm.

completely absent in distal (renal) and peripheral (femoral) arteries. Besides, a subset of CD34+ EPC was seen in the cell harvested from almost all segments; the CD34+ cells peaked in subclavian (26%) and renal (22%) arteries; this observation strengthens the belief that EPCs are resident in the human vascular wall.

Regarding the pericyte phenotype, about 94% of hVW-MSCs derived from the aortic portion and its branches expressed the PDGFR-β surface antigen; this percentage was drastically reduced to about 0.8 and 1.8% in renal and femoral arteries, respectively. As to NG2, a proportional increase, from 33 to 75.2%, was seen in hVW-MSC recovered from aortic branches to the thoracic aorta remaining very low in the other segments (4 and 10%). hVW-MSCs derived from the aorta (aortic arch and thoracic aorta) and its branches (subclavian, brachiocephalic, and common carotid) display a hybrid phenotype, that is, mesenchymal/pericytic, coherent with their presumed origin from pericytes of the adventitial vasa vasorum; on the contrary hVW-MSCs derived from distal and peripheral (renal and femoral) arteries present an almost pure mesenchymal phenotype without significant evidence of pericyte marker expression; this finding suggests that, in these districts, hVW-MSCs may have a different origin; either telocytes or perivascular fibroblasts could be robust candidates.

The analysis of ancestral antigens highlighted that Oct-4 and Notch-1 were constitutively expressed in a high percentage (54.6 and 38.5%, respectively) of hVW-MSC in all arteries and were significantly expressed (until to 88.9% for Oct-4 and 61% for Notch-1) in direction of the thoracic aorta; the same markers were not detected in hVW-MSC derived from renal and femoral arteries. Nestin and αSMA immunostainings were used to explore intermediate and contractile filaments. Few nestin-expressing hVW-MSCs were observed except for brachiocephalic, common carotid, and thoracic arteries where nestin-positive cells increased; a similar trend was seen also for αSMA that was found diffusely expressed in the brachiocephalic artery exclusively. The high density of cells expressing stemness markers, that is, nestin, Oct-4, and Notch-1, in thoracic segments as well as aortic arch may explain why intimal sarcomas, the most undifferentiated tumors of the vascular wall, primarily affect large vascular trunks [58].

To determine the percentage of cycling cells, a single immunofluorescence staining for ki-67 proliferation marker was performed. The semiquantitative analysis revealed that hVW-MSCs were highly proliferating when derived from the thoracic aorta (92.3%); the percentage of ki-67 proliferating cells decreased when hVW-MSCs were recovered from femoral (50.6%), subclavian (40%), renal (37.9%), common carotid (11.9%), and brachiocephalic (6.9%) arteries. All arteries, antigens, and percentage of expression analyzed are listed in **Table 1** and mapped in **Figure 2**.

To prove the multipotency into adipo-osteo-chondrocytes, hVW-MSCs derived from brachiocephalic, thoracic, renal, and femoral vascular segments were stimulated using appropriate experimental conditions; in addition, their angiogenic potential was also investigated considering their vascular origin (**Figure 3**). Results were analyzed using Oil Red O for adipogenic, Alizarin Red for osteogenic, and Alcian blue for chondrogenic differentiation; ultrastructural analysis was used for definitive confirmation. The mesengenic and angiogenic potentials are reported in **Table 2**.

http://dx.doi.org/10.5772/intechopen.68427

	CD90	CD105	CD73	CD44	CD146	STRO-1	CD34	CD31	CD14	CD45	vWF	PDGFR-B	NG2	FSP	Notch-1	Oct-4	Nestina	ASMA	KI-67
Subclavian artery	93.3	99.4	99.5	99.6	45.3	82.1	26.1	4.4	1.3	0.9	26.3	97.7	42.3	Few	38.5	54.6	Few	Few	40
Brachiocephalic artery	90.8	99.6	99.7	99.8	12.2	73.4	6.8	1.3	0	0.3	33.4	96.4	33	Diffuse	33.9	36.4	Diffuse	Diffuse	6.9
Common carotid artery	99	99.5	99.6	99.3	25.5	96.4	9.34	11.2	3.1	1.5	37.2	95	65.1	Few	49.1	72.5	Diffuse	Few	11.9
Aortic arch	95.6	97.9	97.1	98	29.7	82.9	11.1	3.7	1.3	1.3	47	97.8	63.7	Few	58.7	78.1	Diffuse	Few	18.8
Thoracic aorta	99.2	98.8	99	99.5	14.5	82.9	7.1	5.6	1.3	1.2	60.7	85.6	75.2	Few	61	88.9	Diffuse	Few	92.3
Renal artery	81	NA	NA	91	NA	0.8	22.8	0.6	NA	1	NA	0.8	9.8	NA	0.4	NA	NA	Few	37.9
Femoral artery	82.7	99.95	NA	100	0.7	0.7	0.7	NA	NA	0.6	Negative	1.8	4	Few	1	NA	Negative	Few	50.6

Table 1. Phenotypic characterization of MSCs derived from human vascular wall (hVW-MSCs).

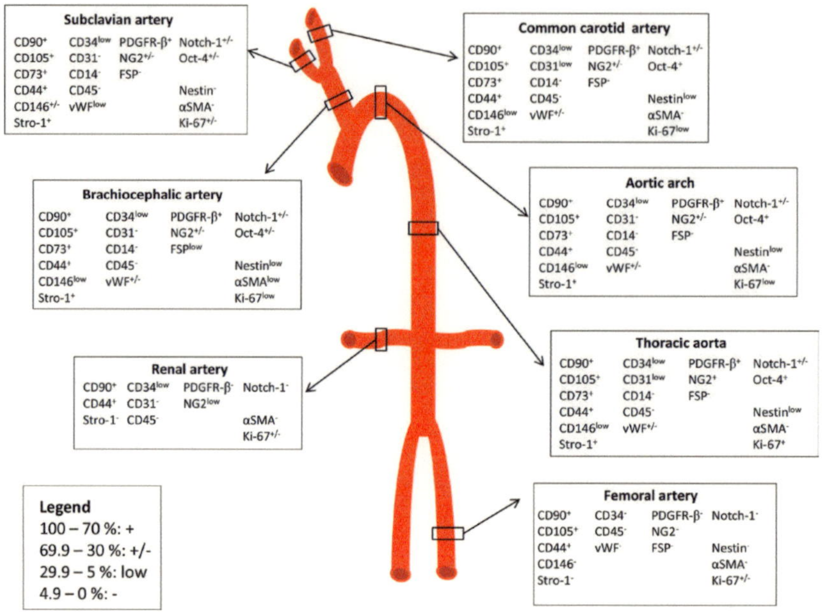

Figure 2. Phenotypic mapping of hVW-MSCs resident in human artery wall.

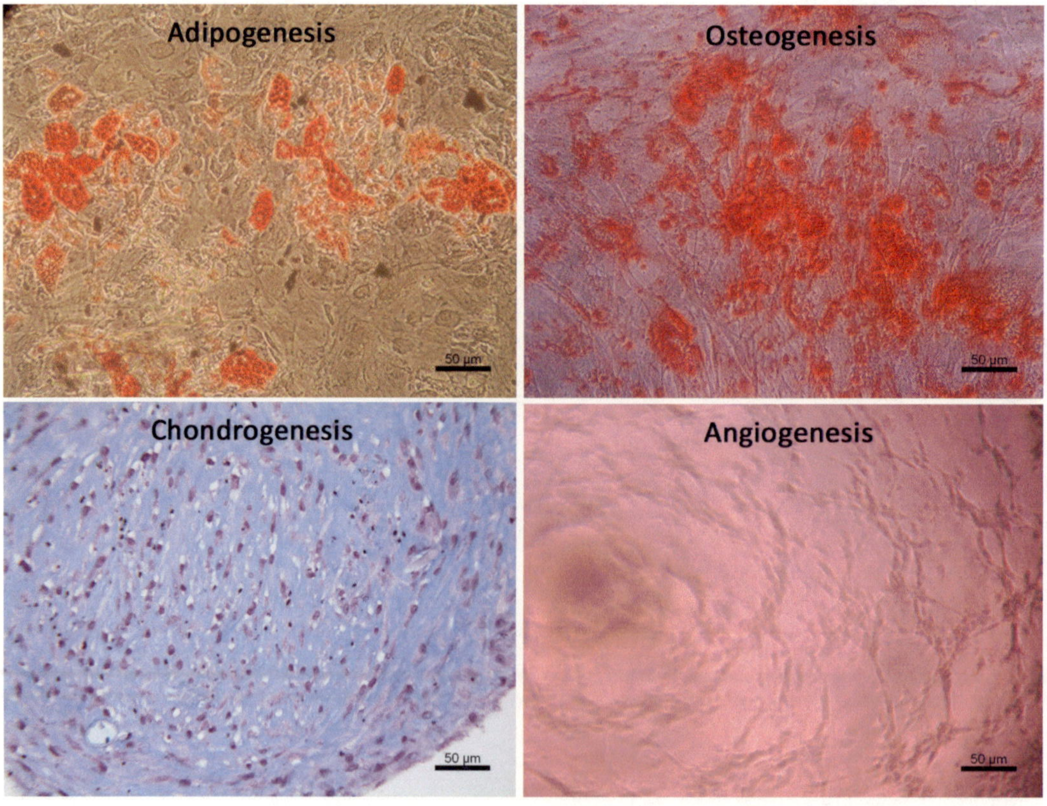

Figure 3. Representative images of hVW-MSCs potential to differentiate into mesodermal and angiogenic lineages. Scale bars: 50 μm.

	Adipogenesis	Osteogenesis	Chondrogenesis	Angiogenesis
Brachiocephalic artery	High and diffuse	High and diffuse	High and diffuse	High
Thoracic aorta	High and diffuse	Moderate and diffuse	High and diffuse	Moderate
Renal artery	NA	High and diffuse	Moderate and diffuse	High
Femoral artery	High and diffuse	Low and patchy	Moderate and diffuse	Low

Table 2. Mesengenic and angiogenic potential of hVW-MSCs derived from human vascular wall.

In particular, adipogenesis had the same high efficiency in each investigated artery. In hVW-MSC cytoplasm, Oil Red O staining revealed the presence of multiple lipid droplets that intensely stained red; the lipid droplets increased in number and size with the time of stimulation; adipocyte formation was confirmed by TEM analysis. Osteogenesis was very intense and diffuse in brachiocephalic and renal arteries, moderate but equally diffuse in the thoracic aorta and almost absent in the femoral artery. Changes in cell morphology as well as progressive deposition of calcium were seen during the induction period and confirmed at the end of treatment by Alizarin Red staining. TEM revealed osteoid matrix and hydroxyapatite crystals in the extracellular space. A successful chondrogenesis was documented using Alcian blue staining and TEM observation. In the brachiocephalic artery, thoracic aorta, and, with lesser intensity, femoral and renal arteries, hVW-MSCs were prone to produce an alcianophilic proteoglycan-rich extracellular matrix accompanied by the presence of clear, glycogen-rich, cytoplasm vacuoles. TEM displayed proteoglycan particles and bodies in the matrix and adherent to the cell plasma membrane. The distinctive features of each mesodermal commitment were not observed in uninduced hVW-MSCs used as controls. These results highlighted quantitative functional differences among hVW-MSCs collected from distinct vascular segments. The different attitudes to differentiate should be helpful for explaining pathological events occurring in specific arterial districts. For example, osteogenesis and chondrogenesis are efficient in hVW-MSCs derived from the common carotid artery and thoracic aorta; this high efficiency may have an impact on the type of calcification seen in atherosclerosis where the calcified plaque represents the result of an active process that involves hVW-MSC reversibly primed by the inflammatory context; on the contrary the inefficient angiogenesis and low osteo-chondrogenesis seen in hVW-MSCs derived from the femoral artery could explain the prevalent occurrence of non-atheromatous calcified arterial lesions seen in this vascular bed and ultimately explain the burden of trophic and ischemic lesions observed in patients with peripheral arterial obstructive disease.

Angiogenesis was assayed using a 3D semisolid model. In brachiocephalic and renal arteries, hVW-MSCs pretreated with VEGF rapidly aligned themselves emitting thin cytoplasmic projections; they formed an intricated and evident capillary-like network when seeded on Matrigel after 6 hours. A similar attitude was seen in hVW-MSCs from the thoracic aorta, while it was decreased in cells from the femoral artery. In untreated hVW-MSCs used as control, most of the cells remained single and dispersed in the culture medium without any hint of tube formation. These differences in angiogenic potential are essential when repair or regenerative cell therapies are to be established; in this case a source of hVW-MSCs capable of responding to the angiogenic stimulus effectively would be preferable.

3. Conclusion

The human vascular wall contains progenitors and stem cells that reside in distinct niches identified in the intima, media, and adventitia. Different anatomic portions of the vascular tree were used to perform a phenotypic and functional sketch of mesenchymal stem cells harvested from the human arterial wall. Although it is well known that the bone marrow remains the best hMSC source, MSCs can be isolated from almost all the arterial districts; subclavian, brachiocephalic, common carotid, aortic arch, thoracic aorta, renal, and femoral arteries are sources of stem cells residing in their wall as the lack of CD45 expression demonstrates consistently. Based on their topographical derivation, hVW-MSCs show a hybrid phenotype (mesenchymal/pericytic) in the aorta and its branches or pure mesenchymal phenotype in distal and peripheral arteries and contain a subset of CD34+ EPCs resident in the vascular wall of all investigated segments and a high cellular density expressing ancestral markers in thoracic segments as well as aortic arch districts. Furthermore, hVW-MSCs show a different predisposition to differentiate in a specific mesodermic lineage rather than another. This aspect should be considered for future clinical applications based on regenerative cell therapies and be helpful to improve the knowledge on pathological events occurring in specific arterial districts.

Abbreviations

MSCs	Mesenchymal stromal/stem cells
hVAFs	Human vascular adventitial fibroblasts
αSMA	Alpha smooth muscle cells
VW-MSCs	Vascular wall resident mesenchymal stem cells
VW-MPSCs	CD44+ multipotent stem cell population
PDGFR-β	Platelet-derived growth factor receptor beta
TGF-β1	Transforming growth factor-beta 1
hC-MSCs	Human cadaveric mesenchymal stem cells
NG2	Neural glial antigen
vWF	von Willebrand factor
3D	Three dimensional
EPCs	Endothelial progenitor cells
HPCs	Hematopoietic progenitor cells
KDR	Vascular endothelial growth factor receptor-2
UEA-1	Ulex europaeus agglutinin-1
LDL	Low-density lipoprotein
MVSC	Multipotent vascular stem cell
bFGF	Basic fibroblast growth factor

PBS	Phosphate-buffered saline
DMEM	Dulbecco's Modified Eagle's Medium
FBS	Fetal bovine serum
moAbs	Monoclonal antibodies
PC5	Phycoerythrin-cyanine 5
PE	Phycoerythrin
FITC	Fluorescein isothiocyanate
APC	Allophycocyanin
rt	Room temperature
BSA	Bovine serum albumin
FSP	Fibroblast surface protein
DAPI	4,6-Diamidino-2-phenylindole
VEGF	Vascular endothelial growth factor
TEM	Transmission electron microscopy
HUVECs	Human umbilical vein endothelial cells

Author details

Valente Sabrina* and Pasquinelli Gianandrea

*Address all correspondence to: sab.valente@gmail.com

Department of Experimental, Diagnostic and Specialty Medicine (DIMES), Alma Mater Studiorum, University of Bologna, Bologna, Italy

References

[1] Deans RJ, Moseley AB. Mesenchymal stem cells: Biology and potential clinical uses. Experimental Hematology. 2000;**28**(8):875-884

[2] Pittenger MF, Mackay AM, Beck SC, Jaiswal RK, Douglas R, Mosca JD, Moorman MA, Simonetti DW, Craig S, Marshak DR. Multilineage potential of adult human mesenchymal stem cells. Science. 1999;**284**(5411):143-147

[3] Friedenstein AJ, Gorskaja JF, Kulagina NN. Fibroblast precursors in normal and irradiated mouse hematopoietic organs. Experimental Hematology. 1976;**4**(5):267-274

[4] Anker PS, Noort WA, Scherjon SA, Kleijburg-van der Keur C, Kruisselbrink AB, van Bezooijen RL, Beekhuizen W, Willemze R, Kanhai HH, Fibbe WE. Mesenchymal stem cells in human second-trimester bone marrow, liver, lung, and spleen exhibit a similar immunophenotype but a heterogeneous multilineage differentiation potential. Haematologica. 2003;**88**(8):845-852

[5] Dowthwaite GP, Bishop JC, Redman SN, Khan IM, Rooney P, Evans DJ, Haughton L, Bayram Z, Boyer S, Thomson B, Wolfe MS, Archer CW. The surface of articular cartilage contains a progenitor cell population. Journal of Cell Science. 2004;**117**(Pt6):889-897. DOI: 10.1242/jcs.00912

[6] Zvaifler NJ, Marinova-Mutafchieva L, Adams G, Edwards CJ, Moss J, Burger JA, Maini RN. Mesenchymal precursor cells in the blood of normal individuals. Arthritis Research. 2000;**2**(6):477-488. DOI: 10.1186/ar130

[7] Kuznetsov SA, Mankani MH, Gronthos S, Satomura K, Bianco P, Robey PG. Circulating skeletal stem cells. The Journal of Cell Biology. 2001;**153**(5):1133-1140

[8] Zuk PA, Zhu M, Mizuno H, Huang J, Futrell JW, Katz AJ, Benhaim P. Multilineage cells from human adipose tissue: Implications for cell-based therapies. Tissue Engineering. 2001;**7**(2):211-228. DOI: 10.1089/107632701300062859

[9] Alhadlaq A, Mao JJ. Mesenchymal stem cells: Isolation and therapeutics. Stem Cells and Development. 2004;**13**(4):436-448. DOI: 10.1089/scd.2004.13.436

[10] Gronthos S, Mankani M, Brahim J, Robey PG, Shi S. Postnatal human dental pulp stem cells (DPSCs) *in vitro* and *in vivo*. Proceedings of the National Academy of Sciences of the United States of America. 2000;**97**(25):13625-13630. DOI: 10.1073/pnas.240309797

[11] Pierdomenico L, Bonsi L, Calvitti M, Rondelli D, Arpinati M, Chirumbolo G, Becchetti E, Marchionni C, Alviano F, Fossati V, Staffolani N, Franchina M, Grossi A, Bagnara GP. Multipotent mesenchymal stem cells with immunosuppressive activity can be easily isolated from dental pulp. Transplantation. 2005;**80**(6):836-842

[12] Alviano F, Fossati V, Marchionni C, Arpinati M, Bonsi L, Franchina M, Lanzoni G, Cantoni S, Cavallini C, Bianchi F, Tazzari PL, Pasquinelli G, Foroni L, Ventura C, Grossi A, Bagnara GP. Term amniotic membrane is a high throughput source for multipotent mesenchymal stem cells with the ability to differentiate into endothelial cells *in vitro*. BMC Developmental Biology. 2007;**7**(11):DOI: 10.1186/1471-213X-7-11

[13] Bieback K, Kern S, Klüter H, Eichler H. Critical parameters for the isolation of mesenchymal stem cells from umbilical cord blood. Stem Cells. 2004;**22**(4):625-634. DOI: 10.1634/stemcells.22-4-625

[14] Kögler G, Sensken S, Airey JA, Trapp T, Müschen M, Feldhahn N, Liedtke S, Sorg RV, Fischer J, Rosenbaum C, Greschat S, Knipper A, Bender J, Degistirici O, Gao J, Caplan AI, Colletti EJ, Almeida-Porada G, Müller HW, Zanjani E, Wernet P. A new human somatic stem cell from placental cord blood with intrinsic pluripotent differentiation potential. The Journal of Experimental Medicine. 2004;**200**(2):123-135. DOI: 10.1084/jem.20040440

[15] Pasquinelli G, Tazzari PL, Vaselli C, Foroni L, Buzzi M, Storci G, Alviano F, Ricci F, Bonafè M, Orrico C, Bagnara GP, Stella A, Conte R. Thoracic aortas from multiorgan donors are suitable for obtaining resident angiogenic mesenchymal stromal cells. Stem Cells. 2007;**25**(7):1627-1634. DOI: 10.1634/stemcells.2006-0731

[16] Pasquinelli G, Pacilli A, Alviano F, Foroni L, Ricci F, Valente S, Orrico C, Lanzoni G, Buzzi M, Luigi Tazzari P, Pagliaro P, Stella A, Paolo Bagnara G. Multidistrict human mesenchymal vascular cells: Pluripotency and stemness characteristics. Cytotherapy. 2010;**12**(3):275-287. DOI: 10.3109/14653241003596679

[17] Hoshino A, Chiba H, Nagai K, Ishii G, Ochiai A. Human vascular adventitial fibroblasts contain mesenchymal stem/progenitor cells. Biochemical and Biophysical Research Communications. 2008;**368**(2):305-310. DOI: 10.1016/j.bbrc.2008.01.090

[18] Zengin E, Chalajour F, Gehling UM, Ito WD, Treede H, Lauke H, Weil J, Reichenspurner H, Kilic N, Ergün S. Vascular wall resident progenitor cells: A source for postnatal vasculogenesis. Development. 2006;**133**(8):1543-1551. DOI: 10.1242/dev.02315

[19] Covas DT, Piccinato CE, Orellana MD, Siufi JL, Silva WA Jr, Proto-Siqueira R, Rizzatti EG, Neder L, Silva AR, Rocha V, Zago MA. Mesenchymal stem cells can be obtained from the human saphena vein. Experimental Cell Research. 2005;**309**(2):340-344. DOI: 10.1016/j.yexcr.2005.06.005

[20] Campagnolo P, Cesselli D, Al Haj Zen A, Beltrami AP, Kränkel N, Katare R, Angelini G, Emanueli C, Madeddu P. Human adult vena saphena contains perivascular progenitor cells endowed with clonogenic and proangiogenic potential. Circulation. 2010;**121**(15):1735-1745. DOI: 10.1161/CIRCULATIONAHA.109.899252

[21] Dominici M, Le Blanc K, Mueller I, Slaper-Cortenbach I, Marini F, Krause D, Deans R, Keating A, Prockop DJ, Horwitz E. Minimal criteria for defining multipotent mesenchymal stromal cells. The International Society for Cellular Therapy position statement. Cytotherapy. 2006;**8**(4):315-317. DOI: 10.1080/14653240600855905

[22] Baksh D, Yao R, Tuan RS. Comparison of proliferative and multilineage differentiation potential of human mesenchymal stem cells derived from umbilical cord and bone marrow. Stem Cells. 2007;**25**(6):1384-1392

[23] Baksh D, Song L, Tuan RS. Adult mesenchymal stem cells: Characterization, differentiation, and application in cell and gene therapy. Journal of Cellular and Molecular Medicine. 2004;**8**(3):301-316

[24] Jiang Y, Jahagirdar BN, Reinhardt RL, Schwartz RE, Keene CD, Ortiz-Gonzalez XR, Reyes M, Lenvik T, Lund T, Blackstad M, Du J, Aldrich S, Lisberg A, Low WC, Largaespada DA, Verfaillie CM. Pluripotency of mesenchymal stem cells derived from adult marrow. Nature. 2002;**418**(6893):41-49

[25] Bartholomew A, Sturgeon C, Siatskas M, Ferrer K, McIntosh K, Patil S, Hardy W, Devine S, Ucker D, Deans R, Moseley A, Hoffman R. Mesenchymal stem cells suppress lymphocyte proliferation *in vitro* and prolong skin graft survival *in vivo*. Experimental Hematology. 2002 ;**30**(1):42-48

[26] Yoo KH, Jang IK, Lee MW, Kim HE, Yang MS, Eom Y, Lee JE, Kim YJ, Yang SK, Jung HL, Sung KW, Kim CW, Koo HH. Comparison of immunomodulatory properties of

mesenchymal stem cells derived from adult human tissues. Cellular Immunology. 2009;**259**(2):150-156. DOI: 10.1016/j.cellimm.2009.06.010

[27] Najar M, Raicevic G, Boufker HI, Fayyad Kazan H, De Bruyn C, Meuleman N, Bron D, Toungouz M, Lagneaux L. Mesenchymal stromal cells use PGE2 to modulate activation and proliferation of lymphocyte subsets: Combined comparison of adipose tissue, Wharton's Jelly and bone marrow sources. Cellular Immunology. 2010;**264**(2):171-179. DOI: 10.1016/j.cellimm.2010.06.006

[28] Valente S, Alviano F, Ciavarella C, Buzzi M, Ricci F, Tazzari PL, Pagliaro P, Pasquinelli G. Human cadaver multipotent stromal/stem cells isolated from arteries stored in liquid nitrogen for 5 years. Stem Cell Research & Therapy. 2014;**5**(1):8. DOI: 10.1186/scrt397

[29] Mariani E, Facchini A. Clinical applications and biosafety of human adult mesenchymal stem cells. Current Pharmaceutical Design. 2012;**18**(13):1821-1845

[30] Wang S, Qu X, Zhao RC. Clinical applications of mesenchymal stem cells. Journal of Hematology & Oncology 2012;**5**:19. DOI: 10.1186/1756-8722-5-19

[31] Klein D, Weisshardt P, Kleff V, Jastrow H, Jakob HG, Ergün S. Vascular wall-resident CD44+ multipotent stem cells give rise to pericytes and smooth muscle cells and contribute to new vessel maturation. PLoS One. 2011;**6**(5):e20540. DOI: 10.1371/journal.pone.0020540

[32] Ciavarella C, Fittipaldi S, Pedrini S, Vasuri F, Gallitto E, Freyrie A, Stella A, Gostjeva E, Pasquinelli G. *In vitro* alteration of physiological parameters do not hamper the growth of human multipotent vascular wall-mesenchymal stem cells. Frontiers in Cell and Development Biology 2015;**3**:36. DOI: 10.3389/fcell.2015.00036

[33] Armulik A, Abramsson A, Betsholtz C. Endothelial/pericyte interactions. Circulation Research. 2005;**97**(6):512-523. DOI: 10.1161/01.RES.0000182903.16652.d7

[34] Hirschi KK, D'Amore PA. Pericytes in the microvasculature. Cardiovascular Research. 1996;**32**(4):687-698

[35] von Tell D, Armulik A, Betsholtz C. Pericytes and vascular stability. Experimental Cell Research. 2006;**312**(5):623-629. DOI: 10.1016/j.yexcr.2005.10.019

[36] Covas DT, Panepucci RA, Fontes AM, Silva WA Jr, Orellana MD, Freitas MC, Neder L, Santos AR, Peres LC, Jamur MC, Zago MA. Multipotent mesenchymal stromal cells obtained from diverse human tissues share functional properties and gene-expression profile with CD146+ perivascular cells and fibroblasts. Experimental Hematology. 2008;**36**(5):642-654. DOI: 10.1016/j.exphem.2007.12.015

[37] Crisan M, Yap S, Casteilla L, Chen CW, Corselli M, Park TS, Andriolo G, Sun B, Zheng B, Zhang L, Norotte C, Teng PN, Traas J, Schugar R, Deasy BM, Badylak S, Buhring HJ, Giacobino JP, Lazzari L, Huard J, Péault B. A perivascular origin for mesenchymal stem cells in multiple human organs. Cell Stem Cell. 2008;**3**(3):301-313. DOI: 10.1016/j.stem.2008.07.003

[38] Crisan M, Chen CW, Corselli M, Andriolo G, Lazzari L, Péault B. Perivascular multipotent progenitor cells in human organs. The Annals of the New York Academy of Sciences 2009;**1176**:118-123. DOI: 10.1111/j.1749-6632.2009.04967.x

[39] Park TS, Gavina M, Chen CW, Sun B, Teng PN, Huard J, Deasy BM, Zimmerlin L, Péault B. Placental perivascular cells for human muscle regeneration. Stem Cells and Development. 2011;**20**(3):451-463. DOI: 10.1089/scd.2010.0354

[40] Domev H, Milkov I, Itskovitz-Eldor J, Dar A. Immunoevasive pericytes from human pluripotent stem cells preferentially modulate induction of allogeneic regulatory T cells. Stem Cells Translational Medicine. 2014;**3**(10):1169-1181. DOI: 10.5966/sctm.2014-0097

[41] Katare R, Riu F, Mitchell K, Gubernator M, Campagnolo P, Cui Y, Fortunato O, Avolio E, Cesselli D, Beltrami AP, Angelini G, Emanueli C, Madeddu P. Transplantation of human pericyte progenitor cells improves the repair of infarcted heart through activation of an angiogenic program involving micro-RNA-132. Circulation Research. 2011;**109**(8):894-906. DOI: 10.1161/CIRCRESAHA.111.251546

[42] Chen CW, Okada M, Proto JD, Gao X, Sekiya N, Beckman SA, Corselli M, Crisan M, Saparov A, Tobita K, Péault B, Huard J. Human pericytes for is chemic heart repair. Stem Cells. 2013;**31**(2):305-316. DOI: 10.1002/stem.1285

[43] Bei Y, Wang F, Yang C, Xiao J. Telocytes in regenerative medicine. Journal of Cellular and Molecular Medicine. 2015;**19**(7):1441-1454. DOI: 10.1111/jcmm.12594

[44] Popescu LM, Faussone-Pellegrini MS. TELOCYTES—a case of serendipity: The winding way from interstitial cells of cajal (ICC), via interstitial cajal-like cells (ICLC) to TELOCYTES. Journal of Cellular and Molecular Medicine. 2010;**14**(4):729-740. DOI: 10.1111/j.1582-4934.2010.01059.x

[45] Cretoiu SM, Popescu LM. Telocytes revisited. Biomolecular Concepts. 2014;**5**(5):353-369. DOI: 10.1515/bmc-2014-0029

[46] Bojin FM, Gavriliuc OI, Cristea MI, Tanasie G, Tatu CS, Panaitescu C, Paunescu V. Telocytes within human skeletal muscle stem cell niche. Journal of Cellular and Molecular Medicine. 2011;**15**(10):2269-2272. DOI: 10.1111/j.1582-4934.2011.01386.x

[47] Gherghiceanu M, Popescu LM. Cardiomyocyte precursors and telocytes in epicardial stem cell niche: Electron microscope images. Journal of Cellular and Molecular Medicine. 2010;**14**(4):871-877. DOI: 10.1111/j.1582-4934.2010.01060.x

[48] Popescu LM, Manole E, Serboiu CS, Manole CG, Suciu LC, Gherghiceanu M, Popescu BO. Identification of telocytes in skeletal muscle interstitium: Implication for muscle regeneration. Journal of Cellular and Molecular Medicine. 2011;**15**(6):1379-1392. DOI: 10.1111/j.1582-4934.2011.01330.x

[49] Asahara T, Murohara T, Sullivan A, Silver M, van der Zee R, Li T, Witzenbichler B, Schatteman G, Isner JM. Isolation of putative progenitor endothelial cells for angiogenesis. Science. 1997;**275**(5302):964-967

[50] Ingram DA, Mead LE, Moore DB, Woodard W, Fenoglio A, Yoder MC. Vessel wall-derived endothelial cells rapidly proliferate because they contain a complete hierarchy of endothelial progenitor cells. Blood. 2005;**105**(7):2783-2786. DOI: 10.1182/blood-2004-08-3057

[51] Bearzi C, Annarosa Leri A, Lo Monaco F, Marcello Rota M, Gonzalez A, Hosoda T, Pepe M, Qanud K, Ojaimi C, Bardelli S, D'Amario D, D'Alessandro DA, Michler RE, Dimmeler S, Zeiher AM, Urbanek K, Hintze TH, Kajstura J, Anversa P. Identification of a coronary vascular progenitor cell in the human heart. Proceedings of the National Academy of Sciences of the United States of America. 2009;**106**(37):15885-15890. DOI: 10.1073/pnas.0907622106

[52] Asahara T, Kawamoto A. Endothelial progenitor cells for postnatal vasculogenesis. American Journal of Physiology. Cell Physiology. 2004;**287**(3):C572-C579. DOI: 10.1152/ajpcell.00330.2003

[53] Hill JM, Zalos G, Halcox JP, Schenke WH, Waclawiw MA, Quyyumi AA, Finkel T. Circulating endothelial progenitor cells, vascular function, and cardiovascular risk. The New England Journal of Medicine. 2003;**348**(7):593-600. DOI: 10.1056/NEJMoa022287

[54] Richardson MR, Yoder MC. Endothelial progenitor cells: Quo vadis?. Journal of Molecular and Cellular Cardiology. 2011;**50**(2):266-272. DOI: 10.1016/j.yjmcc.2010.07.009

[55] Tang Z, Wang A, Yuan F, Yan Z, Liu B, Chu JS, Helms JA, Li S. Differentiation of multipotent vascular stem cells contributes to vascular diseases. Nature Communications. 2012;**3**:875. DOI: 10.1038/ncomms1867

[56] Vasuri F, Fittipaldi S, Buzzi M, Degiovanni A, Stella A, D'Errico-Grigioni A, Pasquinelli G. Nestin and WT1 expression in small-sized vasa vasorum from human normal arteries. Histology and Histopathology. 2012;**27**(9):1195-1202

[57] Del Toro R, Chèvre R, Rodríguez C, Ordóñez A, Martínez-González J, Andrés V, Méndez-Ferrer S. Nestin(+) cells direct inflammatory cell migration in atherosclerosis. Nature Communications 2016;**7**:12706. DOI: 10.1038/ncomms12706

[58] Vasuri F, Resta L, Fittipaldi S, Malvi D, Pasquinelli G. RUNX-1 and CD44 as markers of resident stem cell derivation in undifferentiated intimal sarcoma of pulmonary artery. Histopathology. 2012;**61**(4):737-743. DOI: 10.1111/j.1365-2559.2012.04241.x

Characteristics of Full-Term Amniotic Fluid-Derived Mesenchymal Stem Cells in Different Culture Media

Karuppiah Thilakavathy, Norshariza Nordin,
Rajesh Ramasamy, Peyman Ghoraishizadeh,
Izanwati Mohd Rawi Rohayu and Gurbind Singh

Additional information is available at the end of the chapter

http://dx.doi.org/10.5772/intechopen.69392

Abstract

Amniotic fluid contains precious therapeutic stem cells with ideal features such as they are broadly multipotent, genetically stable, and non-tumorigenic. One of the stem cells that is abundantly found in amniotic fluid is mesenchymal stem cells. Human amniotic fluid mesenchymal stem cells (hAFMSCs) had been successfully isolated from amniotic fluid obtained from second or third trimester amniocentesis. However, studies on hAFMSCs obtained during full-term delivery are still lacking. Furthermore, suitable culture media to propagate hAFMSCs for therapeutic purposes have not been fully established. Basal medium supplemented with fetal bovine serum is commonly used, and unfortunately, this condition has been associated with the risk of transmission of animal pathogens and xenogenic immune reaction. An efficient isolation and expansion method together with suitable culture conditions is essential in establishing a specific homogenous cell population, such as full-term hAFMSCs, of clinical grade. In this chapter we briefly describe the feasibility of generating hAFMSCs from full-term amniotic fluid obtained during cesarean section using serum-free medium as opposed to the conventional serum containing media. These findings would be very useful in utilizing stem cells for bench side application from a source that is accessible and devoid of ethical and safety concerns.

Keywords: full-term amniotic fluid, cesarean, mesenchymal stem cells, postmitotic neurons, serum-free

1. Introduction

The robust development in regenerative medicine, especially the use of stem cells, has opened new treatment modalities in modern medicine. Substantially supported by the scientific evidence, stem cells promise "cures" to chronic diseases and are considered a valuable "substitute" to the conventional therapies. Thus, the dire need for "ready-to-use" or "off-the-shelf" sources of stem cells becomes very apparent because of increased demand for the stem cell therapy. Among the adult stem cells, mesenchymal stem cells (MSCs) have seized the worldwide attention of many clinicians and scientists due to their unique characteristics. The regenerative capability of MSCs with an inherent immunosuppressive ability triggers an excellent outcome in repairing tissue injuries and restoring functions of many organs in the context of inflammatory milieu. However, similar to any other cell therapies, harvesting and acquiring an adequate number of cells for the therapeutical purposes still limit the wide use of MSCs. In line with this, the current study has explored the feasibility of exploiting a human delivery waste, namely, amniotic fluid to generate and propagate fetal-derived MSCs for the potential clinical applications.

Amniotic fluid appears at about 12 days after conception in between the amnion and chorion fetal membranes. Other than nutrients such as proteins, carbohydrates, lipids, phospholipids, and urea to support the growth of the fetus, the fluid also contains heterogenous population of cells that are sloughed off from the fetal skin and the digestive, respiratory, and urogenital tracts and from the amniotic membrane [1, 2]. The total number and proportion of viable cells may vary widely between samples from different pregnancies of the same gestation period [1]. Furthermore, the cell population also found to change with gestation corresponding to the developing fetus [3]. The types of cells found in human amniotic fluid are divided into three main groups—the epithelioid E-type cells, amniotic fluid-specific AF-type cells, and fibroblastic F-type cells—which are classified according to the morphological, biochemical, and growth properties [4, 5]. E-type cells, which are round shaped and slow growing [6], are presumed to derive from the fetal skin and urine, while AF-type cells are from fetal membranes and trophoblast tissue (placenta) because these cells produce estrogen, human chorionic gonadotropin, and progesterone [1, 7]. F-type cells are considered to originate from mesenchymal tissue due to lack of any hormone production, and they do not express human leukocyte class II (HLA-DR) surface antigen [8].

Human amniotic fluid cells have been used to screen for fetal abnormalities for more than 60 years, and only recently, their therapeutic value was discovered. A number of amniotic fluid-derived cells have been identified and examined for their properties. Among all, amniotic epithelial cells have been demonstrated to express glial and neuronal stem cell markers [9]. Midterm amniotic fluid stem cells (AFSCs) have been found to express c-kit as well as mesenchymal stem cell (MSC) surface markers [10]. These AFSCs were shown to have the ability to proliferate and maintain a normal karyotype for more than 250 population doublings [10]. Furthermore, several investigators demonstrated that AFSCs from both mid- and full-term amniotic fluid positively express pluripotent markers, differentiate into derivatives of all the three primary germ layers, and form embryoid bodies under the appropriate conditions; however, they are not tumorigenic [10, 11]. The human amniotic fluid mesenchymal stem cells (hAFMSCs) on the other hand are similar to AFSCs except the c-Kit, SSEA4 and OCT4 expressions are lesser, and their in vitro expansion only could last about 30–50 doublings [12]. Apart from these, amniotic fluid was also found to contain hematopoietic progenitor cells in the first trimester [3].

The characteristics of amniotic fluid cells, particularly AFSCs and hAFMSCs, such as broadly multipotent, high proliferative potential, paracrine secretion activity, and non-tumorigenic in addition to devoid of ethical and safety issues will make them significant candidates in the field of regenerative medicine and drug screening.

2. Human full-term amniotic fluid-derived mesenchymal stem cells

The leftover amniocentesis samples of second-trimester amniotic fluid collected for the routine prenatal diagnosis are usually used for the isolation of amniotic fluid mesenchymal stem cells. At this gestational stage, it is impossible to collect a larger volume of amniotic fluid, and there are increased risks of uterine contamination and miscarriage [13]. Alternatively, stem cells could be isolated from amniotic fluid of full-term pregnancies, specifically during delivery. We attempted to generate hAFMSCs from full-term amniotic fluid obtained during cesarean section. The following sections describe their propagation using serum and serum-free media, phenotypic characterization, and in vitro differentiation potential.

2.1. Culture techniques

Human amniotic fluid samples were obtained under an appropriate Ethical Committee approval and after signed informed consent from 14 women prior to cesarean procedure. The mean term pregnancy duration was 38 ± 1 weeks. About 10 ml of amniotic fluid was collected by puncturing the membranes after the uterine muscle was opened for the cesarean-section delivery. Cells were isolated from the human amniotic fluid samples not more than 4 hours from the time of collection. About 10 ml of human amniotic fluid was centrifuged for 10 minutes at 1200 rpm, and 10,000 cells/cm^2 tumorigenic were seeded in T25 flask and grown to confluence in (1) serum-free MesenCult™-XF (Stem Cell Technologies, Canada) complete medium (SFM) according to the manufacturer protocol, (2) low-glucose DMEM with GlutaMAX™ (Gibco BRL, Invitrogen, USA) supplemented with 15% fetal bovine serum (DMEM-FBS), and (3) low-glucose DMEM with GlutaMAX™ supplemented with 15% human serum (DMEM-HS). The media were added with 1% penicillin/streptomycin (Gibco BRL, Invitrogen, USA). The cells were cultured in respective defined media at 37°C in 5% CO_2 incubator (RS Biotech Galaxy, Irvine, UK). The initial media change was performed at day 5 and subsequently every 3–4 days. Adherent cells achieved approximately 80% confluency around 15th day of the primary culture. The cells were harvested using 0.05% trypsin-EDTA (Gibco BRL, Invitrogen, USA) for 5 minutes at 37°C; cells were reseeded at 3000 cells/cm^2 and subsequently expanded up to P7. Adherent cells with fibroblast-like spindle-shaped morphology (**Figure 1**) were observed at P0 in all respective culture media. However, only SFM medium grown cells were able to expand beyond P7; others, DMEM-FBS and DMEM-HS culture grown cells, were able to attach and proliferate till P3 and P1, respectively. Therefore, the downstream experiments could not be carried out at P3 and P7 for DMEM-HS and P7 for DMEM-FBS culture conditions, respectively.

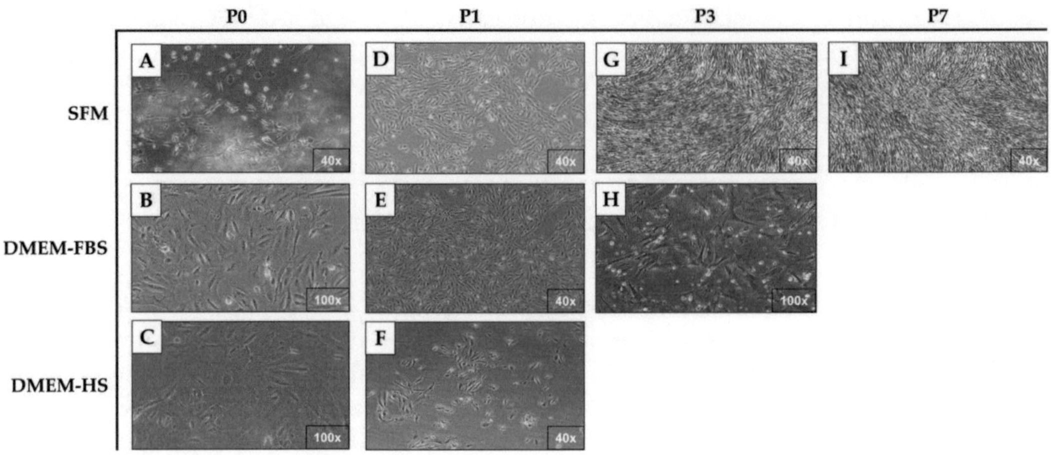

Figure 1. Primary culture of full-term amniotic fluid-derived adherent cells in different culture media. (A, B, and C) Adherent cells were noticed from day 7 onward and took 2 weeks, 25 days, and 38 days to confluence in P0, respectively. (D and E) More spindle-shaped fibroblast-like cells were observed in P1. (F) Cells attached but failed to reach confluency. (G, H, and I) Higher percentage of homogenous population of spindle-shaped fibroblast-like cells was observed in DMEM-FBS and SFM at P3 and P7, respectively.

Morphologically, MSCs derived from human amniotic fluid resembled MSCs from other human tissues, especially the bone marrow. Adherence to the plastic surface and assuming spindle cell morphology are the criteria that minimally define MSCs along with stipulated immunophenotyping and the mesodermal differentiation. Despite the fact that all P0 cells from the respective media acquired an MSC-like morphology, the adherent cells from SFM media showed a much smaller, denser, and defined cell population as compared to other culture systems. These features reflect the nature of the expanding cells, where cells with rapid turnover appear small and with defined spindle morphology (**Figure 1G** and **I**) compared to the slowly growing cells with a broader and polygonal shape (**Figure 1H**). The morphological observation of the SFM-expanded cells confers the high proliferative nature of the culture system (**Figure 1G** and **I**). Although, the content of SFM media is not fully disclosed, it might contain cocktail of growth factors that enhances cell proliferation, preserves the telomere length and prevents early cellular senescence.

2.2. Colony forming unit assay and population doubling time

The proliferations of hAFMSCs in these media were examined by colony forming unit-fibroblast (CFU-F) assay and population doubling time (PDT). The ability to form CFU-F is one of the characteristics of MSCs [14]. PDT was carried out by seeding cells at P3 and P7 in respective T75 flasks with density of 10,000 cells/cm^2 until cultures reach confluency. PDT was calculated using the following formula: PDT = CT/log (Nf/Ni), where CT = culture time, Nf = final number of cells, and Ni = initial number of cells at culture initiation. CFU-F assays were performed by plating 100 cells in the respective culturing media in 100 mm cell culture plate and incubated at 37°C in 5% CO_2. On the 14th day, the cells were washed with

Dulbecco's phosphate buffered saline (DPBS) (pH 7.4) and stained using 10 ml 0.5% crystal violet (Sigma-Aldrich, USA) in methanol for 1 hour at room temperature. Plates were washed with DPBS twice, and colonies containing 50 or more cells were counted.

The hAFMSCs doubled in 36 hours and formed more colonies when cultured in SFM, which is defined for MSCs, while it took 6 days when cultured in DMEM-FBS (**Table 1**). The slow and poor growth of these cells in DMEM-FBS and DMEM-HS most likely because of the media was not supplemented with any additional growth factors. It could be possible that the lack of optimized serum batch selection for FBS and HS that support MSC colony formation and expansion rendered the observed non-conducive proliferation. The selection of serum batch for a particular cell type, especially stem cells, is crucial since the halted cellular expansion is often noticed due to senescence [15].

Medium	Passages	PDT (hours)	CFU-F (no. of colonies)
SFM	3	35.63 ± 0.16	82.11 ± 2.14
SFM	7	37.15 ± 1.22	75.89 ± 2.61
DMEM-FBS	3	146.1 ± 0.97	5.88 ± 1.53

The proliferation of the hAF cells is slower in DMEM-FBS than in SFM. Results are of three independent experiments.

Table 1. Population doubling time (PDT) and colony forming unit-fibroblast (CFU-F) analyses of full-term amniotic fluid-derived mesenchymal stem cells.

2.3. Immunophenotyping by flow cytometry analysis

Second-trimester hAFMSCs were reported to positively express CD44, CD73, CD90, CD105, and CD166 and lack of expression of CD14, CD19, CD34, CD45, and HLA-DR [16, 17]. In order to investigate the expression of MSC surface markers on the full-term hAFMSCs grown in different culture media, fluorescence-activated cell sorting analysis was carried out. Cultured cells were harvested and resuspended in DPBS supplemented with 1% FBS at a cell density of 1.0×10^6 cells/ml. Approximately 1×10^5 cells were incubated with 3 ul of labeled mouse antihuman monoclonal antibodies at 4°C for 30 minutes in dark and then washed with 1 ml 1× DPBS supplemented with 1% FBS. The following antibodies were used: CD44, CD73, CD90, CD105, CD166, CD14, CD19, CD34, CD45, and HLA-DR (BD Pharmingen, San Diego, CA). All analyses were normalized against negative control cells incubated with isotype specific to the respective antibodies. At least 10,000 events were acquired on Guava easyCyte™ flow cytometer, and the results were analyzed using guavaSoft software (Millipore, USA). We found that more than 90% of the cells grown in SFM expressed all MSC-positive markers at P3 and P7 (**Figure 2**). However, in DMEM-FBS, only about 50% of the P3 cultured cells expressed CD105 and CD166. Lower percentage of cells expressing CD105 and CD166 was also reported by several researchers when second- and third-trimester hAFMSCs were cultured in different types of media supplemented with FBS [13, 16, 18].

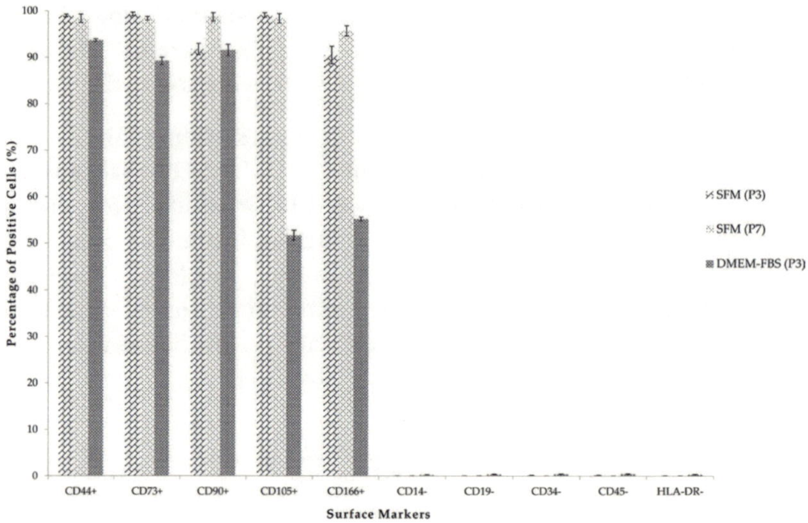

Figure 2. Human full-term amniotic fluid-derived adherent cells grown in SFM and DMEM-FBS showed typical mesenchymal stem cell molecular marker expression. Flow cytometry results showed that the adherent cells were absolutely positive for CD44, CD73, CD90, CD105, and CD166 while negative for CD14, CD19, CD34, CD45, and HLA-DR. Results are of three independent experiments.

2.4. Differentiation into mesodermal lineage

To study the multilineage capacity, hAFMSCs were subjected to differentiate under the conditions that promote osteogenesis, adipogenesis, and chondrogenesis (**Figure 3**). The

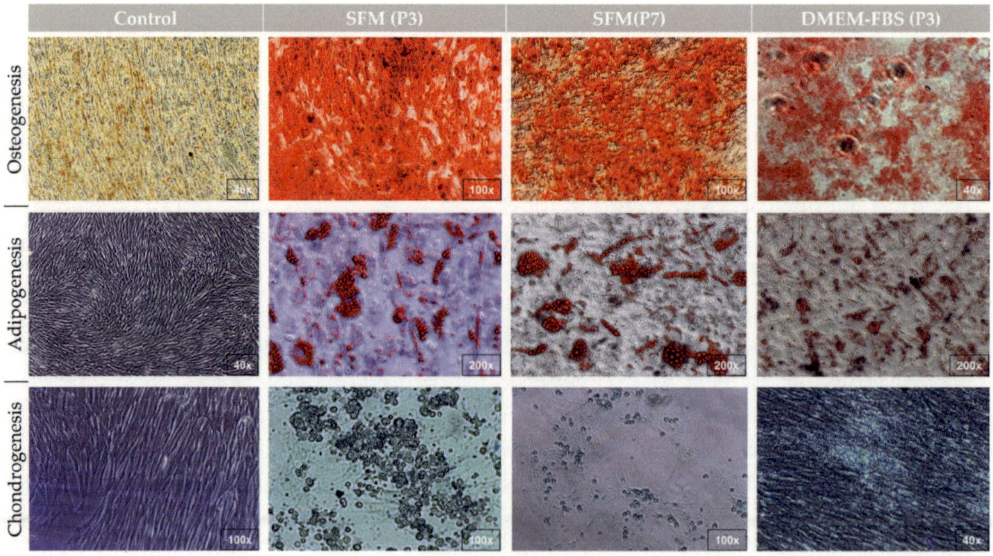

Figure 3. Differentiation potential of SFM and DMEM-FBS cultured human full-term amniotic fluid mesenchymal stem cells into mesodermal lineages. Adherent cells were grown to confluency and subjected to the relevant induction media as per manufacturer's protocol. Osteogenic, adipogenic, and chondrogenic differentiations were evidenced by stained calcium deposits, lipid droplets, and proteoglycan aggregates, respectively.

differentiations were performed using StemPro Adipogenesis, osteogenesis, and chondrogenesis differentiation kits (Gibco, Invitrogen, USA) according to manufacturer's protocol. Differentiation of MSCs into osteoblasts was demonstrated by staining the calcium deposition with Alizarin red. The adipogenic phenotype was determined by staining the cell monolayers with Oil Red O. Multiple intracellular lipid-filled droplets were observed which is consistent with the phenotype of mature adipocytes. Chondrogenic differentiation was detected by staining extracellular proteoglycan aggrecan produced by the differentiated cells with alcian blue stain. Similar to the morphological and proliferation analyses, the mesodermal differentiation of hAFMSCs grown in serum and serum-free media also varied based on the passage numbers and culture media. In agreement with others, the early passage (P3) of SFM-grown hAFMSCs differentiated at the greater extent as compared to the P7 hAFMSCs. Although the degree of differentiation was not quantitatively captured, the distribution and density of the relevant biochemical stainings have indicated the early passaged cells induced into maturation with high magnitudes. When P3 cells from SFM and DMEM-FBS were compared, the degree of differentiation was lesser in the FBS-supplemented medium.

2.5. Spontaneous differentiation into postmitotic neurons

We also observed that the adherent cells derived from full-term amniotic fluid of few samples were growing in atypical manner compared to some of the propagated hAFMSCs when cultured in SFM medium. Morphology of spindle-shaped fibroblast-like cells with neurite-like branching was noticed during expansion at P2 (**Figure 4**) as seen during standard in vitro neural differentiation process.

Preliminary tests were conducted to verify whether these cells had undergone neuronal differentiation. The cells were confirmed positive for class III β-tubulin expression, a specific neuronal marker that can be detected during early neuronal differentiation and neurite outgrowth [19]; however, they were negative for microtubule-associated protein 2 (MAP-2), a marker for mature neurons. Immunocytochemistry (**Figure 5**) and flow cytometry (**Figure 6**) analyses revealed that more than 90% of the cell population expressed the class III β-tubulin protein as well as MSC markers (CD105, CD73, CD44, and GD2), however negative for MAP-2. These results indicate that the cells exhibited postmitotic neuronal cell identity while maintaining the MSC properties.

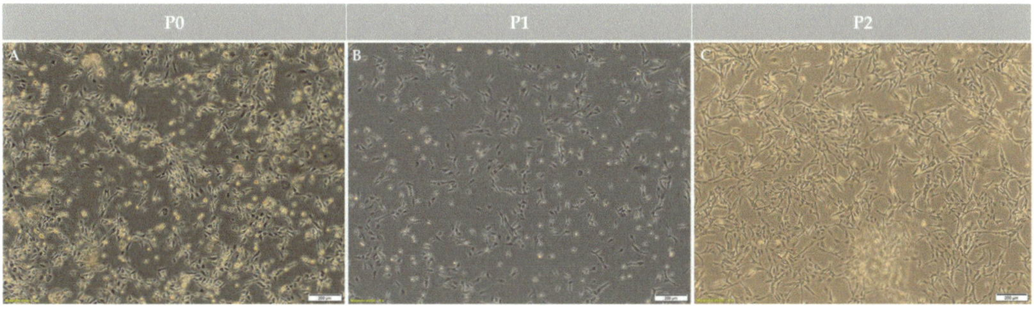

Figure 4. Neurite-like branching in primary culture of human full-term amniotic fluid-derived cells grown in SFM. (A) Heterogenous population of amniotic fluid cells was observed in P0. (B) Spindle-shaped fibroblast-like cells seen in P1. (C) Spindle-shaped fibroblast-like cells with neurite-like branching were noticed in P2.

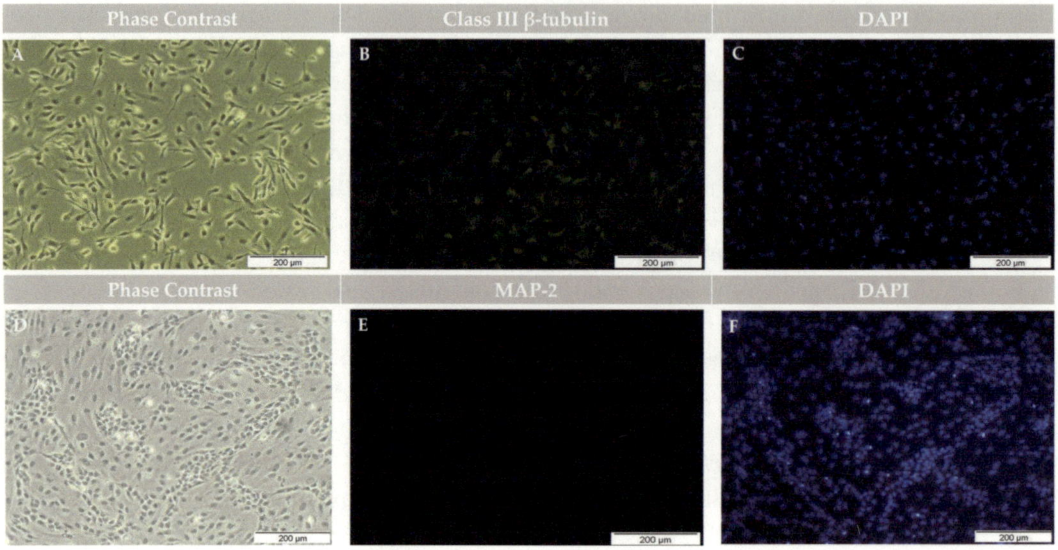

Figure 5. Expression of neuronal protein markers by immunocytochemistry. Passage 2 cells were stained with class III β-tubulin and MAP-2 to confirm the spontaneous neuronal differentiation. (A and D) Stained cells under phase contrast view. (C and F) Cells were counter stained with DAPI for nucleus staining. (B) Cells showing expression of class III β-tubulin indicating these cells were postmitotic neurons. (F) Cells did not express MAP-2 indicating the cells were not matured neurons.

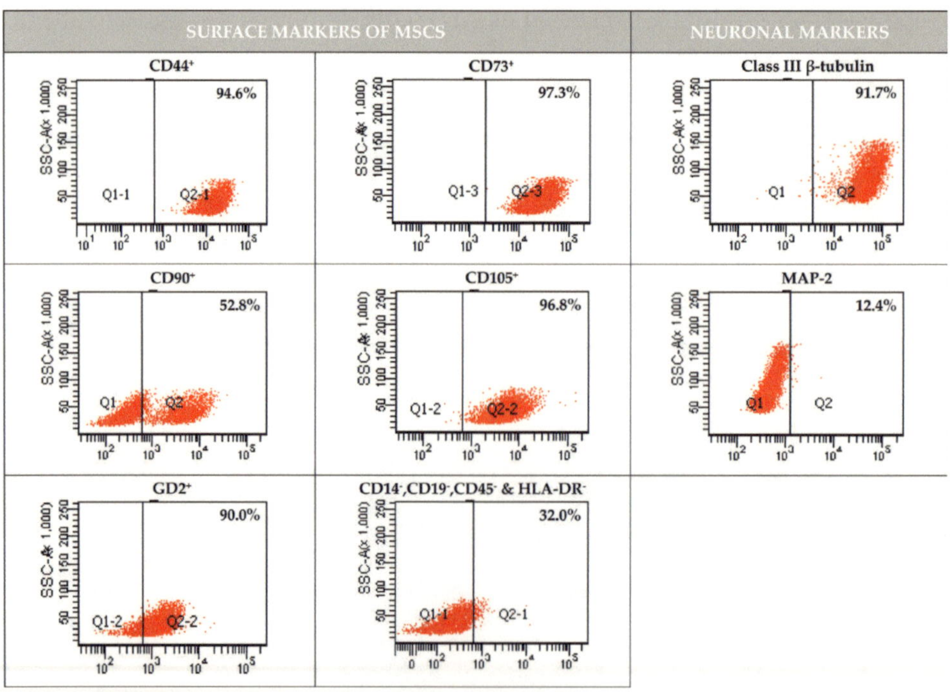

Figure 6. Expression of MSC surface and postmitotic neuron markers by flow cytometry of hAFMSCs grown in SFM. The adherent cells in P2 were positive for CD44, CD73, CD90, CD105, and GD2 while negative for CD14, CD19, CD45, and HLA-DR. About 91% cells exhibited positive class III β-tubulin expression; however, they were negative for MAP-2, suggesting the cells were postmitotic neurons and not matured neurons.

3. Discussion

The present study evaluated the human amniotic fluid-derived MSCs in terms of morphology, expression of cell surface markers, and mesodermal differentiations in various culture conditions. The major limitation that blocks the progress of clinical trials relies on the ability of the cells to propagate in sufficient numbers for transplantation. In many cases, researchers can generate MSCs from various sources but failed to expand these cells beyond certain passages where it jeopardizes the subsequent exploration and transplantation studies. One should bear in mind that a successful use of MSCs from a respective source is only feasible when there is a culture system that allows expansion of particular cell type without compromising its molecular properties. In this study, a commercially available SFM medium was tested along with other conventional media each supplemented with FBS or HS to escalate the expansion of hAFMSCs.

The standard practice for expansion of MSCs is mainly the use of FBS as it is the basic source of growth factors and low-molecular-weight bioactive compounds [20]. FBS is simple to use and supports undifferentiated MSC expansion; however, the drawbacks of using FBS-containing medium in therapeutic application are lack of experimental reproducibility and may cause immunogenic reactions in patients [21]. In order to replace FBS, human serum has been used; nonetheless similar to FBS, there are issues such as variability between lots due to genetic diversity and lack of assurance that it will not transmit new and emerging infectious disease viruses [22]. Recently, for therapeutic purposes stem cells are being cultured in xeno-free or animal component-free media formulations to circumvent transmission of xenogenic proteins and pathogens and to improve the outcome of cell transplantation studies. Xeno-free media may contain material derived from the human plasma, while animal component-free media do not contain serum and other materials derived from animal or human sources.

In our work, commercial xeno-free, serum-free culture medium was used in conjunction with serum-free attachment substrate to support cell adhesion. The cells were also cultured in FBS and HS-containing media to evaluate the capacity of expansion and differentiation of hAFMSCs. The growth rate and morphology observations of our study were similar with other MSC studies carried out using serum-free and serum-supplemented media [21, 23, 24]. We found that the proliferation rate of the full-term amniotic fluid cells was higher in SFM, where the doubling time remained short and relatively consistent till seven passages compared to serum-containing media. Besides, we also noticed that the cells grown in SFM exhibited more elongated, spindle-shaped morphology and grow in distinct bundles of cells when continually expanded. In contrast, cells cultured in serum media displayed more flattened, fibroblast-like morphology and even monolayer of cells. Human MSCs undergo replicative senescence with decreasing proliferation and changes in cell morphology, which were observed in early passage of cells cultured in both DMEM-FBS and DMEM-HS. This could be due to undefined factors in the serum; alternatively the rapid proliferation and longer life span of hAFMSCs in SFM are possibly contributed by the higher concentrations of growth factors present in this medium. In addition to robust proliferation rate, SFM also enhanced the clonogenic potential of the full-term hAFMSCs which is one of the properties of bona fide MSCs.

Immunophenotyping is one of the main criteria for characterizing MSCs. International Society for Cellular Therapy (ISCT) proposed that multipotent human MSCs must express CD105, CD73, and CD90 and lack of expression of CD45, CD34, CD14 or CD11b, CD79α, CD19, and HLA-DR surface antigen [25]. Other positive markers currently included to define MSCs are CD166, CD44, CD29, and CD9 [26]. In our study, all the MSC-positive markers were highly expressed in full-term hAFMSCs expanded in SFM. The hAFMSCs cultured in DMEM-FBS moderately expressed CD105 and CD166. A similar expression pattern of CD105 and CD166 was also observed by other researchers during generation of MSCs derived from second- and third-trimester amniotic fluid using FBS-supplemented media [13, 16, 18]. CD105 plays a role in chondrogenic differentiation [27], while CD166 is involved in neurite extension [28]. Probably due to lower expression of CD105, hAFMSCs grown in DMEM-FBS differentiated poorly into chondrocytes. hAFMSCs isolated in SFM were found to have higher trilineage mesoderm differentiation capacity. Martinez et al. [29] studied on neural ganglioside GD2 surface antigen on bone marrow MSCs after several researchers reported on neural antigen expression on MSCs. They found that GD2 was consistently expressed at a high level on all freshly isolated or ex vivo expanded bone marrow MSCs but was not expressed in all other cells within the marrow. Likewise, Xu et al. [30] found that umbilical cord-derived MSCs were the only cells within umbilical cord that expressed this marker. These findings suggested that GD2 can be a unique marker for MSCs. It is also noteworthy that GD2 is one of the major gangliosides of the postmitotic neurons [31] and was found to increase during neurite outgrowth [32]. When spindle-shaped cells with neurite-like outgrowth were noted in most of the SFM cultures, we investigated the MSC surface antigen expressions and included GD2 marker to confirm that the cells were indeed MSCs and have neurogenic potential. Class III β-tubulin and MAP-2 expression were also analyzed to examine whether the cells were postmitotic neurons or matured neurons. As suspected, the cells had differentiated spontaneously into postmitotic neurons. Interestingly, this morphology was not seen in DMEM-FBS cultures, probably due to lower expression of CD166 on these cells. Chase et al. [24] had reported that bone marrow MSCs grown in serum-free medium significantly enhanced the expression of the intermediate filament nestin when compared to cells expanded in serum-containing medium. MSCs at the earliest developmental stage were found to harbor stronger neurodifferentiation capacity than postnatal MSCs, acquiring characteristics of postmitotic neurons [33]. It was suggested that MSCs exhibit both stem cell and precursor functions allowing neuronal differentiation through both mitotic stem cells and nonmitotic precursor pathways [34]. Many studies demonstrated that the morphological and molecular modifications of MSCs were probably due to stress response, rather than to a real differentiation into neuronal cells; however, some recent studies had demonstrated that MSC-derived cells not only showed morphological features of neurons, but that they also demonstrate functional properties of neurons [35]. There is evidence that MSCs from different sources may not have the same biological and genetic properties; probably for these reasons and neurotrophic cytokines that might be present in the microenvironment, the full-term hAFMSCs spontaneously differentiated into postmitotic neurons in our hands. These postmitotic neurons might turn into mature neurons upon further expansion. These findings suggest the possibility that the full-term amniotic fluid-derived mesenchymal stem cells expanded in serum-free medium may serve as a source for stem cell-based regenerative medicine to a variety of therapeutic scenarios including treating the neurological disorders.

4. Conclusion

Our findings clearly demonstrated the feasibility of generating MSCs from full-term human amniotic fluid. SFM media were found to be most efficient in isolation and expansion of full-term hAFMSCs. It is very interesting to note that the characteristics and behavior of the established hAFMSCs change under the influence of different culture media. Keeping these observations in mind, further work needs to be done to understand differentiation potential of full-term hAFMSCs before they can be applied in bedside settings.

Acknowledgements

This research was supported by Universiti Putra Malaysia (RUGS/04-04-11-1491RU and GP-IPS/2017/9523900) and Ministry of Education, Malaysia (FRGS: 04-01-14-1454FR). We thank Dr. Tan Boon Chong and Dr. Zulida Rejali for sample collection and Dr. Anjan Kumar Das for the permission to use the Stempeutics Research Malaysia facility. We also would like to thank Ms. Nur Izzati Mansor for assistance with immunocytochemistry technique.

Abbreviations

CFU-F	Colony forming unit-fibroblast
DMEM-FBS	Low-glucose DMEM with GlutaMAX™ supplemented with 15% fetal bovine serum
DMEM-HS	Low-glucose DMEM with GlutaMax™ supplemented with 15% human serum
DPBS	Dulbeccos's phosphate buffered saline
hAFMSC	Human amniotic fluid mesenchymal stem cell
HLA-DR	Human leukocyte antigen class II
MAP-2	Microtubule-associated protein 2
MSC	Mesenchymal stem cell
PDT	Population doubling time
SFM	Serum-free MesenCult™-XF complete medium

Author details

Karuppiah Thilakavathy[1,2*], Norshariza Nordin[1,2], Rajesh Ramasamy[2,3], Peyman Ghoraishizadeh[1], Izanwati Mohd Rawi Rohayu[1,4] and Gurbind Singh[5]

*Address all correspondence to: thilathy@upm.edu.my

1 Department of Biomedical Science, Faculty of Medicine and Health Sciences, Universiti Putra Malaysia, Selangor, Malaysia

2 Genetics and Regenerative Medicine Research Centre, Faculty of Medicine and Health Sciences, Universiti Putra Malaysia, Selangor, Malaysia

3 Department of Pathology, Faculty of Medicine and Health Sciences, Universiti Putra Malaysia, Selangor, Malaysia

4 Faculty of Health Sciences, Universiti Sultan Zainal Abidin (UniSZA), Kampus Gong Badak, Kuala Terengganu, Terengganu, Malaysia

5 Department of Life Sciences, International Medical University, Kuala Lumpur, Malaysia

References

[1] Gosden CM. Amniotic fluid cell types and culture. British Medical Bulletin. 1983;**39**:348-354. DOI: 10.1093/oxfordjournals.bmb.a071847

[2] Fauza D. Amniotic fluid and placental stem cells. Best Practice & Research Clinical Obstetrics & Gynaecology. 2004;**18**:877-891. DOI: 10.1016/j.bpobgyn.2004.07.001

[3] Torricelli F, Brizzi L, Bernabei PA, Gheri G, Di Lollo S, Nutini L, Lisi E, Di Tommaso M, Cariati E. Identification of hematopoietic progenitor cells in human amniotic fluid before the 12th week of gestation. Italian Journal of Anatomy and Embryology. 1993;**98**:119-126

[4] Priest JH. Prenatal chromosomal diagnosis and cell culture. In: Barch MJ, editor. The ACT Cytogenetics Laboratory Manual. 2nd ed. New York: Raven Press; 1991. pp. 149-203

[5] Prusa AR, Hengstschlager M. Amniotic fluid cells and human stem cell research: A new connection. Medical Science Monitor. 2002;**08**:RA253-RA257

[6] Bai J, Wang Y, Liu L, Chen J, Yang W, Gao L, Wang Y. Human amniotic fluid-derived c-kit$^+$ and c-kit$^-$ stem cells: Growth characteristics and some differentiation potential capacities comparison. Cytotechnology. 2012;**64**:577-589. DOI: 10.1007/s10616-012-9441-6

[7] Thakar N, Priest RE, Priest JH. Estrogen production by cultured amniotic fluid cells. Clinical Research. 1982;**30**:888A

[8] Whitsett CF, Priest JH, Priest RE, Marion J. HLA-typing of cultured amniotic fluid cells. American Journal of Clinical Pathology. 1983;**79**:186-194. DOI: 10.1093/ajcp/79.2.186

[9] Sakuragawa N, Thangavel R, Mizuguchi M, Hirasawa M, Kamo I. Expression of markers for both neuronal and glial cells in human amniotic epithelial cells. Neuroscience Letters. 1996;**209**:9-12. DOI: 10.1016/0304-3940(96)12599-4

[10] De Coppi P, Bartsch G Jr, Siddiqui MM, Xu T, Santos CC, Perin L, Mostoslavsky G, Serre AC, Snyder EY, Yoo JJ, Furth ME, Soker S, Atala A. Isolation of amniotic stem cell lines with potential for therapy. Nature Biotechnology. 2007;**25**:100-106. DOI: 10.1038/nbt1274

[11] Kim J, Lee Y, Kim H, Hwang KJ, Kwon HC, Kim SK, Cho DJ, Kang SG, You J. Human amniotic fluid-derived stem cells have characteristics of multipotent stem cells. Cell Proliferation. 2007;**40**:75-90. DOI: 10.1111/j.1365-2184.2007.00414.x

[12] Bernado JD, Kunisaki SM. Amniotic fluid stem cell populations. In: Fauza DO, Bani M, editors. Fetal Stem Cells in Regenerative Medicine. Principles and Translational Strategies. New York: Springer; 2016. pp. 167-179. DOI: 10.1007/978-1-4939-3483-6_9

[13] Savickiene J, Treigyte G, Baronaite S, Valiuliene G, Kaupinis A, Valius M, Arlauskiene A, Navakauskiene R. Human amniotic fluid mesenchymal stem cells from second- and third-trimester amniocentesis: Differentiation potential, molecular signature, and proteome analysis. Stem Cells International. 2015;**2015**:319238. DOI: 10.1155/2015/319238

[14] Caplan AI. Mesenchymal stem cells. Journal of Orthopaedic Research. 1991;**9**:641-650. DOI: 10.1002/jor.1100090504

[15] Ridzuan N, Abbar AA, Yip WK, Maqbool M, Ramasamy R. Characterization and expression of senescence marker in prolonged passages of rat bone marrow-derived mesenchymal stem cells. Stem Cells International. 2016;**2016**:8487264. DOI: 10.1155/2016/8487264

[16] Tsai MS, Lee JL, Chang YJ, Hwang SM. Isolation of human multipotent mesenchymal stem cells from second-trimester amniotic fluid using a novel two-stage culture protocol. Human Reproduction. 2004;**19**:1450-1456. DOI: 10.1093/humrep/deh279

[17] Roubelakis MG, Pappa KI, Bitsika V, Zagoura D, Vlahou A, Papadaki HA, Antsaklis A, Anagnou NP. Molecular and proteomic characterization of human mesenchymal stem cells derived from amniotic fluid: Comparison to bone marrow mesenchymal stem cells. Stem Cells and Development. 2007;**16**:931-952. DOI: 10.1089/scd.2007.0036

[18] Gucciardo L, Ochsenbein-Kölble N, Ozog Y, Verbist G, Van Duppen V, Fryns JP, Lories R, Deprest J. A comparative study on culture conditions and routine expansion of amniotic fluid-derived mesenchymal progenitor cells. Fetal Diagnosis and Therapy. 2013;**34**:225-235. DOI: 10.1159/000354895

[19] Tropel P, Platet N, Platel JC, Noel D, Albrieux M, Benabid AL, Berger F. Functional neuronal differentiation of bone marrow-derived mesenchymal stem cells. Stem Cells. 2006;**24**:2868-2876. DOI: 10.1634/stemcells.2005-0636

[20] Kume T, Taguchi R, Katsuki H, Akao M, Sugimoto H, Kaneko S, Akaike A. Serofendic acid, a neuroprotective substance derived from fetal calf serum, inhibits mitochondrial membrane depolarization and caspase-3 activation. European Journal of Pharmacology. 2006;**542**:69-76. DOI: 10.1016/j.ejphar.2006.04.038

[21] Dolley-Sonneville PJ, Romeo LE, Melkoumian ZK. Synthetic surface for expansion of human mesenchymal stem cells in xeno-free, chemically defined culture conditions. PLoS One. 2013;**8**:e70263. DOI: 10.1371/journal.pone.0070263

[22] Bsargent. What Does Xeno-Free Really Mean, and Why Does It Matter to Cell Culture Scientists Today? [Internet]. 10 March 2016. Available from: http://cellculturedish.com/2016/03/xeno-free-really-mean-matter-cell-culture-scientists-today/ [Accessed: 1 March 2016]

[23] Sohn MY, Seo JM, Cho HJ, Suh JS, Seong WJ, Park IS, Atala A, Yoo JJ, Shon YH. Effective in vitro expansion of amniotic fluid-derived stem cells in serum free medium. Tissue Engineering and Regenerative Medicine. 2012;**9**:48-54

[24] Chase LG, Lakshmipathy U, Solchaga LA, Rao MS, Vemuri MC. A novel serum-free medium for the expansion of human mesenchymal stem cells. Stem Cell Research & Therapy. 2010;**1**:8. DOI: 10.1186/scrt8

[25] Dominici M, Le Blanc K, Mueller I, Slaper-Cortenbach I, Marini F, Krause D, Deans R, Keating A, Prockop D, Horwitz E. Minimal criteria for defining multipotent mesenchymal stromal cells. The International Society for Cellular Therapy position statement. Cytotherapy. 2006;**8**:315-317. DOI: 10.1080/14653240600855905

[26] Halfon S, Abramov N, Grinblat B, Ginis I. Markers distinguishing mesenchymal stem cells from fibroblasts are downregulated with passaging. Stem Cells and Development. 2011;**20**:53-66. DOI: 10.1089/scd.2010.0040

[27] Barry FP, Boynton RE, Haynesworth S, Murphy JM, Zaia J. The monoclonal antibody SH-2, raised against human mesenchymal stem cells, recognizes an epitope on endoglin (CD105). Biochemical and Biophysical Research Communications. 1999;**265**:134-139. DOI: 10.1006/bbrc.1999.1620

[28] Sato C, Matsuda T, Kitajima K. Neuronal differentiation-dependent expression of the disialic acid epitope on CD166 and its involvement in neurite formation in Neuro2A cells. Journal of Biological Chemistry. 2002;**277**:45299-45305. DOI: 10.1074/jbc.M206046200

[29] Martinez C, Hofmann TJ, Marino R, Dominici M, Horwitz EM. Human bone marrow mesenchymal stromal cells express the neural ganglioside GD2: A novel surface marker for the identification of MSCs. Blood. 2007;**109**:4245-4248. DOI: 10.1182/blood-2006-08-039347

[30] Xu J, Liao W, Gu D, Liang L, Liu M, Du W, Liu P, Zhang L, Lu S, Dong C, Zhou B, Han Z. Neural ganglioside GD2 identifies a subpopulation of MSC in human umbilical cord. Cellular Physiology and Biochemistry. 2009;**23**:415-424. DOI: 10.1159/000218188

[31] Rosenberg A, Noble EP. EGF-induced neuritogenesis and correlated synthesis of plasma membrane gangliosides in cultured embryonic chick CNS neurons. Journal of Neuroscience Research. 1989;**24**:531-536. DOI: 10.1002/jnr.490240411

[32] Seifer W and Fink HJ. In-vitro and in-vivo studies on gangliosides in the developing and regenerating hippocampus of the rat. In: Ledeen R, Yu RK, Rapport MM, Suzuki K, editors. Ganglioside Structure, Function, and Biomedical Potential. New York: Plenum Press; 1984. pp. 535-545. DOI: 10.1007/978-1-4684-1200-0_45

[33] Peng KY, Lee YW, Hsu PJ, Wang HH, Wang Y, Liou JY, Hsu SH, Wu KK, Yen BL. Human pluripotent stem cell (PSC)-derived mesenchymal stem cells (MSCs) show potent neurogenic capacity which is enhanced with cytoskeletal rearrangement. Oncotarget. 2016;**7**:43949-43959. DOI: 10.18632/oncotarget.9947

[34] Muñoz-Elías G, Woodbury D and Black IB. Marrow stromal cells, mitosis, and neuronal differentiation: Stem cell and precursor functions. Stem Cells. 2003;**21**:437-448. DOI: 10.1634/stemcells.21-4-437

[35] Scuteri A, Miloso M, Foudah D, Orciani M, Cavaletti G, Tredici G. Mesenchymal stem cells neuronal differentiation ability: A real perspective for nervous system repair?. Current Stem Cell Research & Therapy. 2011;**6**:82-92. DOI: 10.2174/157488811795495486

Chapter 4

Stem Cells from Human Exfoliated Deciduous Teeth: Biology and Therapeutic Potential

Waleerat Sukarawan and Thanaphum Osathanon

Additional information is available at the end of the chapter

http://dx.doi.org/10.5772/intechopen.68173

Abstract

Stem cells isolated from human exfoliated deciduous teeth (SHEDs) are a type of mesenchymal stem cells (MSCs), widely investigated for regenerative treatment. They are isolated from dental pulp tissues remaining in physiologically shedding human deciduous teeth. Thus, SHEDs are easy to access and not required invasive procedure to obtain cells. SHEDs are multipotent mesenchymal stem cells; however, they possess distinct properties when compared to other MSCs. In this regard, SHEDs exhibit higher proliferative rate than bone marrow-derived MSCs and greater osteogenic differentiation potency than human dental pulp stem cells. This chapter reviews the isolation technique and basic characteristics of SHEDs. Moreover, the intracellular signalling involved in the stemness regulation and differentiation ability of SHEDs is discussed, particularly on fibroblast growth factor, Notch, and Wnt signalling. Finally, the potential regenerative therapeutic application of SHEDs is also described.

Keywords: stem cells, deciduous teeth, basic fibroblast growth factor, Wnt signalling, Notch signalling, mechanical stress

1. Introduction

Dental pulp is a loose connective tissue residing in pulp chamber inside both deciduous and permanent teeth. It surrounds by hard tissues called dentin. Nutrients and oxygen supply are acquired from blood vessels passing through apical and accessory foramen of the teeth's root. Dental pulp originates from cranial neural crest cells [1]. Dental pulp tissues are composed of extracellular matrix and various cell types, e.g. fibroblasts, odontoblasts, endothelial cells, pericytes, immune cells and stem cells. When injured, cells in dental pulp tissues are capable of differentiating odontoblasts or odontoblast-like cells, leading to the promotion of tertiary dentin formation. The formation of tertiary dentin is a mechanism which can protect the tooth vitality. Dental pulp tissues remaining in physiological shedding of deciduous teeth are the alternative source of mesenchymal stem cells, due to the ease of accessibility and minimally invasive technique to obtain tissues [2]. Stem cells from human exfoliated deciduous teeth (SHEDs) are firstly identified by Miura et al. in 2003 [2]. SHEDs have high proliferation potency and are multipotent mesenchymal stem cells. These cells are able to differentiate into, not only, dental pulp-related cells, but also, other cell lineages, for example osteoblasts, adipocytes, neuronal-like cells and endothelial cells [2–8]. Taking these advantageous properties together, SHEDs are one of the candidate cell types for tissue regeneration study.

2. SHEDs' characteristics

SHEDs are heterogeneous population of cells isolated from dental pulp tissues remained in exfoliated deciduous teeth. Similar to those mesenchymal stem cells (MSCs), SHEDs exhibit fibroblast-like morphology, adhere on plastic tissue culture surface, express mesenchymal stem cell surface marker and have multipotential differentiation ability (**Figure 1**). SHEDs have higher proliferation rate compared to dental pulp stem cells (DPSCs) and bone marrow-derived mesenchymal stem cells (BMMSCs) [2, 9]. This could be due to the high expression of genes related to cell proliferation and extracellular matrix in SHEDs comparing with DPSCs [9]. First, a study by Miura et al. demonstrated that SHEDs express mesenchymal surface markers, STRO-1 and CD146 [2], though, the percentage of positive cells is low [2]. Later studies utilized various surface markers for SHEDs characterization protocol. SHEDs expressed CD44, CD73, CD90, CD105 and STRO-1 [6]. In addition, these cells lack of CD45 expression [6]. Besides these markers described above, SHEDs also express other surface markers for example, CD166 and SSEA4. Lack of CD34 is also reported [10]. There is no specific surface marker to precisely identify SHEDs population.

Up to date, MSCs can be isolated from many tissue types. Though, there is no specific marker to clearly identify these cells. According to the Mesenchymal and Tissue Stem Cell Committee of the International Society for Cellular Therapy, the minimum criteria to identify MSCs are as follow [11]. First, the isolated MSCs should adhere to plastic tissue culture plate [11]. Second, MSCs must express several specific surface markers, namely CD105, CD73 and CD90 [11]. They also should not express CD45, CD34, CD14 or CD11b, CD79α or CD19, and HLA-DR [11]. Finally, MSCs have to be able to differentiate into osteoblasts, adipocytes and chondroblasts *in vitro* [11]. The following section describes general SHEDs' characteristics and addresses MSCs' characteristics of SHEDs according to these criteria.

2.1. Isolation technique

Two methods have been utilized for SHEDs isolation, namely an enzymatic digestion and a tissue explant. The enzymatic digestion is performed by digesting minced remaining pulp tissues from deciduous teeth, normally with type I collagenase and dispase mixed enzyme solution

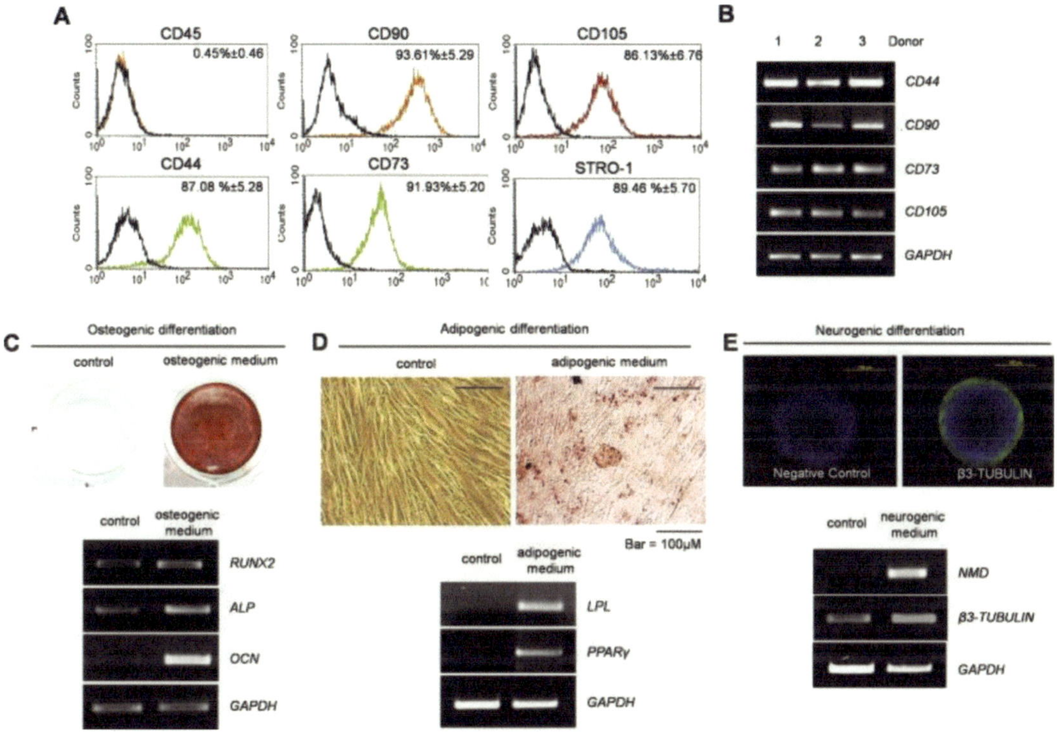

Figure 1. Characteristics of stem cells isolated from human exfoliated deciduous teeth. The expression of stem cells markers was evaluated using flow cytometry and conventional semi-quantitative PCR (A and B). The osteogenic, adipogenic and neurogenic differentiation were evaluated at day 14, 16 and 7 after induction, respectively (C–E). The mineral deposition and intracellular lipid accumulation were determined using alizarin red and oil red O staining, respectively (C and D). The β3-tubulin protein expression was evaluated by immunocytochemistry staining (E). The expression of differentiation marker was examined using conventional semi-quantitative PCR. Reprinted from Archives of Oral Biology, 60(3), Nunthawan Nowwarote, Prasit Pavasant, Thanaphum Osathanon, Role of endogeneous basic fibroblast growth factor in stem cells isolated from human exfoliated deciduous teeth, 408–15, Copyright (2015), with permission from Elsevier [6].

[12–14]. For tissue explant, minced pulp tissues are placed on the tissue culture dishes, allowing the outgrowth of the cells from the tissues [12]. Enzymatic digestion technique leads to more heterogeneous population of isolated cells than those obtained from tissue outgrowth protocol [14]. A study illustrated that there is no significant difference regarding cell morphology and proliferation between cells isolated using enzymatic digestion and tissue outgrowth [14]. Enzymatic digestion-derived SHEDs had higher mineralization ability *in vitro* [14]. However, another study demonstrated that SHEDs isolated using enzymatic digestion exhibited higher cell proliferation and colony forming unit ability as well as adipogenic differentiation potency [13]. However, tissue explant-derived SHEDs had higher osteogenic differentiation ability than enzymatic digestion-derived SHEDs *in vitro* and *in vivo* [13]. The difference between these two studies, especially in osteogenic potential, may be due to the dissimilar osteogenic medium supplementation. The first study employed 0.01 μM dexamethasone disodium phosphate, 1.8 mM monopotassium phosphate (KH_2PO_4) and 5 mM β-glycerophosphate [14]. However, the osteogenic medium of later study was supplemented with 0.1 μM dexamethasone, 2 mM β-glycerophosphate and 50 μM ascorbic acid-2 phosphate [13]. Together, the different isolation technique resulted in the different population of SHEDs. Further comparison is needed to determine a suitable protocol for specific application of these cells.

2.2. Differentiation potential of SHEDs

Studies have shown that SHEDs possess multi-differentiation potency similar to MSCs. Those lineages include odontogenic/osteoblastic, adipogenic, neurogenic and angiogenic differentiation [2].

2.2.1. Odontogenic/osteoblastic differentiation potential

The ability of SHEDs to differentiate into odontoblastic lineage is widely known [2, 15, 16]. Primitively, SHEDs were characterized by their *in vivo* osteoinductivity [2] and follow by several *in vitro* studies to confirm their odontogenic/osteoblastic differentiation potential [5, 10]. SHEDs showed similar osteogenic potency when comparing with BMMSCs, exhibiting significantly elevate levels of ALP activity after 1 week of induction. In addition, several osteogenic markers such as RUNX2, DSP and OCN are also upregulated [10]. When cultured in an osteogenic medium, SHEDs formed mineralized nodules after 4 weeks of induction which indicate calcium deposition *in vitro* [2]. Transplantation of *ex vivo* expanded-SHEDs with hydroxyapatite/tricalcium phosphate (HA/TCP) into immunocompromised mice also induced mineralized tissue formation [2]. Recently, osteoinductivity of SHEDs has been shown in mice. SHEDs formed an osteoinductive template in immunocompromised mice and induced the recruitment of native osteogenic cells to repair calvarial defects [16]. The osteogenic potential of SHEDs in regenerating bone defects in maxillofacial region was also investigated by Zheng et al., the results found that autologous graft using stem cells from miniature pig primary teeth has the ability to regenerate and repair mandibular defects [15]. SHEDs were able to regenerate bone tissues with blood vessels around dental implants in dog model when mixed with platelet-rich plasma (PRP) [17].

Evidence suggested that SHEDs might have the preference towards the odontoblastic lineage due to its origin. SHEDs can be induced to become functional odontoblasts *in vitro* [2]. SHEDs can differentiate to become the odontoblast-like cells and regenerate the tissue with architecture and cellularity similar to the physiologic dental pulp when cultured in scaffolds prepared within human tooth slices and transplanted into immunodeficient mice [18]. It has been recently shown that SHEDs can generate functional dental pulp when injected with PuraMatrix or Collagen into root canals [19]. However, majority of the studies focusing on regenerating bone or dentin-pulp complex *in vivo* were performed in ectopic implantation models, mostly in skin or renal capsule, in mice or rats [20–22] which might not close to real clinical situation. More studies in the clinical relevance area such as tooth socket or jaw bone in larger animals such as pig or dog should be considered to make the results more valuable for application.

2.2.2. Neurogenic differentiation potential

Neurogenic potential of SHEDs is expecting due to their neural crest embryonic origin. Several research studies focusing on differentiating dental stem cells to be used for neurodegenerative disease therapy. These cells are prone to undergo neurogenic differentiation both *in vitro* and *in vivo*. Under the undifferentiating condition, SHEDs and other dental stem cells expressed the neural progenitor markers, nestin and the glial marker, glial fibrillary acidic protein (GFAP), at both the mRNA and protein levels [2, 23]. SHEDs can be induced to become a variety of specialized cells in neural lineage including dopaminergic neuron like cells and glial cells [24, 25]. When induced, SHEDs could form neural-like spheres *in vitro*. Further incubation with a combination of cytokines including sonic hedgehog, fibroblast growth factor 8, glial cell line-derived neurotrophic factor and forskolin can drive these neural spheres into the dopaminergic like neurons [25]. A similar trend was observed where SHEDs showed positive expression of both glial and neuronal markers after 21 days of neurogenic induction. Deposition of antimyelin basic protein was seen and the differentiated cells showed positive expression for neuronal markers such as βIII-tubulin, apolipoprotein E (ApoE), intermediate filament peripherin and Brn3a [26]. SHEDs are able to differentiate into dopaminergic neuronal like cells *in vitro* [27]. However, SHEDs exhibited inferior differentiation ability towards dopaminergic neurons as compared with DPSCs [28]. In this regard, DPSCs upregulated dopaminergic neuron markers (Nurr1, Engrailed1 and Pitx3) higher than SHEDs after treated with sonic hedgehog, fibroblast growth factor 8 and basic fibroblast growth factor [28].

In vivo studies also show the promising results for generating the specialized cells in the neural system. Transplantation of neural-like spheres derived from SHEDs into the striatum of parkinsonian rats significantly improved the apomorphine-evoked rotation of behavioural impairment compared to transplantation of control SHEDs [25]. The results were in line with another study showing the partially recovery after inducing neural maturation of SHEDs into dopaminergic neuron-like cells and transplantation in parkinsonian rats [27]. Moreover, a complete recovery of hindlimb motor function was observed after implantation of neural-induced SHEDs in a rat spinal cord injury [29]. These results suggested that pre-induction of the undifferentiated SHEDs into the neural-like cells before implantation might improve the efficiency of SHEDs in regenerating specialized neural cells and potentially improve the treatment outcome.

2.2.3. *Angiogenic differentiation potential*

Angiogenic potential of SHEDs is another aspect of interest for the benefit of connective tissue regeneration. The rapid and effective induction of vasculation is required for sufficiently supply of oxygen and nutrients as well as removing the toxic waste from the newly synthesized tissues. Unstimulated SHEDs expressed VEGFR1 and NP-1, the known important receptors in angiogenesis and VEGFR1 signalling play an important role in VEGF-induced capillary tube formation by SHEDs as shown by VEGFR1 gene silencing [30]. SHEDs cultured in the tooth slice/scaffolds in combine with VEGF expressed several endothelial differentiation markers such as VEGFR1, VEGFR2, platelet endothelial cell adhesion molecule-1 (PECAM-1) and vascular endothelial cadherin (VE-Cadherin). When transplanted in immunodeficient mice, SHEDs actually lined the new blood vessels within the tooth slice/scaffolds close to the blood vessels of host [3]. Similar results were observed when SHEDs seeded in human tooth slice/scaffolds and transplanted into immunodeficient mice differentiate into human blood vessels that anastomosed with the mouse vasculature and VEGF induced the angiogenic differentiation of SHEDs through Wnt/β-catenin signalling [31]. Another study also showed that SHEDs can differentiate into VEGFR2-positive and CD31-positive endothelial cells *in vitro*. This phenomenon occurred via VEGF/MEK-1/ERK signalling pathway [30]. In addition to *in vitro* data, an *in vivo* study also showed that SHEDs differentiate into endothelial cells when seeded in biodegradable scaffolds and transplanted into immunodeficient mice [18], confirming the plasticity of SHEDs.

2.2.4. Adipogenic differentiation potential

Several studies have reported that SHEDs can be induced into adipogenic lineage [6, 32–34]. After cultured in an adipogenic medium, SHEDs' morphology changed from spindle-like to polygonal shapes and lipid vacuoles were observed, along with the increased in PPARγ2 and LPL mRNA [32]. However, the studies evaluated the adipogenic potential of SHEDs *in vivo* are sparse and the clinical application may not come in the near future.

2.3. Immunomodulatory property

Like other MSCs, SHEDs exhibit immunomodulatory properties. Though, the potency and mechanism are not exact the same to those of BMMSCs [10, 35]. SHEDs significantly reduced the percentage of $IL17^{+}IFN\gamma$ cells population in $CD4^{+}$ T cells *in vitro* [10]. In addition, IL17 expression was decreased compared with the naïve T cell culture alone [10]. SHEDs were also able to rescue the systemic lupus erythematosus-associated symptoms in mice by increasing the ratio of regulatory T cells [10]. It has also been shown that acetylsalicylic acid treatment could improve the immunomodulation of SHEDs [36]. In this regard, acetylsalicylic acid-treated SHEDs enhanced apoptosis of T cells and reduction of $IL17^{+}IFN\gamma$ cells via TERT/FASL pathway [36]. SHEDs also modulate dendritic cell maturation. When co-culture with SHEDs, mature dendritic cells decreased CD40, CD80, CD83 and CD86 expression [37]. SHEDs treating monocyte-derived dendritic cells reduced CD4+ and CD8+ cell proliferation when co-culture with peripheral blood lymphocyte as compared to the control [37]. These immunomodulatory functions of SHEDs encourage them as an interesting MSCs source for regenerative therapy.

3. Basic fibroblast growth factor signalling in SHEDs

Basic fibroblast growth factor (bFGF) is a member in fibroblast growth factor family [38]. It binds to fibroblast growth factor receptors (FGFR) and further initiates intracellular signalling [39]. bFGF has been shown to participate in the regulation of stemness maintenance and cellular differentiation. In human DPSCs, bFGF promotes pluripotent stem cell marker expression, corresponding with the increase of colony-forming unit [40]. Furthermore, bFGF inhibits osteogenic differentiation by SHEDs, human DPSCs and human periodontal ligament stem cells (PDLSCs) when supplemented in osteogenic induction medium (**Figure 2**) [5, 40]. In this regard, alkaline phosphatase enzymatic activity and mineralization are markedly decreased under bFGF-treated condition compared with the control [5, 40]. On the contrary,

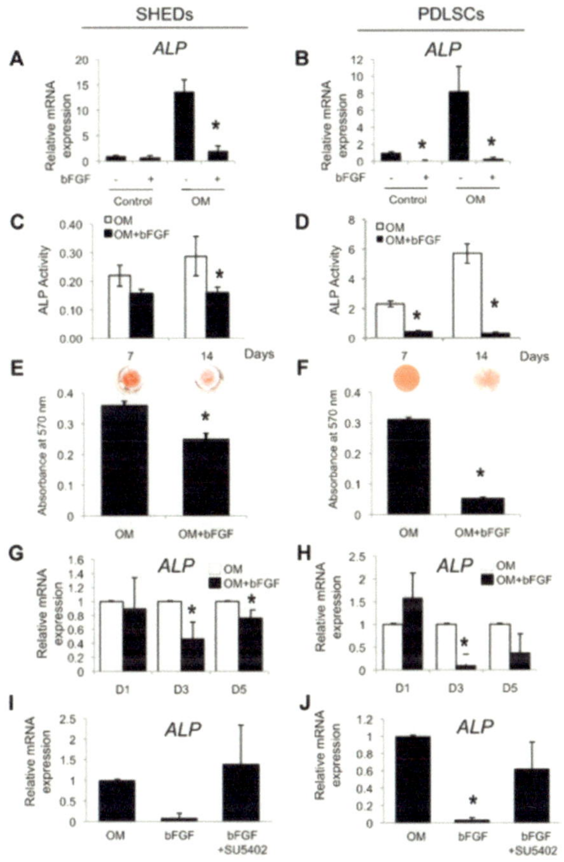

Figure 2. bFGF inhibited *ALP* expression and mineral deposition. The attenuation of *ALP* mRNA expression by bFGF at 7 days in normal and osteogenic medium was illustrated (A and B). The ALP enzymatic activity in osteogenic medium was shown at 7 and 14 days (C and D). Mineral deposition was determined at 14 days after maintaining in osteogenic medium (E and F). The time course experiments illustrated that bFGF attenuated *ALP* mRNA expression at 1, 3 and 5 days after cultured under osteogenic induction condition (G and H). The influence of bFGF was confirmed using FGFR inhibitor (SU5402). The *ALP* mRNA expression at 3 days was evaluated in osteogenic condition (OM), osteogenic condition supplemented with bFGF (bFGF) and osteogenic condition supplemented with bFGF and SU5402 (bFGF + SU5402) (I and J). The asterisk indicated the statistical significance compared to the control. Reprinted from Journal of Cellular Biochemistry, 114(11), Thanaphum Osathanon, Nunthawan Nowwarote, Jeeranan Manokawinchoke, Prasit Pavasant, bFGF and JAGGED1 regulated alkaline phosphatase expression and mineralization in dental tissue-derived mesenchymal stem cells, 2551–61, Copyright (2013), with permission from John Wiley & Sons Inc. [5].

bFGF enhances the expression of neurogenic marker, βIII-tubulin, via FGFR and PLCγ when human DPSCs are cultured in a neurogenic induction medium supplemented with bFGF [40].

In SHEDs, long-term culture *in vitro* leads to the decrease of stemness as determined by the reduction of pluripotent stem cell markers, i.e. *OCT4*, *NANOG* and *REX1* [41]. In addition, the reduction of colony-forming unit ability is observed in high passage (passage 10) of cells [41]. Interestingly, bFGF enhanced *OCT4*, *NANOG* and *REX-1* mRNA levels in both short- and long-term maintaining *in vitro* [41]. bFGF also increased colony-forming unit in passage 10 [41]. Similarly, an attenuation of endogenous bFGF expression or blocking FGFR results in the reduction of colony-forming number by SHEDs [6]. Further, bFGF promotes colony-forming unit ability in SHEDs isolated from inflamed dental pulp tissues [42]. For the regulatory mechanism, it has been demonstrated that bFGF regulated REX-1 expression in SHEDs via FGFR and Akt pathway [34]. IL-6 is also shown to involve in bFGF induced REX-1 expression as pre-treatment with antibody against IL-6 attenuates REX-1 expression [34].

Regarding osteogenic differentiation, bFGF attenuated osteogenic differentiation. In this regard, bFGF attenuated alkaline phosphatase enzymatic activity and mineralization in SHEDs after osteogenic induction [5, 43]. The inhibition of endogenous bFGF in SHEDs either by a chemical inhibitor for FGFR or lentiviral shRNA against bFGF resulted in the enhancement of osteogenic differentiation [6]. It was also demonstrated that bFGF attenuated alkaline phosphatase mRNA expression and mineral deposition via FGFR and MEK signalling pathway [5].

Several possible mechanisms were reported. Firstly, bFGF might attenuate osteogenic differentiation in SHEDs via decreasing Notch signalling [5]. Notch signalling activation led to the enhancement of mineralization in SHEDs [7]. Treatment with bFGF attenuated Notch receptor, ligand and target gene expression which may participate in bFGF attenuated osteogenic differentiation in SHEDs [5]. Secondly, bFGF inhibited matrix metalloproteinase (MMP) expression, for example *MMP-2*, *MMP-13* and *MT1-MMP* [5]. It has been demonstrated that MMP2 influenced odontogenic differentiation by DPSCs [44]. In this regard, MMP2 cleaved dentin matrix protein 1 (DMP1), resulting in the release of bioactive peptide that could promote odontogenic differentiation of DPSCs [44]. Finally, it has been shown that bFGF might inhibit canonical Wnt signalling pathway via the activation of ERK1/2 signalling [43]. ERK attenuation rescued bFGF inhibiting osteogenic differentiation by SHEDs both *in vitro* and *in vivo* [43]. In addition, ERK inhibitor increase β-catenin levels in bFGF-treated SHEDs [43]. Besides odonto/osteogenic differentiation, it has been shown that bFGF participated in angiogenesis induction properties of SHEDs. Priming with bFGF promoted SHEDs-induced angiogenesis *in vivo* [45]. This could be due to the increase production of VEGF and HGF by bFGF-treated SHEDs [45]. Together, these data denote the crucial influence of bFGF in the regulation of SHEDs stemness and differentiation mechanisms.

4. Wnt signalling in SHEDs

Canonical Wnt signalling also has a significant role in tooth development and stem cells self-renewal through β-catenin [46, 47]. Inactivation of β-catenin in the mesenchyme of developing tooth results in arrested tooth developmental at the bud stage [48]. Various studies established the influence of canonical Wnt signalling pathway to promote the osteogenic differentiation of dental stem cells, i.e. DPSCs, PDLSCs, stem cells from apical papilla (SCAPs) and dental follicle stem cells (DFSCs) [49–52]. However, the effect of the canonical Wnt/ β-catenin on SHEDs is very limited. The involvement of Wnt/β-catenin on SHEDs-mediated mineralized tissue regeneration was investigated with the addition of basic fibroblast growth factor (bFGF) [43]. Treatment with bFGF attenuated SHEDs-mediated mineralized tissue regeneration via activation of ERK 1/2 pathway and consequently inhibited Wnt/β-catenin pathway, leading to osteogenic deficiency of SHEDs [43].

A recent *in vitro* and *in vivo* study reported that an activation of the canonical Wnt signalling pathway induced by Wnt3A can promote osteogenic differentiation of DPSCs [52]. Similar to previous study that activated Wnt signalling by using various concentrations of lithium chloride (LiCl), the result showed that Wnt/β-catenin strongly upregulated expression of dentin sialophosphoprotein (DSPP), OCN and ALP in time- and dose-dependent manner [50, 51]. LiCl also upregulated protein expression of osteogenic transcription factors, including RUNX2, MSX2 and OSX. Whereas cells treated with Dickkopf-1 (DDK1), an inhibitor for canonical Wnt signalling, resulted in the inhibition of osteogenic mRNA expression and reduction the alkaline phosphatase enzymatic activity and matrix mineralization [50]. On the other hand, the contradictory evidence demonstrates that the canonical Wnt signalling can inhibit osteogenic differentiation, alkaline phosphatase enzymatic activity and formation of mineralized nodules in DPSCs [53]. Canonical Wnt signalling inhibited the odontoblast-like differentiation of DPSCs was first reported by Scheller et al. in 2008 [53]. This study showed that Wnt-1 inhibited alkaline phosphatase enzymatic activity and the formation of mineralized nodules in DPSCs when transduced with canonical Wnt-1 or the active form of β-catenin, with retrovirus-mediated infection. Moreover, overexpression of β-catenin was also sufficient to suppress the differentiation and mineralization of DPSCs [53]. Another study was established using Wnt3A and LiCl to examine the possible involvement of canonical Wnt signalling in regulating cementoblast behaviours. Activation of endogenous canonical Wnt signalling with both Wnt3A and LiCl suppressed alkaline phosphatase enzymatic activity and expression of genes associated with cementum function: *Alp*, *Bsp* and *Ocn*. This effect was accompanied by decreased gene expression of *Runx2* and *Osx* and by increased gene expression of *Lef-1*. In conclusion, these observations suggest that Wnt signalling inhibits cementoblast differentiation and promotes cell proliferation [54].

Activation of β-catenin by LiCl in SHEDs led to the significant decrease of colony formation by SHEDs [55]. In addition, LiCl enhanced subG0 population in SHEDs [55]. *OSX* and *DMP1* mRNA expression was markedly decreased in LiCl-treated SHEDs. These results imply the influence of Wnt signalling in SHED behaviours [55]. The canonical Wnt/β-catenin pathway also implicates in angiogenic differentiation of SHEDs. Transplantation of SHEDs in human tooth slice/scaffolds into immunodeficient mice differentiates into new blood vessels that anastomose with the host vasculature. *In vitro* data showed that VEGF induced the vasculogenic differentiation of SHEDs via potent activation of Wnt/β-catenin signalling while Wnt inhibition blocked this process [31]. Moreover, the study has been shown that the Wnt/β-catenin pathway also participates in immunomodulatory properties of SHEDs [36]. Acetylsalicylic acid treatment enhances immunomodulatory properties of SHEDs as indicated by increased in SHED-mediated T-cell apoptosis and the decreased levels of T helper 17. Moreover, acetylsalicylic acid significantly improves SHED-based bone formation and these effects of acetylsalicylic acid treatment on SHEDs occurred via the regulation of the telomerase reverse transcriptase/Wnt and TERT/FASL pathways [36].

5. Notch signalling in SHEDs

Notch signalling controls various function of stem cells, ranging from stemness maintenance to cell-specific differentiation [56]. It is a highly conserved pathway, firstly identified in Drosophila. Notch signalling is initiated by the binding between membrane-bound Notch receptors and ligands of neighbouring cells [56–58]. Further, Notch receptors are cleaved by a γ-secretase enzyme, leading to the release of Notch intracellular domain (NICD) [56–58]. Subsequently, NICD translocates into nucleus and forms complex with other transcriptional molecules, resulting in the activation of Notch target genes [56–58]. Common Notch signalling target genes are Hes and Hey families [56–58]. In the canonical Notch signalling pathway, four receptors and five ligands are identified [56–58]. The four types of Notch receptors are Notch1, Notch2, Notch3 and Notch4. Five ligands are Delta-like-1 (Dll-1), Delta-like-3 (Dll-3), Delta-like-4 (Dll-4), Jagged1 and Jagged2 [56–58].

Notch signalling participates in odontogenesis, dental pulp repair and regeneration. Mice lacking of Jagged2 expression exhibited defective enamel formation of incisors and malformation of molars [59]. The expression of Notch receptors and ligands was upregulated in response to calcium hydroxide, a material for direct pulp capping treatment [60]. Human DPSCs over-expressing Jagged1 exhibited the reduction of osteogenic differentiation ability and mineralization *in vitro* and *in vivo* [61]. Dll-1 over-expressing human DPSCs had higher proliferative rate than the control and knock down Dll-1 expression in human DPSCs led to significantly enhancement of osteo/odontogenic differentiation [62, 63]. These evidence support the role of Notch signalling in the regulation of human DPSCs' behaviours and dental pulp tissue homeostasis of permanent teeth. Though, knowledge of Notch signalling in SHEDs is yet limited.

Studies illustrated that indirectly immobilized Notch ligands, Jagged1 or Dll-1, on tissue culture surface increased *HES1* and *HEY1* mRNA levels in SHEDs, implying the successful activation of intracellular Notch signalling [7]. The activation of Notch signalling in SHEDs led to the enhancement of osteogenic differentiation [5, 7]. However, Jagged1 exhibited higher potency to promote alkaline phosphatase enzymatic activity and mineralization than Dll-1 (**Figure 3**) [7]. Corresponding to study in primary human dental pulp cells isolated from deciduous teeth, Jagged1, but not Dll-1, attenuated cell proliferation [64]. The influence of Jagged1 on alkaline phosphatase enzymatic activity and mineralization *in vitro* could be attenuated by pretreatment with a γ-secretase inhibitor, DAPT, confirming the involvement of Notch signalling pathway [7]. Jagged1 significantly enhanced osteogenic marker gene expression, namely *ALP* and *COL1* [7]. In addition, Jagged1 downregulated a negative regulator of osteogenic differentiation, *TWIST2,* in SHEDs [7].

It has been shown that bFGF inhibited the mRNA expression of Notch signalling components. In this regard, bFGF significantly reduced the mRNA levels of *NOTCH1, NOTCH2, JAGGED1, DLL1* and *HES1* in SHEDs cultured in osteogenic induction medium [5]. In addition, bFGF was able to attenuate Jagged1-induced alkaline phosphatase mRNA expression and mineralization when SHEDs were maintained in osteogenic medium for 7 and 14 days, respectively [5]. bFGF significantly reduced alkaline phosphatase mRNA expression as early as 1 day in culture, corresponding to the significant reduction of *HES1* [5]. Taken all evidence together, bFGF and Notch signalling possibly interact and regulate mineralization process in SHEDs.

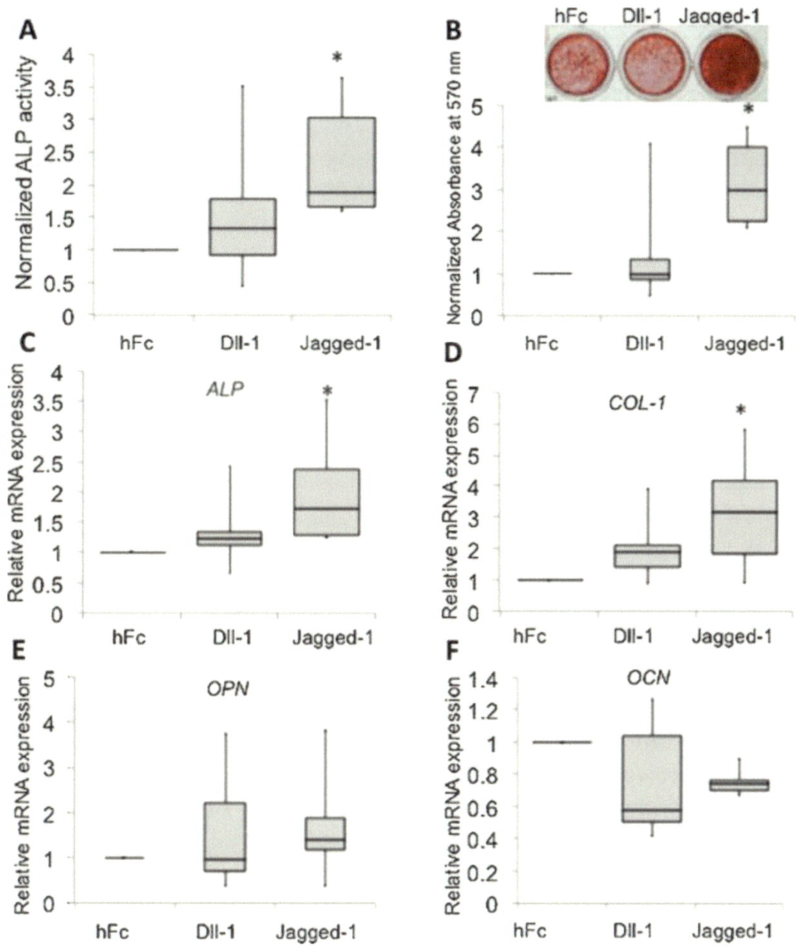

Figure 3. Effects of Dll-1 and Jagged1 on osteogenic differentiation. The alkaline phosphatase enzymatic activity (A) and mineralization (B) were evaluated at day 7 and 14 after osteogenic induction, respectively. For osteoblast marker gene expression, cells were cultured on Dll-1, Jagged1 or hFc treated surface for 7 days after osteogenic differentiation. The graphs demonstrated the relative mRNA expression of *ALP, COL1, OPN* and *OCN* upon seeding cells on Dll-1 or Jagged1 immobilized surface and normalized to the hFc control (D–F). Asterisks indicated statistical significance compared to the hFc control. Reprinted from Archives of Oral Biology, 65, Waleerat Sukarawan, Kannapas Peetiakarawach, Prasit Pavasant, Thanaphum Osathanon, Effect of Jagged1 and Dll-1 on osteogenic differentiation by stem cells from human exfoliated deciduous teeth, 1–8, Copyright (2016), with permission from Elsevier [7].

6. Mechanical stress influences SHEDs' behaviours

Dental pulp tissues are surrounded by hard tissues, namely dentin. During inflammation, an interstitial fluid pressure increases [65, 66], causing biological changes in local cells and tissues. In addition, fluid movement in dentin-pulp complex during normal occlusal force may expose cells to mechanical stimuli [67]. Mechanical forces are shown to regulate biological functions in many cell types, for example osteoblasts, osteocytes, periodontal ligament cells and dental pulp cells. Different types and magnitude of force lead to different cell responses.

In human DPSCs, uniaxial cycle stretching inhibited odonto/osteogenic differentiation but increased cell proliferation [68, 69], while cyclic hydrostatic pressure synergistically enhanced BMP-2-induced DSPP expression by human DPSCs *in vitro* and increased hard tissue formation *in vivo* [70]. Studies in SHEDs demonstrate that mechanical force may regulate stemness maintenance. In this respect, static compressive force upregulated pluripotent marker mRNA expression in SHEDs [8]. *REX-1, SOX2, OCT4* and *NANOG* mRNA levels increased in a magnitude-dependent manner [8]. Mechanical stress-induced REX-1 expression is partly controlled via IL-6/JAK and ATP-P2Y1 signalling pathways [8, 71]. Though, the clinical significance of these phenomena requires further investigation.

7. Potential application of SHEDs in regenerative therapy

SHEDs are the good candidate for the stem cells used in regenerative therapy due to their high plasticity as well as ability to cross lineage boundaries and differentiate into several specialized cells. Current progresses have been made for tissue engineering-based therapies involving a large number of tissues. However, dentin-pulp complex and neuronal tissue seem to be the most promising aspects for the application of SHEDs in regenerative therapy.

The first evidence to show that SHEDs can differentiate to become the functional odontoblasts with the ability to generate the mineralized tissue resemble to dentin was shown in mice [3]. SHEDs were seeded within a scaffold in a tooth slice and implanted into the dorsum of mice. Dental pulp-like tissue was observed in the central area of the pulp chamber of the tooth slice [3]. The expression of odontoblastic differentiation markers such as DSPP and DMP-1 was detected [3]. Remarkably, the newly deposited dentin was observed and suggested that SHEDs can differentiate into fully functional odontoblasts *in vivo* [3]. Later in 2013, the regeneration of the dental pulp within the full length of the root canal was reported [19]. SHEDs were transplanted into the root canals with the scaffold and were observed for 28 days *in vitro*. The transplanted SHEDs were able to proliferate and inside the root canal [19]. The expression of odontoblastic differentiation marker such as DSPP, DMP-1 and MEPE was observed [19]. Interestingly, when the roots with SHEDs were implanted in the subcutaneous space of mice, a dental pulp-like tissue was formed in the majority of space in the root canal [19]. This *de novo* dental pulp-like tissue was capable of depositing new dentin [19]. However, this model is still considered as the ectopic transplantation model. Another concern for clinical translation is that most of the results interpretation was made from histological evaluation with the lack of functional testing. Therefore, the regeneration of dentin-pulp complex by SHEDs still needs further study in the more related oral environment and the additional functional of nerve innervation or vascularization should be performed before clinical application. Current possible experimental approaches for dentin pulp complex regeneration are summarized in **Figure 4**.

In addition to dentin-pulp complex regeneration, SHEDs also show the potential to be used in neuroregeneration. Stem cell therapy is the promising therapeutic options for treating the neurodegenerative diseases due to the limited regenerative capacity of the specialized cells in the nervous system. The neural crest cell in origin makes SHEDs the candidate cell model for neuron tissue regeneration. These cells are prone to undergo neurogenic differentiation

http://dx.doi.org/10.5772/intechopen.68173

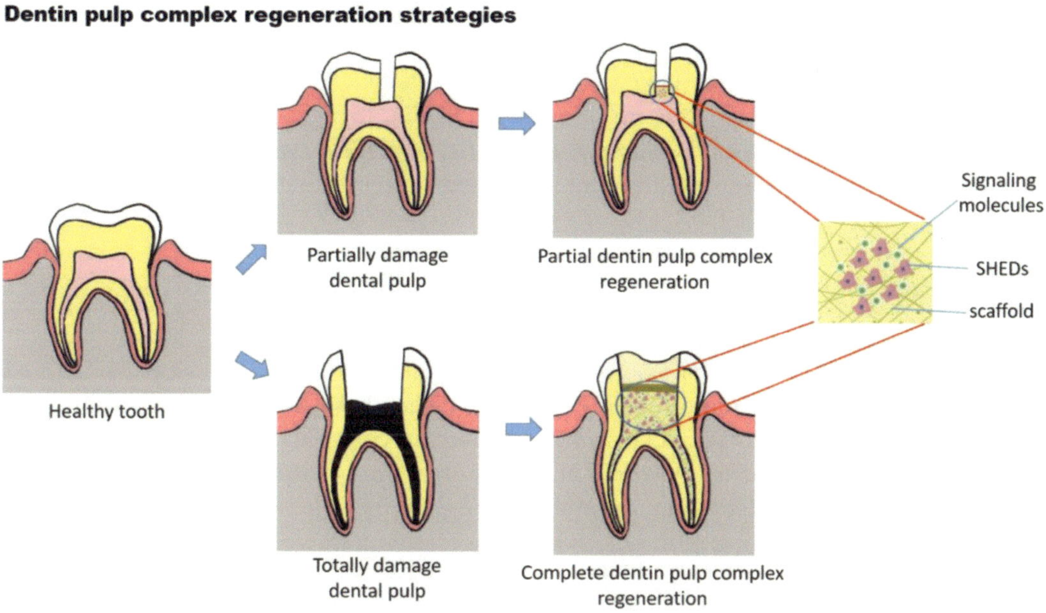

Figure 4. Current possible experimental approaches for dentin pulp complex regeneration using SHEDs.

both *in vitro* [2, 24–26] and *in vivo* [25, 27, 72]. Promising results from several *in vivo* studies lay the spotlight on SHEDs for their use as a stem cell source for treating neurodegenerative disease and other neuron-related conditions such as Parkinson disease, Alzheimer's disease, focal cerebral ischemia and spinal cord injuries [27, 29, 72–74]. Transplantation of neurogenic induced SHEDs into the parkinsonian rat model significantly improved the recovery behavioural impairment compared to transplantation of control SHEDs [25, 27].

In a focal cerebral ischemia rat model induced by permanent middle cerebral artery occlusion, intranasal administration of supernatants from the medium used to culture SHEDs significant decreased in the motor disability score and significantly reduced in the infarct volume [72]. Moreover, positive signals for neuronal nucleus, neurofilament H, doublecortin and rat endothelial cell antigen in the peri-infarct area were observed in the rats treated with SHEDs conditioned media compared to the DMEM control from approximately 140 mm^3 in DMEM control to 50 mm^3 in SHEDs conditioned medium [72]. These results suggest that SHEDs might secrete some compounds that positively influence the recovery of the brain lesion in focal cerebral ischemia [72].

Studies have shown that SHEDs have remarkable neuroregenerative activity and promote functional recovery in a spinal cord injury animal model [29, 75]. Rats that received SHEDs transplantation within the lesion created at the 9th–11th thoracic vertebral levels exhibited higher scores in the locomotor rating scale compared to the bone marrow stromal cells or fibroblasts transplantation control [75]. In addition, the rescue of hindlimb locomotor function was prominent in the rats that received SHEDs. These animals were able to move hindlimb coordinately and walk, while the bone marrow stromal cells transplantation exhibited only subtle movements

[75]. A similar trend was observed in another study, a complete recovery of hindlimb motor function was observed after implantation of neural-induced SHEDs in a rat spinal cord injury [29] which suggested that preinduction of the undifferentiated SHEDs into the neural-like cells before implantation might improve the efficiency of SHEDs in regenerating specialized neural cells. Taken together, these high neurogenic potential of SHEDs especially in animal models makes them the favourable source for stem cell regeneration treatment for neural diseases.

8. Conclusion

Dental stem cells, including SHEDs, have been extensively studied in the past decades leading to the better understanding in their unique biological properties and therapeutic potential. As SHEDs can be easily obtained with limited ethical concern, their multi-differentiation potentials have been demonstrated, which creates great opportunities for the application in the regenerative therapy. However, despite the intriguing results, we still need further study to deepen the understanding of the mechanisms underlying the differentiation processes to attain clinical reality. Also, the potential risks for the clinically use of SHEDs or other dental stem cells should be thoroughly studied for the safety of the patients who will greatly benefit from their regenerative ability.

Acknowledgements

The authors thank for support of the Faculty of Dentistry Research Fund, Chulalongkorn University. We would like to thank Dr. Pattarin Potisomporn for the illustration in **Figure 4**.

Abbreviations

Akt	Protein kinase B
ALP	Alkaline phosphatase
ApoE	apolipoprotein E
ATP	Adenosine triphosphate
bFGF	Basic fibroblast growth factor
BMMSCs	Bone marrow-derived mesenchymal stem cells
BSP	Bone sialoprotein
CD	Cluster of differentiation
COL1	Collagen type 1
DAPT	N-[N-(3,5-Difluorophenacetyl)-L-alanyl]-S-phenylglycine t-butyl ester
DDK	Dickkopf
DFSCs	Dental follicle stem cells

Dll	Delta-like
DMEM	Dulbecco's Modified Eagle Medium
DMP-1	Dentin matrix acidic phosphoprotein 1
DMP	Dentin matrix protein
DPSCs	Dental pulp stem cells
DSP	Dentin sialoprotein
DSPP	Dentin phosphoprotein
ERK	Extracellular signal-regulated kinase
FASL	Fas ligand
FGFR	Fibroblast growth factor receptor
GFAP	Glial fibrillary acidic protein
HA	Hydroxyapatite
Hes	Hairy and enhancer of split
Hey	Hairy and enhancer of split related with YRPW motif protein
HGF	Hepatocyte growth factor
HLA-DR	Human leukocyte antigen-antigen D related
IFN	Interferron
IL	Interleukin
JAK	Janus kinase
LEF-1	Lymphoid enhancer binding factor 1
LiCl	Lithium chloride
LPL	Lipoprotein
MEK	Mitogen-activated protein kinase kinase
MEPE	Matrix extracellular phosphoglycoprotein
MMP	Matrix metalloproteinase
MSCs	Mesenchymal stem cells
MSX2	Msh homeobox 2
MT1-MMP	Membrane type1- matrix metalloproteinase
NICD	Notch intracellular domain
Nurr1	Nuclear receptor related 1 protein
OCN	Osteocalcin
OCT4	Octamer-binding transcription factor 4
OPN	Osteopontin
OSX	Osterix
P2Y1	Purinergic receptor P2Y1

PCR	Polymerase chain reaction
PDLSCs	Periodontal ligament stem cells
PECAM-1	Platelet endothelial cell adhesion molecule 1
Pitx3	Paired like homeodomain 3
PLCγ	Phospholipase C gamma
PPARγ2	Peroxisome proliferator-activated receptor-gamma 2
PRP	Platelet-rich plasma
REX1	Reduced Expression Protein 1
RUNX2	Runt-related transcription factor 2
SCAPs	Stem cells from apical papilla
SHEDs	Stem cells isolated from human exfoliated deciduous teeth
shRNA	Short hairpin ribonucleic acid
SOX2	Sex determining region Y-box 2
SSEA4	Stage-specific embryonic antigen-4
TCP	Tricalcium phosphate
TERT	Telomerase reverse transcriptase
TWIST	Twist Family BHLH Transcription Factor
VE-Cadherin	Vascular endothelial cadherin
VEGF	Vascular endothelial growth factor
VEGFR	Vascular endothelial growth factor receptor

Author details

Waleerat Sukarawan[1] and Thanaphum Osathanon[2*]

*Address all correspondence to: thanaphum.o@chula.ac.th

1 Department of Pediatric Dentistry, Faculty of Dentistry, Chulalongkorn University, Bangkok, Thailand

2 Department of Anatomy, Faculty of Dentistry, Chulalongkorn University, Bangkok, Thailand

References

[1] Chai Y, Jiang X, Ito Y, Bringas P Jr., Han J, Rowitch DH, Soriano P, McMahon AP, Sucov HM. Fate of the mammalian cranial neural crest during tooth and mandibular morphogenesis. Development. 2000;**127**:1671-9.

[2] Miura M, Gronthos S, Zhao M, Lu B, Fisher LW, Robey PG, Shi S. SHED: stem cells from human exfoliated deciduous teeth. Proc Natl Acad Sci U S A. 2003;**100**:5807-12. doi: 10.1073/pnas.0937635100.

[3] Sakai VT, Zhang Z, Dong Z, Neiva KG, Machado MA, Shi S, Santos CF, Nor JE. SHED differentiate into functional odontoblasts and endothelium. J Dent Res. 2010;**89**:791-6. doi: 10.1177/0022034510368647.

[4] Chadipiralla K, Yochim JM, Bahuleyan B, Huang CY, Garcia-Godoy F, Murray PE, Stelnicki EJ. Osteogenic differentiation of stem cells derived from human periodontal ligaments and pulp of human exfoliated deciduous teeth. Cell Tissue Res. 2010;**340**:323-33. doi: 10.1007/s00441-010-0953-0.

[5] Osathanon T, Nowwarote N, Manokawinchoke J, Pavasant P. bFGF and JAGGED1 regulate alkaline phosphatase expression and mineralization in dental tissue-derived mesenchymal stem cells. J Cell Biochem. 2013;**114**:2551-61. doi: 10.1002/jcb.24602.

[6] Nowwarote N, Pavasant P, Osathanon T. Role of endogenous basic fibroblast growth factor in stem cells isolated from human exfoliated deciduous teeth. Arch Oral Biol. 2015;**60**:408-15. doi: 10.1016/j.archoralbio.2014.11.017.

[7] Sukarawan W, Peetiakarawach K, Pavasant P, Osathanon T. Effect of Jagged-1 and Dll-1 on osteogenic differentiation by stem cells from human exfoliated deciduous teeth. Arch Oral Biol. 2016;**65**:1-8. doi: 10.1016/j.archoralbio.2016.01.010.

[8] Govitvattana N, Osathanon T, Taebunpakul S, Pavasant P. IL-6 regulated stress-induced Rex-1 expression in stem cells from human exfoliated deciduous teeth. Oral Dis. 2013;**19**:673-82. doi: 10.1111/odi.12052.

[9] Nakamura S, Yamada Y, Katagiri W, Sugito T, Ito K, Ueda M. Stem cell proliferation pathways comparison between human exfoliated deciduous teeth and dental pulp stem cells by gene expression profile from promising dental pulp. J Endod. 2009;**35**:1536-42. doi: 10.1016/j.joen.2009.07.024.

[10] Yamaza T, Kentaro A, Chen C, Liu Y, Shi Y, Gronthos S, Wang S, Shi S. Immunomodulatory properties of stem cells from human exfoliated deciduous teeth. Stem Cell Res Ther. 2010;**1**:5. doi: 10.1186/scrt5.

[11] Dominici M, Le Blanc K, Mueller I, Slaper-Cortenbach I, Marini F, Krause D, Deans R, Keating A, Prockop D, Horwitz E. Minimal criteria for defining multipotent mesenchymal stromal cells. The international society for cellular therapy position statement. Cytotherapy. 2006;**8**:315-7. doi: 10.1080/14653240600855905.

[12] Kerkis I, Caplan AI. Stem cells in dental pulp of deciduous teeth. Tissue Eng Part B Rev. 2012;**18**:129-38. doi: 10.1089/ten.TEB.2011.0327.

[13] Jeon M, Song JS, Choi BJ, Choi HJ, Shin DM, Jung HS, Kim SO. *In vitro* and *in vivo* characteristics of stem cells from human exfoliated deciduous teeth obtained by enzymatic disaggregation and outgrowth. Arch Oral Biol. 2014;**59**:1013-23. doi: 10.1016/j.archoralbio.2014.06.002.

[14] Bakopoulou A, Leyhausen G, Volk J, Tsiftsoglou A, Garefis P, Koidis P, Geurtsen W. Assessment of the impact of two different isolation methods on the osteo/odontogenic differentiation potential of human dental stem cells derived from deciduous teeth. Calcif Tissue Int. 2011;**88**:130-41. doi: 10.1007/s00223-010-9438-0.

[15] Zheng Y, Liu Y, Zhang CM, Zhang HY, Li WH, Shi S, Le AD, Wang SL. Stem cells from deciduous tooth repair mandibular defect in swine. J Dent Res. 2009;**88**:249-54. doi: 10.1177/0022034509333804.

[16] Seo BM, Sonoyama W, Yamaza T, Coppe C, Kikuiri T, Akiyama K, Lee JS, Shi S. SHED repair critical-size calvarial defects in mice. Oral Dis. 2008;**14**:428-34.

[17] Yamada Y, Nakamura S, Ito K, Sugito T, Yoshimi R, Nagasaka T, Ueda M. A feasibility of useful cell-based therapy by bone regeneration with deciduous tooth stem cells, dental pulp stem cells, or bone-marrow-derived mesenchymal stem cells for clinical study using tissue engineering technology. Tissue Eng Part A. 2010;**16**:1891-900. doi: 10.1089/ten.TEA.2009.0732.

[18] Cordeiro MM, Dong Z, Kaneko T, Zhang Z, Miyazawa M, Shi S, Smith AJ, Nor JE. Dental pulp tissue engineering with stem cells from exfoliated deciduous teeth. J Endod. 2008;**34**:962-9. doi: 10.1016/j.joen.2008.04.009.

[19] Rosa V, Zhang Z, Grande RH, Nor JE. Dental pulp tissue engineering in full-length human root canals. J Dent Res. 2013;**92**:970-5. doi: 10.1177/0022034513505772.

[20] Kim S, Shin SJ, Song Y, Kim E. *In vivo* experiments with dental pulp stem cells for pulp-dentin complex regeneration. Mediators Inflamm. 2015;**2015**:409347. doi: 10.1155/2015/409347.

[21] Shi S, Bartold PM, Miura M, Seo BM, Robey PG, Gronthos S. The efficacy of mesenchymal stem cells to regenerate and repair dental structures. Orthod Craniofac Res. 2005;**8**:191-9. doi: 10.1111/j.1601-6343.2005.00331.x.

[22] Huang GT, Yamaza T, Shea LD, Djouad F, Kuhn NZ, Tuan RS, Shi S. Stem/progenitor cell-mediated de novo regeneration of dental pulp with newly deposited continuous layer of dentin in an *in vivo* model. Tissue Eng Part A. 2010;**16**:605-15. doi: 10.1089/ten.TEA.2009.0518.

[23] Gronthos S, Brahim J, Li W, Fisher LW, Cherman N, Boyde A, DenBesten P, Robey PG, Shi S. Stem cell properties of human dental pulp stem cells. J Dent Res. 2002;**81**:531-5.

[24] Wang J, Wang X, Sun Z, Wang X, Yang H, Shi S, Wang S. Stem cells from human-exfoliated deciduous teeth can differentiate into dopaminergic neuron-like cells. Stem Cells Dev. 2010;**19**:1375-83. doi: 10.1089/scd.2009.0258.

[25] Heng BC, Lim LW, Wu W, Zhang C. An overview of protocols for the neural induction of dental and oral stem cells *in vitro*. Tissue Eng Part B Rev. 2016;**22**:220-50. doi: 10.1089/ten.TEB.2015.0488.

[26] Jarmalaviciute A, Tunaitis V, Strainiene E, Aldonyte R, Ramanavicius A, Venalis A, Magnusson KE, Pivoriunas A. A new experimental model for neuronal and glial differentiation using stem cells derived from human exfoliated deciduous teeth. J Mol Neurosci. 2013;**51**:307. doi: 10.1007/s12031-013-0046-0

[27] Fujii H, Matsubara K, Sakai K, Ito M, Ohno K, Ueda M, Yamamoto A. Dopaminergic differentiation of stem cells from human deciduous teeth and their therapeutic benefits for Parkinsonian rats. Brain Res. 2015;**1613**:59-72. doi: 10.1016/j.brainres.2015.04.001.

[28] Majumdar D, Kanafi M, Bhonde R, Gupta P, Datta I. Differential neuronal plasticity of dental pulp stem cells from exfoliated deciduous and permanent teeth towards dopaminergic neurons. J Cell Physiol. 2016;**231**:2048-63. doi: 10.1002/jcp.25314.

[29] Taghipour Z, Karbalaie K, Kiani A, Niapour A, Bahramian H, Nasr-Esfahani MH, Baharvand H. Transplantation of undifferentiated and induced human exfoliated deciduous teeth-derived stem cells promote functional recovery of rat spinal cord contusion injury model. Stem Cells Dev. 2012;**21**:1794-802. doi: 10.1089/scd.2011.0408.

[30] Bento LW, Zhang Z, Imai A, Nor F, Dong Z, Shi S, Araujo FB, Nor JE. Endothelial differentiation of SHED requires MEK1/ERK signaling. J Dent Res. 2013;**92**:51-7. doi: 10.1177/0022034512466263.

[31] Zhang Z, Nor F, Oh M, Cucco C, Shi S, Nor JE. Wnt/beta-catenin signaling determines the vasculogenic fate of postnatal mesenchymal stem cells. Stem Cells. 2016;**34**:1576-87. doi: 10.1002/stem.2334.

[32] Zhang N, Chen B, Wang W, Chen C, Kang J, Deng SQ, Zhang B, Liu S, Han F. Isolation, characterization and multi-lineage differentiation of stem cells from human exfoliated deciduous teeth. Mol Med Rep. 2016;**14**:95-102. doi: 10.3892/mmr.2016.5214.

[33] Kushnerev E, Shawcross SG, Hillarby MC, Yates JM. High-plasticity mesenchymal stem cells isolated from adult-retained primary teeth and autogenous adult tooth pulp—a potential source for regenerative therapies? Arch Oral Biol. 2016;**62**:43-8. doi: 10.1016/j.archoralbio.2015.11.009.

[34] Nowwarote N, Sukarawan W, Pavasant P, Osathanon T. Basic fibroblast growth factor regulates rex1 expression via il-6 in stem cells isolated from human exfoliated deciduous teeth. J Cell Biochem. 2016. doi: 10.1002/jcb.25807.

[35] Alipour R, Adib M, Masoumi Karimi M, Hashemi-Beni B, Sereshki N. Comparing the immunoregulatory effects of stem cells from human exfoliated deciduous teeth and bone marrow-derived mesenchymal stem cells. Iran J Allergy Asthma Immunol. 2013;**12**:331-44.

[36] Liu Y, Chen C, Liu S, Liu D, Xu X, Chen X, Shi S. Acetylsalicylic acid treatment improves differentiation and immunomodulation of SHED. J Dent Res. 2015;**94**:209-18. doi: 10.1177/0022034514557672.

[37] Silva Fde S, Ramos RN, de Almeida DC, Bassi EJ, Gonzales RP, Miyagi SP, Maranduba CP, Sant'Anna OA, Marques MM, Barbuto JA, Camara NO, da Costa Maranduba CM. Mesenchymal stem cells derived from human exfoliated deciduous teeth (SHEDs) induce immune modulatory profile in monocyte-derived dendritic cells. PLoS One. 2014;**9**:e98050. doi: 10.1371/journal.pone.0098050.

[38] Ornitz DM, Itoh N. Fibroblast growth factors. Genome Biol. 2001;**2**:Reviews 3005.

[39] Dvorak P, Hampl A. Basic fibroblast growth factor and its receptors in human embryonic stem cells. Folia Histochem Cytobiol. 2005;**43**:203-8.

[40] Osathanon T, Nowwarote N, Pavasant P. Basic fibroblast growth factor inhibits mineralization but induces neuronal differentiation by human dental pulp stem cells through a FGFR and PLCgamma signaling pathway. J Cell Biochem. 2011;**112**:1807-16. doi: 10.1002/jcb.23097.

[41] Sukarawan W, Nowwarote N, Kerdpon P, Pavasant P, Osathanon T. Effect of basic fibroblast growth factor on pluripotent marker expression and colony forming unit capacity of stem cells isolated from human exfoliated deciduous teeth. Odontology. 2014;**102**:160-6. doi: 10.1007/s10266-013-0124-3.

[42] Kim J, Park JC, Kim SH, Im GI, Kim BS, Lee JB, Choi EY, Song JS, Cho KS, Kim CS. Treatment of FGF-2 on stem cells from inflamed dental pulp tissue from human deciduous teeth. Oral Dis. 2014;**20**:191-204. doi: 10.1111/odi.12089.

[43] Li B, Qu C, Chen C, Liu Y, Akiyama K, Yang R, Chen F, Zhao Y, Shi S. Basic fibroblast growth factor inhibits osteogenic differentiation of stem cells from human exfoliated deciduous teeth through ERK signaling. Oral Dis. 2012;**18**:285-92. doi: 10.1111/j.1601-0825.2011.01878.x.

[44] Chaussain C, Eapen AS, Huet E, Floris C, Ravindran S, Hao J, Menashi S, George A. MMP2-cleavage of DMP1 generates a bioactive peptide promoting differentiation of dental pulp stem/progenitor cell. Eur Cell Mater. 2009;**18**:84-95.

[45] Gorin C, Rochefort GY, Bascetin R, Ying H, Lesieur J, Sadoine J, Beckouche N, Berndt S, Novais A, Lesage M, Hosten B, Vercellino L, Merlet P, Le-Denmat D, Marchiol C, Letourneur D, Nicoletti A, Vital SO, Poliard A, Salmon B, Muller L, Chaussain C, Germain S. Priming dental pulp stem cells with fibroblast growth factor-2 increases angiogenesis of implanted tissue-engineered constructs through hepatocyte growth factor and vascular endothelial growth factor secretion. Stem Cells Transl Med. 2016;**5**:392-404. doi: 10.5966/sctm.2015-0166.

[46] Jarvinen E, Salazar-Ciudad I, Birchmeier W, Taketo MM, Jernvall J, Thesleff I. Continuous tooth generation in mouse is induced by activated epithelial Wnt/beta-catenin signaling. Proc Natl Acad Sci U S A. 2006;**103**:18627-32. doi: 10.1073/pnas.0607289103.

[47] Liu F, Chu EY, Watt B, Zhang Y, Gallant NM, Andl T, Yang SH, Lu MM, Piccolo S, Schmidt-Ullrich R, Taketo MM, Morrisey EE, Atit R, Dlugosz AA, Millar SE. Wnt/beta-catenin signaling directs multiple stages of tooth morphogenesis. Dev Biol. 2008;**313**:210-24. doi: 10.1016/j.ydbio.2007.10.016.

[48] Liu F, Millar SE. Wnt/beta-catenin signaling in oral tissue development and disease. J Dent Res. 2010;**89**:318-30. doi: 10.1177/0022034510363373.

[49] Han N, Zheng Y, Li R, Li X, Zhou M, Niu Y, Zhang Q. beta-catenin enhances odontoblastic differentiation of dental pulp cells through activation of Runx2. PLoS One. 2014;**9**:e88890. doi: 10.1371/journal.pone.0088890.

[50] Heo JS, Lee SY, Lee JC. Wnt/beta-catenin signaling enhances osteoblastogenic differentiation from human periodontal ligament fibroblasts. Mol Cells. 2010;**30**:449-54. doi: 10.1007/s10059-010-0139-3.

[51] Wang J, Liu B, Gu S, Liang J. Effects of Wnt/beta-catenin signalling on proliferation and differentiation of apical papilla stem cells. Cell Prolif. 2012;**45**:121-31. doi: 10.1111/j.1365-2184.2012.00806.x.

[52] Yang Y, Ge Y, Chen G, Yan Z, Yu M, Feng L, Jiang Z, Guo W, Tian W. Hertwig's epithelial root sheath cells regulate osteogenic differentiation of dental follicle cells through the Wnt pathway. Bone. 2014;**63**:158-65. doi: 10.1016/j.bone.2014.03.006.

[53] Scheller EL, Chang J, Wang CY. Wnt/beta-catenin inhibits dental pulp stem cell differentiation. J Dent Res. 2008;**87**:126-30.

[54] Nemoto E, Koshikawa Y, Kanaya S, Tsuchiya M, Tamura M, Somerman MJ, Shimauchi H. Wnt signaling inhibits cementoblast differentiation and promotes proliferation. Bone. 2009;**44**:805-12. doi: 10.1016/j.bone.2008.12.029.

[55] Rattanawarawipa P, Pavasant P, Osathanon T, Sukarawan W. Effect of lithium chloride on cell proliferation and osteogenic differentiation in stem cells from human exfoliated deciduous teeth. Tissue Cell. 2016;**48**:425-31. doi: 10.1016/j.tice.2016.08.005.

[56] Osathanon T, Pavasant P, Giachelli CM. Notch signaling biomaterials and tissue regeneration. In: Zhang LG, Khademhosseini A, Webster TJ, editors. Tissue and organ regeneration: advances in micro- and nanotechnology. Pan Stanford Publishing; Singapore. 2014. p. 535-63. doi: 10.1201/b15595-18.

[57] Zanotti S, Canalis E. Notch and the skeleton. Mol Cell Biol. 2010;**30**:886-96. doi: 10.1128/MCB.01285-09.

[58] Tien AC, Rajan A, Bellen HJ. A Notch updated. J Cell Biol. 2009;**184**:621-9. doi: 10.1083/jcb.200811141.

[59] Mitsiadis TA, Graf D, Luder H, Gridley T, Bluteau G. BMPs and FGFs target Notch signalling via jagged 2 to regulate tooth morphogenesis and cytodifferentiation. Development. 2010;**137**:3025-35. doi: 10.1242/dev.049528.

[60] Lovschall H, Tummers M, Thesleff I, Fuchtbauer EM, Poulsen K. Activation of the Notch signaling pathway in response to pulp capping of rat molars. Eur J Oral Sci. 2005;**113**:312-7. doi: 10.1111/j.1600-0722.2005.00221.x.

[61] Zhang C, Chang J, Sonoyama W, Shi S, Wang CY. Inhibition of human dental pulp stem cell differentiation by Notch signaling. J Dent Res. 2008;**87**:250-5.

[62] Wang X, He F, Tan Y, Tian W, Qiu S. Inhibition of Delta1 promotes differentiation of odontoblasts and inhibits proliferation of human dental pulp stem cell *in vitro*. Arch Oral Biol. 2011;**56**:837-45. doi: 10.1016/j.archoralbio.2011.02.006.

[63] He F, Yang Z, Tan Y, Yu N, Wang X, Yao N, Zhao J. Effects of Notch ligand Delta1 on the proliferation and differentiation of human dental pulp stem cells *in vitro*. Arch Oral Biol. 2009;**54**:216-22. doi: 10.1016/j.archoralbio.2008.10.003.

[64] Peetiakarawach P, Osathanon, T., Pavasant, P., Sukarawan, W. Effect of Jagged-1 and Delta-like-1 on the proliferation of primary deciduous pulp cells. SWU Dent J. 2014; **7**(Suppl):58-64.

[65] Tonder KJ, Kvinnsland I. Micropuncture measurements of interstitial fluid pressure in normal and inflamed dental pulp in cats. J Endod. 1983;**9**:105-9. doi: 10.1016/S0099-2399(83)80106-X.

[66] Heyeraas KJ, Berggreen E. Interstitial fluid pressure in normal and inflamed pulp. Crit Rev Oral Biol Med. 1999;**10**:328-36.

[67] Paphangkorakit J, Osborn JW. The effect of normal occlusal forces on fluid movement through human dentine *in vitro*. Arch Oral Biol. 2000;**45**:1033-41.

[68] Hata M, Naruse K, Ozawa S, Kobayashi Y, Nakamura N, Kojima N, Omi M, Katanosaka Y, Nishikawa T, Naruse K, Tanaka Y, Matsubara T. Mechanical stretch increases the proliferation while inhibiting the osteogenic differentiation in dental pulp stem cells. Tissue Eng Part A. 2013;**19**:625-33. doi: 10.1089/ten.tea.2012.0099.

[69] Cai X, Zhang Y, Yang X, Grottkau BE, Lin Y. Uniaxial cyclic tensile stretch inhibits osteogenic and odontogenic differentiation of human dental pulp stem cells. J Tissue Eng Regen Med. 2011;**5**:347-53. doi: 10.1002/term.319.

[70] Yu V, Damek-Poprawa M, Nicoll SB, Akintoye SO. Dynamic hydrostatic pressure promotes differentiation of human dental pulp stem cells. Biochem Biophys Res Commun. 2009;**386**:661-5. doi: 10.1016/j.bbrc.2009.06.106.

[71] Govitvattana N, Osathanon T, Toemthong T, Pavasant P. IL-6 regulates stress-induced REX-1 expression via ATP-P2Y1 signalling in stem cells isolated from human exfoliated deciduous teeth. Arch Oral Biol. 2015;**60**:160-6. doi: 10.1016/j.archoralbio.2014.09.008.

[72] Inoue T, Sugiyama M, Hattori H, Wakita H, Wakabayashi T, Ueda M. Stem cells from human exfoliated deciduous tooth-derived conditioned medium enhance recovery of focal cerebral ischemia in rats. Tissue Eng Part A. 2013;**19**:24-9. doi: 10.1089/ten.TEA.2011.0385.

[73] Yamamoto A, Sakai K, Matsubara K, Kano F, Ueda M. Multifaceted neuro-regenerative activities of human dental pulp stem cells for functional recovery after spinal cord injury. Neurosci Res. 2014;**78**:16-20. doi: 10.1016/j.neures.2013.10.010.

[74] Mita T, Furukawa-Hibi Y, Takeuchi H, Hattori H, Yamada K, Hibi H, Ueda M, Yamamoto A. Conditioned medium from the stem cells of human dental pulp improves cognitive function in a mouse model of Alzheimer's disease. Behav Brain Res. 2015;**293**:189-97. doi: 10.1016/j.bbr.2015.07.043.

[75] Sakai K, Yamamoto A, Matsubara K, Nakamura S, Naruse M, Yamagata M, Sakamoto K, Tauchi R, Wakao N, Imagama S, Hibi H, Kadomatsu K, Ishiguro N, Ueda M. Human dental pulp-derived stem cells promote locomotor recovery after complete transection of the rat spinal cord by multiple neuro-regenerative mechanisms. J Clin Invest. 2012;**122**:80-90. doi: 10.1172/JCI59251.

Chapter 5

Olfactory Mucosa Mesenchymal Stem Cells and Biomaterials: A New Combination to Regenerative Therapies after Peripheral Nerve Injury

Rui Damásio Alvites, Ana Rita Caseiro Santos,
Artur Severo Proença Varejão and
Ana Colette Pereira de Castro Osório Maurício

Additional information is available at the end of the chapter

http://dx.doi.org/10.5772/intechopen.68174

Abstract

The peripheral nerve injury after trauma is a common occurrence in both human and veterinary medicine and has severe consequences for the survival and quality of life of the patients. Despite the continuous efforts and the creation of diverse medical and surgical techniques, the harmful effects of this type of injury are far from being overcome. Regenerative medicine has been growing in the scientific milieu as a new therapeutic approach for different situations. Among the cell-based therapies explored, the mesenchymal stem cells are evidenced by their features, versatility and potential applications. The olfactory mucosa mesenchymal stem cells, components of the olfactory system and identified in the *lamina propria,* were newly identified and are still undergoing characterization, appearing as a new promise in the regenerative therapy of several tissues but with special emphasis on the nervous system in general and the peripheral nervous system in particular, for which they appear to have special regenerative aptitude.

Keywords: stem cells, mesenchymal stem cells, olfactory mucosa mesenchymal stem cells, peripheral nerve injury, regenerative medicine, biomaterials

1. Introduction

Peripheral nerve injuries (PNI) lead to serious consequences in the life of the injured, impairing the performance of physiological functions and occupational activities [1]. The causes of

PNI are multiple and varied and may include traumatic events and iatrogenic interventions. In this second group, the peripheral nerve can be injured during manipulation of the nerve with different surgical instruments, due to poor nerve exposure during the procedure, during removal of tumours or lymph nodes, due to the inexperience of the surgeon and even during removal of osteosynthetic devices [2]. The main consequences of this type of injury are the loss of motor, sensory and autonomic function in the denervated body segments, resulting in substantial functional deficits [3].

When compared to the central nervous system (CNS), the peripheral nervous system (PNS) presents a higher reparative and regenerative activity. This contrast in the regenerative capacity depends on the intrinsic characteristics of the injured neurons of CNS and PNS and also on the physiology and functional environment of the two systems [4]. The ability of regeneration also depends on the age of the injured, the mechanism of injury and, particularly, survival and functional status of neural cell bodies [5]. Nevertheless, poor functional recovery is common due to chronic Schwann cell denervation, chronic neuronal axotomy and misdirection of regenerating axons into wrong endoneurial tubes. The effect of muscle denervation atrophy is secondary to the nerve injured and most of the times, implies fibrosis and neurogenic atrophy of the muscle [6].

Peripheral nerve damage not only removes a source of sensory input from the somatosensory system but also triggers a set of modifications in the neural circuits that lead to long-term changes in spinal somatosensory functions [7]. In fact, one of the worst consequences of PNI is the development of neuropathic pain characterized by allodynia and pain hypersensitivity in the partially denervated region [7, 8].

Peripheral nerve lesions can vary widely in severity and in most cases do not show complete recovery with the injured individual suffering from chronic lifelong disabilities. Satisfactory outcomes are usually limited to relatively minor injuries [9]. Even rapidly intervened patients are likely to undergo prolonged denervation in the distal segment of the injured nerve due to the slow rate of regeneration [10].

Achieving better outcomes depends on the advancements in microsurgical performances, introduction of new techniques into clinical practice and improvement of the therapeutics options already in use [11]. Tissue cell therapy and mesenchymal stem cells (MSCs) has been proposed as a promising alternative to treat a variety of neurologic injuries. The use of MSCs that can differentiate into appropriate cell types in the affected area or can secrete important growth factors that promote the regeneration process and positively modulate the local inflammatory response has developed rapidly in the last years [12]. Although MSCs' functional mechanisms are still poorly understood, nasal olfactory mucosa mesenchymal stem cells (OM-MSCs) stand as a promising competitor for therapeutic application due to its advantages [13].

This chapter will focus on the phenomena of PNI and its nuances, on the characteristics of OM-MSCs, their secretome and current applications. Finally, the potential use of these MSCs associated with biomaterials in cases of nerve damage, a tissue engineering technique that has not been applied until today, will be explored.

2. Peripheral nerve injury

2.1. Nerve functional anatomy

The peripheral nerve is composed by sensory and motor neurons whose long axons communicate with distant target organs [14]. The cell bodies of sensory neurons are located in the dorsal root ganglion while those of the motor neurons are found within the CNS, into the spinal cord or brainstem [15]. Its coating is complex and consists of three distinct layers (**Figure 1**). The axons are directly involved by a connective tissue sheath named *endoneurium* whose mechanical load is reduced. A fine network of capillaries exists in association with the *endoneurium*. Groups of axons involved by *endoneurium,* which together form the nervous fascicles, are covered by *perineurium,* a thin but dense epithelial layer. The perineurium offers strength in tension, and also maintains the blood-nerve barrier and endoneurial homeostasis [16]. Groups of fascicles are contained within the peripheral nerve surrounded by a connective tissue layer called *epineurium* that comprises 50% of the total cross-sectional area of the peripheral nerve. The inner epineurial layer separates fascicles, contains the vessels supplying and coursing through the nerve and a small amount of adipose tissue. The external layer surrounds all the fascicles, protects and defines the nerve anatomically [14, 17]. The *endoneurium* is longitudinally oriented while the *perineurium* and *epineurium* are circumferential (**Figure 1**) [18].

2.2. Nerve injury: pathophysiology

Mechanisms of PNI can be divided into three categories: mechanical (traumatic), vascular (ischemic) or chemical (neurotoxic) [19]. Mechanical processes can occur due to a sufficiently aggressive trauma, iatrogenic or not, due to perforating injuries with needles or due to administrations,

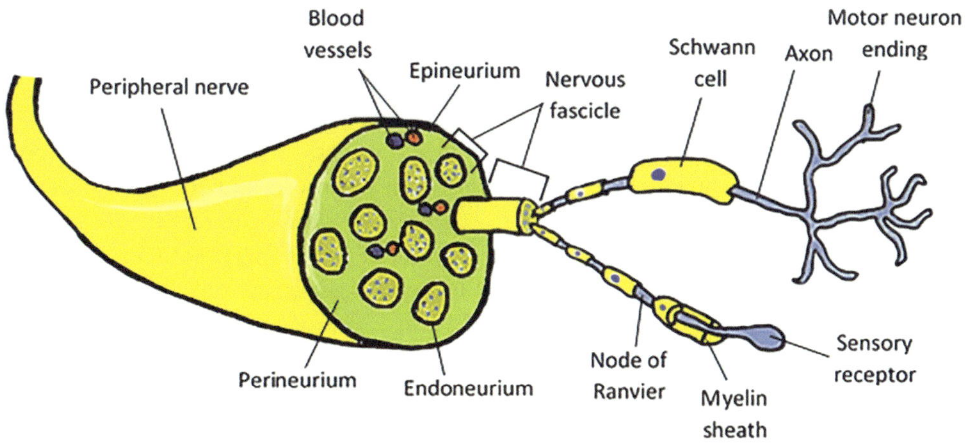

Figure 1. Schematic representation of the peripheral nerve structure and anatomical overview of the PNS. Axons, surrounded by myelinating Schwann cell sheaths, are enclosed by *endonerium*. Next, the *perineurium* binds individual axons together to form fascicles. Several axons are contained in each fascicle. Lastly, *epineurium* groups fascicles to one another, forming the nerve cable. *Endoneurium, perineurium* and *epineurium* present a tubular shape.

such as anaesthesia, on the periphery or into the nerve itself [20]. Nerve compression may trigger a blockage in nerve conductivity and, if prolonged, cause focal demyelination of the axons, ischemic phenomena, increased neuropeptide production [21] and increased spinal dorsal horn circuits activity that are involved in sensory information processing, including pain perception [22]. Stretch lesions are generally associated with intense exercise and fractures in the extremities where there is a close contact between the bone and nerves. Peripheral nerves exhibit intrinsic elasticity due to its collagen content at the *endoneurium*, but a strong enough force can lead to stretch injuries, resulting in a complete loss of continuity with nerve avulsion. Despite this, in some cases, continuity is maintained [21, 23]. Nervous lacerations or transections, caused by sharp objects, are common and represent about 30% of the identified cases. These lesions can result in complete transections or maintenance of structural continuity [24].

Vascular damage during nerve injury can lead to local or diffuse ischemia, occlusion of the arteries from which the *vasa nervorum* is derived or haemorrhage occurring within the nerve sheats. This vascular dysfunction and consequent hypoxia, contributes to the manifestation of neuropathic pain [25]. The epineural circulation is constituted by plexuses of microvessels running longitudinally in the *epineurium* that sends transverse branches through the *perineurium* to form a vascular network consisting primarily of capillaries in the *endoneurium*. It is of central importance and any alterations can reduce the nervous blood supply to residual levels. The connective tissue of the internal epineural layer makes the vessels less susceptible to compression since the forces are not directly transmitted to the epineural vessels. Sufficiently intense traumatic forces increase the permeability of epineural vessels and even larger forces or prolonged compressions can also injure endoneurial vessels, leading to intrafascicular oedema and secondary nerve damages [21]. Local anaesthetics and adjuvants also reduce blood flow, depending on both the agent used and its concentration [26].

Chemical lesions originate in the toxicity of solutions injected directly into the nerve or adjacent tissues, with development of acute inflammatory reactions and chronic fibrosis involving the nerve [27]. The site of administration (extraneural, intraneural, interfascicular or intrafascicular) determines the degree of toxicity and the same substance administered at different sites or portions of the nerve can cause different toxicity and lesions [28].

Before regeneration begins, a series of degenerative processes must occur, a direct prelude to the regenerative process. Regenerative success depends on the severity of the lesion and subsequent degenerative changes [21]. Any structural change or defect in the axon or its phospholipid bilayer leads to a programmed cascade of cell death that is interrupted only if there is rapid repair. Axonal degeneration follows a sequence of events that proliferate both proximally and distally to the site of injury. Axons disconnected from their cell bodies undergo degeneration through phenomena of chromatolysis [29]. Once the nerve is injured, its distal portion begins to degenerate due to the activity of proteases and the functional disruption of metabolic resources of the nervous cell body, in a calcium-mediated process known as Wallerian degeneration that involves invasion by myelomonocytic cells and results in the destruction of myelin and the onset of mitosis in Schwann cells (**Figure 2**). The degeneration of the distal axonal endings occurs due to autolytic mechanisms. The proximal end of the nerve swells but suffers minimal damage associated with retrograde degradation. The cytoskeleton starts to breakdown, followed by the dissolution of the cell membrane [30]. Once the cytoskeleton and membrane degrade, the Schwann cells that surround the distal portion of the axon shed their myelinated lipids. Axonal and myelinic debris are then removed by cells with phagocytic activity such as macrophages and Schwann cells which also release interleukins-6 to stimulate other Schwann cell and fibroblast proliferation (**Figure 2**) [31].

After removal of the debris, the regenerative process begins at the proximal end of the injured nerve and extends to the distal end. The new axonal buds (50–100) emanate from the most distal Ranvier nodes, the non-myelinated areas of the axon localized between the Schwann cells; these, in turn, guide the new cytoplasmic axonal extensions between basal membranes of the two nerve ends [32]. Proteases are also released from the growth cone to aid axonal regeneration through tissue. Several axonal extensions develop from the growth cone to contact the receptor at the distal end. The remaining neurites are abraded; otherwise, they continue to grow disorganized and may lead to neuromas that manifest clinically as painful nodules [14]. This process, however, is not free of complications. Uncontrolled branching or misdirecting of growing axons and dysfunctional innervation of target organs are common occurrences (**Figure 2**) [8]. Disruption of motor or peripheral targets secondary to PNI decreases the cortical representation of this zone in the ipsilateral cerebral hemisphere. Thus, adjacent regions in the ipsilateral hemisphere and regions of the contralateral hemisphere overgrow to compensate for deficits. The interpretation of the stimuli becomes unpredictable between regions associated with the lesion and healthy regions, which may lead to phenomena of neuropathic pain and phantom limbs [33]. In humans and rodents, axon regeneration occurs at a slow rate of 1–2 mm/day. Thus, significant injuries can take months to heal [34].

2.3. Nerve injury grading system

Success in regenerative processes after PNI depends directly on the severity of the lesion. Grading systems were developed in order to correlate the microscopic changes of the injured nerve with the clinical manifestations and prognosis. The first classification system of nerve injuries in three categories was proposed by Seddon in 1943 [35]. In this system, neuropraxia

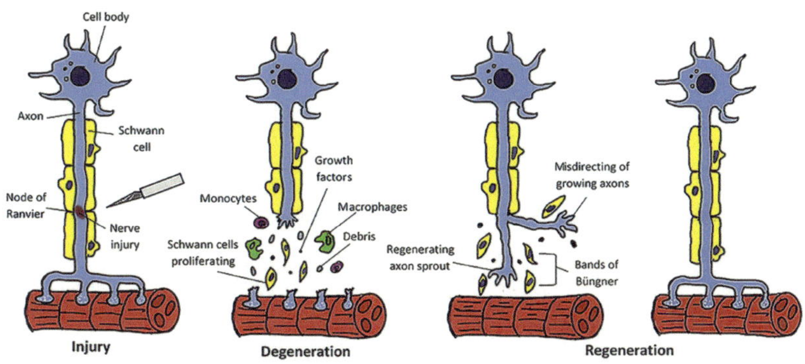

Figure 2. Schematic representation of the Wallerian degeneration.

is the least severe injury, without loss of nerve continuity: axons are anatomically intact, but nonfunctional. Since the affected nerve is unable to transmit impulses, the corresponding body regions become paralyzed. The lesion is followed by a temporary paralysis due to a local ion-induced conduction block and ischemia at the injury site, with a consequent recovery. Usually, no signs of Wallerian degeneration or regeneration are identified, but subtle alterations in myelin structure can be found and lead to motor and sensory loss due to segmental demyelination. Functional changes remain until re-myelination occurs. If decompression disappears, complete recovery is expected, without any intervention, within 3–6 months. Axonotmesis, the second level of injury, is characterized by a complete disruption of the axon and surrounding myelin while structures of supporting connective tissue, namely the *perineurium* and *epineurium*, remain intact. Axon and myelin degeneration occur distal to the point of injury by Wallerian degeneration, causing complete denervation (**Figure 2**). Despite this, once the integrity of the collagenous structures involving the nerve and that function as guides to growth of new axonal buds is maintained, the prognosis of recovery is excellent with a recovery rate of 1 inch per month [35, 36]. Neurotmesis results in a total disconnection between the two ends of the injured nerve. The functional loss is complete and recovery without surgical intervention or any other alternative is unlikely due to the intense scarring phenomena and loss of the collagen coatings and their guide function to axonal regrowth [35, 37].

In 1951, Sunderland proposed the existence of five categories in PNI according to its severity (**Figure 3**) [38, 39]. First- and second-degree injuries are equivalent to Seddon's neurapraxia and axonotmesis respectively. Third-degree lesions refer to a total disruption of the axon (axonotmesis) but are associated with partial lesions of the *endoneurium*. This category is placed between axonotmesis and neurotmesis in Seddon´s classification. The recovery prognosis depends on the extension of the endoneurial lesion and usually occurs over many months with conservative treatment or surgery to release the entrapment sites over the swollen nerve with or without limited neurolysis. Sunderland further divides neurotmesis into fourth or fifth-degree lesions. In fourth-degree lesions, all portions of the nerve undergo disruption with the exception of *epineurium* and internal haemorrhage and fibrous tissue imprisons the growing nerve sprouts due to fascicular discontinuity, inhibiting the axonal growth and originating neuromas-in-continuity. In fifth-degree lesions, the *epineurium* is also injured and the formation of end-bulb neuromas is observed. In both cases, recovery without surgical or similar intervention is impossible (**Figure 3**) [8, 36, 39].

Finally, Mackinnon and Dellon described a sixth-degree for mixed lesions. This degree is based on the evidence that a single trauma can affect different regions of the nerve transverse session variably, causing different degrees of injury at different points of the same nerve. Presumably, this is the most common type of injury, especially in perforating lesions, and is associated with bone fractures. Recovery and treatment vary according to the degree of injuries observed [17, 39].

2.4. Nerve repair and therapeutic options

Primary nerve repair through micro sutures is still the standard method in cases of axonotmesis and neurotmesis. The procedure should be performed immediately after the injury

http://dx.doi.org/10.5772/intechopen.68174

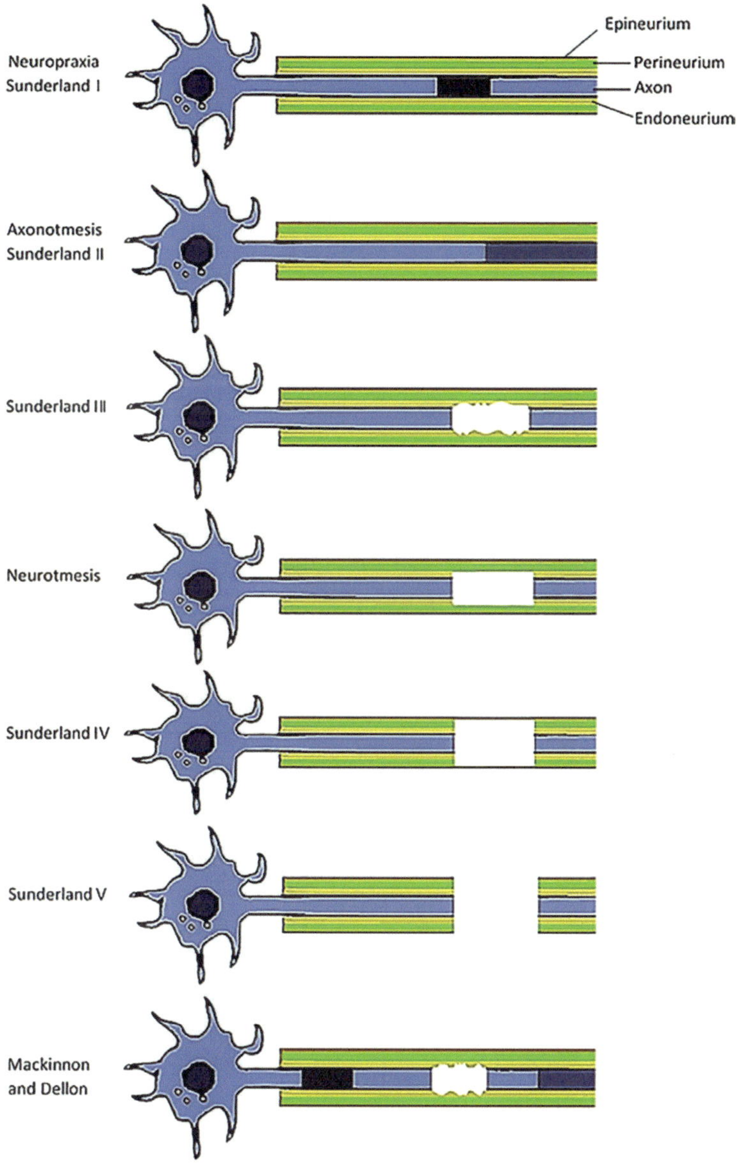

Figure 3. Schematic representation of the Sunderland classification of peripheral nerve injuries. More recently, Mackinnon and Dellon have proposed a sixth-degree for mixed lesions, based on the evidence that a single trauma can affect different regions of the nerve transverse session variably, causing different degrees of injury at different points of the same nerve.

or within a short period of time. Epineural repair is performed when a tension free coaptation in a well-vascularized bed is achieved. The surgical procedure may be divided into distinct phases. In the preparation phase, the nerve ends are prepared to get viable nerves without necrotic tissue. In the approach phase, the nerve ends are coaptated in order to leave a minimum gap between them. This gap will be rapidly filled with blood clots, and a fibrin matrix containing macrophages serves as a transport medium for Schwann cells between the proximal and distal nerve segments. Axons of the proximal segment grow in association with Schwann cells. In order to maintain the coopted position, interrupted microstructures are performed (neurorrhaphy) in the *epineurium*, always ensuring the physiological position of the segments after suturing and avoiding rotation of the nerve ends [40]. Another surgical technique, more suitable for larger nerves includes an intranerve dissection and suture of fascicular groups. This technique allows a better fascicular alignment but causes more trauma and intranerve scarring due to the permanent presence of the suture [41].

Procedures connecting uninjured nerves to the distal portion of an injured nerve (neurotisation) are sometimes employed if direct repair of the injured nerve is not possible. A healthy nerve transfer and coaptation in cases of nerve root avulsions, as well as reimplantation of an avulsed nerve root, are also possible techniques [11].

When the injury originates a gap too large to perform a neurorrhaphy, a graft or nerve conduit may be used to provide a guidewire for the growing axons. Autologous nerve grafts are most indicated since they have all the microstructural components that facilitate axonal migration and have no antigenic components, but the collection of nerves with adequate diameter and the consequences of sacrificing a healthy nerve are important limitations [42]. Allografts, generally from cadaveric donors, despite providing the necessary cellular structures, require immunosuppressive treatments for long periods of time in order to prevent rejections in the receptor [43]. The allografts can be enzymatically processed to become acellular, thus alleviating the need for immunosuppression, presenting high success rates compared to other techniques. Even so, inflammatory reactions can, even rarely, lead to scarring phenomena that preclude normal nerve regeneration [44, 45].

Nerve guidance conduits (NGCs) have been used as viable alternatives to the grafts in a technique called entubulation or tubulisation [45] that allows the entrapment of the fibrous tissue around the injury site and the local maintenance of the neurotrophic and neurotropic factors secreted in the damaged nerve ends [44]. Since they do not have the microstructure of the nerves, can only be used in defects with no more than 10 mm if these tube-guides are not associated with cell-based therapies or growth factors local delivery. Due to this fact, more attention has recently been paid to its effectiveness in assisting coaptation than to its direct function in repairing the gap. They are usually applied to smaller nerves and overcome the disadvantages of the organic options [45]. To ensure its functionality, the characteristics of the NGCs used must comply with all the criteria established for this type of biomaterials: the material used must be (i) biocompatible with the regenerating tissue where it will be applied and should never trigger any local or systemic inflammatory response [46]; (ii) biodegradable, while ensuring mechanical and architectural stability during the regenerative process and resisting to the application of sutures and to inflammatory tissue reaction [47]; (iii) flexible and resistant in a balanced way in order to avoid compression of the regenerating axons and to limit tissue inflammation [48]; (iv) capable of preventing the growth of excessive fibrous tissue associated with the site of injury and reducing the loss of neurotrophic factors secreted by damaged nerve endings [44]; (v) capable to provide an orientation line to the growth cone through a 3-D tubular structure, thereby diminishing misdirection phenomena

[49]; (vi) semi-permeable and with pores of adequate diameter that allow the influx of oxygen and interstitial nutrients to nourish the growing axon that simultaneously prevent the entry of inflammatory cells and the loss of growth factors [50]; technically efficient, ensuring requirements related to production, sterilization, storage and handling [51]. Adapted in each case, these biomaterials must have appropriate dimensions that allow the connection of the nerve defects without tension, and the diameter of the conduit and the thickness of the wall should be sufficient to accommodate the two stumps at the nerve ends without any compression being exerted. In fact, these dimensional variations seem to have an influence on the rate of nerve regeneration [52]. Various materials can be used in nerve conducts, which can be divided into non-resorbable devices, natural resorbable devices and synthetic resorbable devices.

Non-resorbable devices with synthetic origin, or polyvinyl alcohol hydrogels, consist of water in proportions identical to those observed in biological tissues and in polyvinyl alcohol (PVA) that guarantees mechanical structural stability and facilitates sterilization [53]. In contrast, the nature of these materials creates problems related to compression and tension at the suture lines, even after nerve regeneration has occurred. In addition, there are still few clinical studies evaluating the efficacy of these materials in controlled and randomized models [44].

Resorbable devices with natural origin include type I collagen based devices, a natural and abundant organic component that can easily be isolated and purified to reduce its antigenicity [54]. Its reabsorbability can be defined to varying degrees and its adhesiveness allows cell survival and proliferation for long periods of time [55]. With proven biocompatibility and ability to support and guide tissue regeneration *in vivo*, these devices have already demonstrated efficacy in large gaps as recorded in the literature [56]. The main disadvantages related to the use of these materials are discrepancies observed between the different products available in the market with respect to the months needed to complete biodegradation and the observation of immune responses that require the use of immunosuppressive drugs. [57]. Furthermore, there is evidence that materials of different lots can lead to different results, hindering reproducibility, degeneration intervals can be increased and the regenerative supporting ability of the stored nerve may be compromised [58].

Synthetic scaffolds have recently been developed with cellular guidance channels that facilitate propagation of Schwann cell processes, which may improve the chances of successful nerve regeneration (**Figure 4**) [11]. Poly(L-lactide): poly (glycolide) (PLGA) and poly(DL-lactide-ε-caprolactone) (PLC) subgroups may be included in this group of resorbable devices with synthetic origin. The biomaterials of the first group are characterized by good levels of degradability, mechanical properties and associated cellular viability, having good performance in clinical trials with gaps of considerable dimensions. In contrast, high rates of acidic degradation and their products, rapid changes in mechanical properties and low solubility are important limitations [59]. PCL biomaterials are characterized by being transparent, which brings great clinical advantages during surgical application across the nerve gap defect [60], besides demonstrating efficacy in cases of large gaps [44]. Among the limitations, it is worth mentioning its high rigidity, which requires immersion in saline solution prior to use. In addition, it is necessary to use needles of larger dimensions and more resistant during the

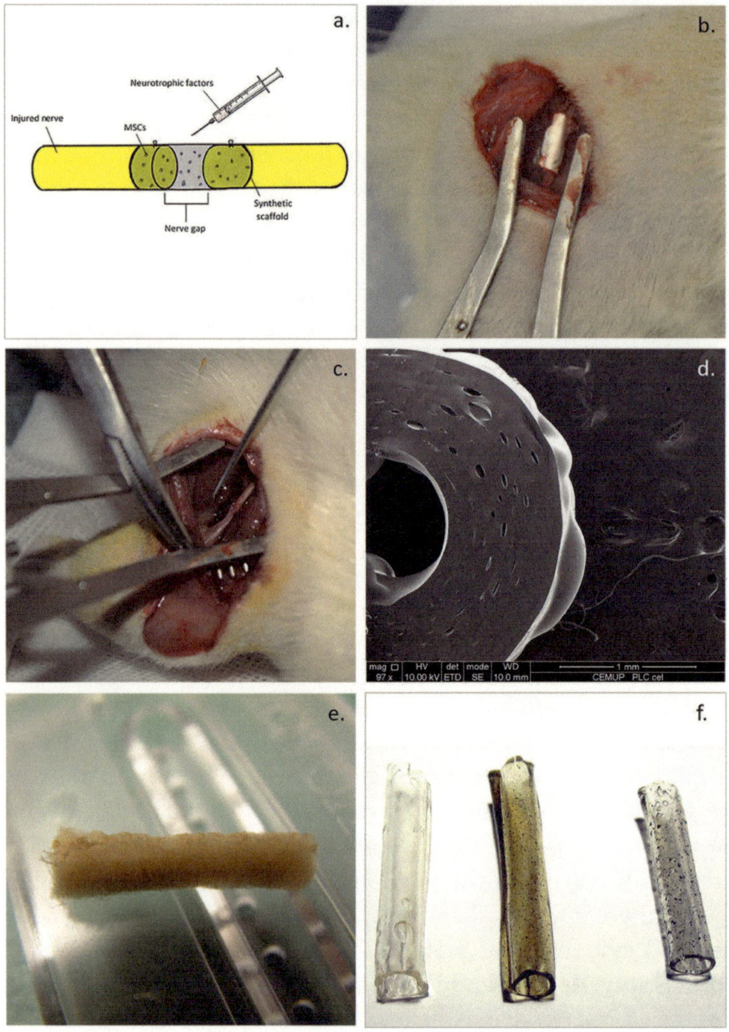

Figure 4. Schematic representation of a tube-guide that can be implanted in neurotmesis injuries (a). Tube-guide made of poly(lactic acid) (PLA)-gelatin piezoelectric material implanted the sciatic nerve of a rat, used as animal model for studying the nerve regeneration process (b). The sciatic nerve of the rat where the neurotmesis injury was reconstructed with an end-to-end suture (c). Scanning Electron Microscope (SEM) image of the inner and outer diameter of a tube-guide made of PLC (transversal section; 1500×) (d). A tube-guide made of PLGA (e). PVA tube-guide; PVA tube-guide loaded with carbon nanotubes (CNTs), and PVA tube-guide loaded with poly-pyrrole (PPy) (f).

application of sutures [50]. Other complications include foreign body reactions, severe swelling with possible device lumen obstruction, fragmentation due to incomplete degradation, early collapse of the device with possible formation of neuromas, and reduced number of myelinated nerve fibres connecting the gap defect [61, 62].

Nerve sheaths constituted by collagen extracellular matrices, acellular and animal-derived, namely with origin in swine intestinal submucosa, have already been used with relative success in the regeneration of different tissues and as nerve guidance channel for regeneration of the peripheral nerve [63]. This technique supports early neovascularization and acts as scaffold in axonal regeneration without immunogenicity problems. In addition, its use guarantees the presence of several growth factors and cytokines that also aid in neuronal survival and growth [64].

Neurotrophic factors, secreted by neuronal or non-neuronal cells in the proximal and distal nodes of the injured nerve, are essential in the conduction of the regenerative process. The addition of these factors to the wall or lumen of the conduits and their slow release by diffusion at the lesion site are techniques currently applied and without which the synthetic conduits may fail to aid the regenerative process over longer graft lengths [65]. Among the neurotrophic factors identified and commonly used are transforming growth factor, beta superfamily, nerve growth factor, insulin-like growth factors, neurotrophins 3, 4, and 5, ciliary neurotrophic factor, neuregulin-1, brain-derived neurotrophic factor, glial cell line-derived neurotrophic factor [63] and platelet-rich plasma [66].

Cell-based therapies have been proposed as a promising alternative to treat a variety of neurologic injuries and the use of stem cells that can differentiate into appropriate cell types in the affected area has developed rapidly in the last years [67]. Stem cells are undifferentiated cells capable to proliferate and produce both new stem cells and different types of cells and tissues [68]. More specifically, MSCs are multipotent, heterogenic stromal cells derived from the mesoderm [69], one of the two populations of bone marrow progenitors (bone marrow stromal progenitors), and were initially characterized as presenting adherence to plastic culture dishes, fibroblast-like morphology and a unique ability to differentiate into multi-lineage MSCs, phenotypes and specialized tissues [70]. Due to the attention that these cells have received in recent years, it became necessary to create a more precise definition that unified the basic characteristics of the MSCs, which emerged in 2006 by the International Society for Cellular Therapy. Thus, MSCs [71]:

- are plastic-adherent under standard culture conditions (α minimal essential medium plus 20% fetal bovine serum);
- express non-specific markers CD105, CD90 CD73 and CD44, and lack the expression of hematopoietic lineage markers CD45, CD34, CD14 or CD11b, CD79α or CD19 and major histocompatibility complex- (MHC-) II/human leukocyte antigen- (HLA-) DR
- are capable to in vitro differentiate into at least three different cell types, like osteoblasts, adipocytes and chondroblasts.

While this initial definition is broad enough to cover the most obvious features of MSCs, several studies over the years have shown that these cells are able to differentiate not only in osteoblasts, adipocytes and chondrocytes but also in other cells and tissues with mesodermal origin (ligaments and tendons, cardiomyocytes, muscle) and also ectodermal and endodermal origins (skin, retinal epithelial pigment, lungs, hepatocytes, renal tubules, pancreatic islets, sebaceous gland ducts and neural cells) [68]. MSCs can be obtained from a vast array of tissues that include adipose tissue, lungs, bone marrow, umbilical cord (Wharton's jelly and umbilical cord blood), *synovium*, amniotic fluid, fetal blood, dental pulp, skeletal muscle, circulatory system [68, 69] and olfactory mucosa [13]. Applied to regenerative medicine, MSCs present exceptional features that make them great options, such as easy expansion, differentiation into different cell types, immune-privileges and immune modulation, tropism to injured sites, trophic stimulation and modulation of tissues functions [46]. In addition to being able to secrete neurotrophic factors and provide an environment conducive to neurogenesis and proliferation of Schwann cells in nerve injury sites, they can themselves differentiate into cells with Schwann cell phenotype and modulate the local inflammatory process and the Wallerian degeneration [72], being a precious addition to the use of biomaterials and growth factors in therapeutic techniques after PNI.

3. Nasal olfactory mucosa mesenchymal stem cells

3.1. General features

The olfactory mucosa (OM), as a component of the olfactory system, consists of different types of cells. Among these are olfactory neurons (ON) that are able to regenerate continuously throughout adult life. This regenerative capacity is attributed to olfactory stem cells and supporting cells of OM, together promoting axonal regeneration [73]. In addition to the bipolar ON or olfactory neurosensory cells, several cell types can be identified in the OM: horizontal basal stem cells (HBCs) and globular basal stem cells (GBCs) in the olfactory epithelium, MSCs in the lamina propria, olfactory ensheathing cells (OECs) and support cells (**Figure 5**). New nerves generate from GBS in the olfactory epithelium, which are guided to their correct position in the olfactory bulb by OECs in the olfactory mucosa. GBCs, derived from HBCs, were initially thought to be the exclusive source of ON and support cells [74] but it is currently known that MSCs are also capable of producing neurons *in vitro* [75].

The OM-MSCs were initially identified from an embryonic rat OM culture [76] and the first characterization studies evidenced the expression of mesenchymal-specific markers such as CD90, CD105, STRO-1 and differentiation into the three main cell lines [74, 77]. OM-MSCs have important characteristics such as neural crest origin, high versatility, vast distribution, advantageous localization and are not susceptible to chromosomal abnormalities or tumorigenicity [78]. They exhibit high mitotic activity when compared to the bone marrow MSCs (BM-MSCs) (nearly three times higher) and are able to self-renew in long-term cultures (over 15 weeks) by maintaining telomerase activity and lack apoptotic activity [79]. The olfactory mucosa itself is a great cell source since its renewal continues throughout life and OM-MSC potency is not even affected by the age of the donor [80]. Different from BM-MSCs, OM-MSCs promote CNS myelination and induce the differentiation of neural stem cells into oligodendrocytes and oligodendrocyte maturation. All that suggest an easy and rapid propagation to sufficient levels that allow transplantation and that MSCs from a more neurogenic niche may have different properties to the classical BM-MSCs [81, 82].

Due to its origin from ectoderm (resulting from the interaction between cranial neural crest and olfactory placodes) and its high expression of neural cell-related genes, it was proposed that OM-MSCs be renamed as ectomesenchymal stem cells [81]. Even its origin highlights the predisposition for OM-MSCs to differentiate into neural lineage cells.

3.2. Isolation, characterization and differentiation

The properties of the OM-MSCs are still far from being fully understood. Located in the olfactory region of the nasal cavity, OM-MSCs are primarily derived from neural crest cells, have

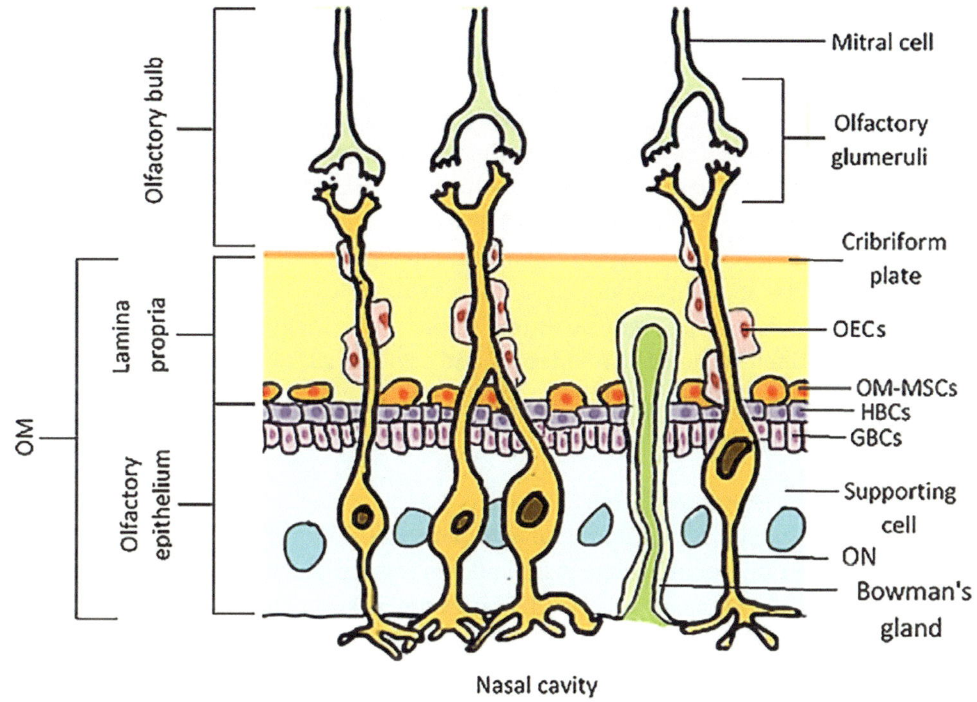

Figure 5. Schematic representation of the olfactory mucosa and relative location of its cells.

a high proliferation rate, self-renewal ability and multiple lineage differentiation capabilities [83]. These cells have already been isolated from human and mice [84], rat [85], rabbit [86] and dog [87], both with *in vivo* [84] and post-mortem protocols [86].

The described collection methods include a delicate discarding of the turbinates [82, 88] or complete collection of the olfactory bulb [80], being the harvested material taken in balanced solutions containing antibiotics and antifungals [82]. The olfactory epithelium or olfactory bulb tissue sample is then cut into small pieces, digested with collagenase and DNase [80, 82] or dispase [78] to separate the cells from their extracellular matrix. Then, they are cultured in flasks containing appropriate medium like Dulbecco's modified Eagle's medium or HAM´s F-12 supplemented with fetal calf serum or fetal bovine serum, antibiotics, antifungals and growth factors [75]. After removal of non-adherent cells, the remaining cells can be trypsinized, expanded [82, 88] and then banked in liquid nitrogen or induced to form for differentiation. [89]. Multilineage differentiation can be achieved by culturing under induction conditions and staining with specific dyes [88].

Determination of the phenotypic markers is achieved through flow cytometric analysis where single cell suspensions are stained with specific fluorochrome-conjugated antibodies [78, 80], RT-PCR [80] or by immunocytochemistry [82]. In addition, the identification of specific proteins of these cells can be performed using immunofluorescence microscopy [88].

In colonies formed after culture, OM-MSCs exhibit mostly fibroblast-like morphology [78, 80, 88, 90]. Phenotypic analysis reveals the presence of Stro-1 [75–77], CD29 [80, 88], CD44 [80, 88], CD49b

[90], CD54 [13, 77, 82], CD73 [77, 82, 90], CD90 [13, 75–78, 80, 82, 88, 90], CD105 [13, 75–78, 80, 82], CD106 [77, 78] and CD166 [77, 78, 80, 82, 90] markers and the absence of haematopoietic stem cell markers such as CD34 [13, 75–78, 80, 88, 90], CD45 [13, 75–78, 80, 88] and CD11b [88]. Strangely, one work indicates lack of Stro-1 [82]. Immunofluorescence microscopy provides evidence for the expression of neural stem cell–related proteins such as vimentin [88], nestin [75–77, 81, 82, 88] and also NG2 [77, 81], a marker whose presence indicates the ability to form neurospheres and to generate neurons in vivo and in vitro [81]. They also express p75-NTR and SMA [77]. Consistent with the characteristics of multilineage differentiation capacity of MSCs, OM-MSCs can be differentiated into osteocytes and adipocytes [13, 82, 88], chondrocytes [78, 90], neuron like cells [13, 78, 81] and myocardial-like cells [80] when cultured under appropriate conditions. They, therefore, present abilities to differentiate into mesodermal and ectodermal cell types.

3.3. Secretome and metabolome

It is known that the regenerative effect of MSCs, as well as OM-MSCs, is not exclusively related to their differentiating ability but also to paracrine factors. These are important in the creation of a support microenvironment that allows cellular survival, differentiation, activation of endogenous neural stem cells, reduction of the inflammatory reaction and induction of angiogenesis [91]. MSCs have also the ability to produce potent protective factors that promote tissue repair and immunomodulation, reducing fibrosis and cell death [92]. The determination of the characteristics and components of the secretome and metabolome of a given cell is essential to uncover the essential needs for its success in regenerative processes [93]. Similar cell types with different secretome profile can reflect their cellular niches and local function. The only study of the determination of the secretome and metabolome of OM-MSCs allowed to identify 274 proteins in OM-MSCs conditioned medium and the identification of some processes that are usually associated to transplantation processes such as biological regulation, cellular processes, metabolic processes, development processes and response to stimuli [13]. These processes promote repairing by facilitating migration to injured sites, remodelling the extracellular matrix and increasing metabolism and cellular activity. In the OM-MSCs, genes related to cell growth and migration, angiogenesis and blood circulation, inflammatory and immune regulation and neurotrophy, the major components of transplantation and regenerative promotion, were identified and they can also produce cytokines that promote hematopoietic stem cell survival, proliferation and differentiation [90]. In addition, it has also been identified the secretion of important molecules in neural differentiation such as Dystroglycan that can organize axon guidance cue location which is critical for nervous system development and plays important roles in perisynaptic and axonal matrix formations, contributing to synaptic homeostatic plasticity [94]. Proteins like Dermcidin, retinoic acid induced 1 and cadherin 13 that contribute to cell cycle related events and play roles in neural differentiation were also identified. Dermcidin acts as a neural growth factor [95]; retinoic acid is involved in neurobehavioural disorders and plays a role in normal neural development [96] and Cadherin acts as a regulator of neural cell growth [97]. OM-MSCs were identified as secreting high levels of the chemokine CXCL12 [98] that is known to be important in the promotion of endogenous myelinations [99]. Evidences that OM-MSCs are capable to alter the biological properties of the precursors of OECs and oligodendrocytes and of increasing the myelination of the oligodendrocytes in the CNS [63], in conjunction with the other findings referred, demonstrate the enormous potential of these cells to be applied in regenerative medicine of the nervous system in general and in the PNS in particular.

3.4. Applications

The studies and characterizations carried out to date suggest that the OM-MSCs have phenotypes and differentiation characteristics similar to other MSCs and can efficiently proliferate in culture. However, studies of direct clinical application of these cells in regenerative therapies are still few and the results obtained need more research and deeper approaches.

Several studies have demonstrated the improvement of locomotor function in animal and human patients with spinal cord injury after implantation of entire OM grafts and OECs [100] but isolated OM-MSCs have never been used singly in this type of cell therapies [73]. In these cases, the effect of OM-MSCs may be masked by the use of OM as a whole, making it impossible to determine the direct effect of MSCs on motor recovery. Despite this, it has been already demonstrated that OM-MSCs promote rat CNS myelination *in vitro* [77]. A study on the use of cell-based therapy in cases of deafness revealed that OM-MSCs present repair efficiency in the spiral ganglion neuron after lesion induction in the cochlea [101]. Transplantation of OM-MSCs into a brain after injury led to a partial reconstitution of the hippocampus, with observation of important phenomena such as migration of stem cells to the inflamed region, *in situ* neuronal differentiation and local stimulus to neurogenesis. The injured individual presented reversal of learning deficits, recovered memorization capabilities and enhanced physiological function. All these events are also observed if the OM-MSCs were transplanted directly into the cerebrospinal fluid. These results open precedents for the use of OM-MSCs in patients with post-traumatic memory loss [102]. Another study has shown that OM-MSCs generated dopaminergic cells and reduced the asymmetries resulting from the ablation of dopaminergic neurons when transplanted into a rat brain model of Parkinson's disease, also opening precedents [103] for its use in neurodegenerative diseases [104]. Recently, immunoregulatory properties of OM-MSCs have been identified, which can exert immunosupressive functions and modulate T-cell responses. These findings indicate a potential use of these cells targeting autoimmune diseases [88, 105].

It is possible that in some studies in which olfactory mucosa cells were used without an exact determination of the cell types present in the heterogeneous cell mix, OM-MSCs were part of the group used and performed specific actions. In these cases where there is no detailed determination of the composition of the cell matrix used, it is impossible to know which type of cell has dominant function or effectively reparative properties. For instance, a study in which transplantation of OM to rat hearts after infarction led to differentiation into cells resembling cardiomyocytes, but failure to specify the types of cells included in the transplantation makes it impossible to attribute regenerative function to OM-MSCs and makes it difficult to optimize future procedures [106]. The use of unpurified olfactory mucosal cells cannot be directly compared to purified OM-MSC and it is essential that each type of transplanted cell is characterized prior to the procedure. Even the preliminary study of the use of OM-MSCs in the treatment of Parkinson's disease resulted from the use of the OM as a whole and not from the MSCs isolated [103]. The combined use of cells is not, obviously, a wrong procedure and it has already been shown that the combined use of OECs and OM-MSCs has beneficial effects of inter-stimulation between cells with respect to their functionality and secretome [106, 107].

It is, however, important to define exactly the role of each of them in the regenerative process before its use on a larger scale. At this point, and given the existing knowledge about the functional characteristics of OM-MSCs, the experience obtained in the few realized studies and the potential that has been established for these cells, is important to set specifc goals and start more focused works to determine the importance of these cells in the future of regenerative medicine, in the approach to lesions of the peripheral nerve as well as to nervous system in general and to other organic systems.

4. Conclusions and further directions

Although we have known for some time that there are multipotent cells in the olfactory epithelium and the olfactory bulb, only recently OM-MSC was identified at the *lamina propria* of OM. Since its discovery, most of the studies that addressed OM-MSC focused on its complete characterization and few studies have yet applied these cells in order to determine their regenerative capacity. The use of these cells presents clear technical advantages due to their location and their collection can be easily made in the donors under anaesthesia and practically without any side effect. In addition, besides its high versatility and clonogenic activity, OM-MSC may be used for autologous transplants, circumventing possible rejections at the site of application and ethical issues. Since MS-MSCs are submitted to an environment with continuous regenerative activity, it is understandable that they secrete higher levels of neurotrophic and myelinating factors when compared to MSCs with other origins. Thus, OM-MSCs appear as a robust candidate for the approach to PNI cases when compared to other MSCs that do not achieve significant success in promoting nerve growth through the glial scar. The regeneration of the peripheral nerve has been the subject of multiple studies and there are many therapeutic techniques currently under development. Despite this, and considering the restrictions still observed in these approaches, regenerative medicine is one of the options with higher potential to achieve success in this type of lesions. Although several types of stem cells and MSCs are already under study, the OM-MSCs, their identified characteristics and preliminary results observed after their application, make them a promise in regenerative medicine in general and, specifically, a revolutionary approach to PNI lesions or demyelinating diseases and in situations in which neuroprotection or neurite outgrowth is important for repair. In this moment of rapid expansion of the knowledge we have about OM-MSCs, the next steps will have to include a complete and unambiguous characterization of these cells, their secretome and their metabolome and a precise determination of their regenerative potential in different tissues, specifically in the peripheral nerve, after isolation from the remaining OM cells. Finally, it will be necessary to explore their use associated with different biomaterials (something that has not yet been done), growth factors and even other cells whose associations have already been shown to be effective, such as OECS.

Acknowledgements

This research was supported by Programa Operacional Regional do Norte (ON.2 – O Novo Norte), QREN, FEDER with the project 'iBone Therapies: Terapias inovadoras para a regeneração óssea', ref. NORTE-01-0247-FEDER-003262, and by the program COMPETE—Programa Operacional Factores de Competitividade, Projects PEst-OE/AGR/UI0211/2011 and PEst-C/EME/UI0285/2013 funding from FCT. This research was also supported by Programa Operacional Competitividade e Internacionalização (P2020), Fundos Europeus Estruturais e de Investimento (FEEI) and FCT with the project 'BioMate—A novel bio-manufacturing system to produce bioactive scaffolds for tissue engineering' with reference PTDC/EMS-SIS/7032/2014 and by COMPETE 2020, from ANI—Projectos ID&T Empresas em Copromoção, Programas Operacionais POCI, by the project 'insitu.Biomas—Reinvent biomanufacturing systems by using an usability approach for in situ clinic temporary implants fabrication' with the reference POCI-01-0247-FEDER-017771. The research was also supported by the research project 'BEPIM III—Microdispositivos médicos com capacidades osteintegradoras por micoPIM', with the reference POCI-01-0247-FEDER-017935, from Fundo Europeu de Desenvolvimento Regional (FEDER), by the Programa Operacional da Competitividade & Internacionalização. Ana Rita Caseiro (SFRH/BD/101174/2014) acknowledges FCT, for financial support.

Conflict of Interests

The authors declare that there is no conflict of interests regarding the publication of this chapter.

Abbreviations

BM-MSCs	Bone marrow MSCs
CNS	Central nervous system
CNTs	Carbon nanotubes
GBCs	Globular basal cells
HBCs	Horizontal basal cells
MSCs	Mesenchymal stem cells
NGCs	Nerve guidance conduits
OECs	Olfactory ensheathing cells
OM	The olfactory mucosa
OM-MSCs	Olfactory mucosa mesenchymal stem cells
ON	Olfactory neurons
PLA	Poly(lactic acid)
PLC	Poly(DL-lactide-ε-caprolactone)
PLGA	Poly(L-lactide):poly(glycolide)
PVA	Poly(vinyl alcohol)
PNI	Peripheral nerve injuries
PNS	Peripheral nervous system
PPY	Poly-pyrrole
SEM	Scanning Electron Microscope

Author details

Rui Damásio Alvites[1,2], Ana Rita Caseiro Santos[1,2,3], Artur Severo Proença Varejão[4] and Ana Colette Pereira de Castro Osório Maurício[1,2*]

*Address all correspondence to: ana.colette@hotmail.com

1 Centro de Estudos de Ciência Animal (CECA), Instituto de Ciências, Tecnologias e Agroambiente da Universidade do Porto (ICETA), Praça Gomes Teixeira, Apartado, Porto, Portugal

2 Departamento de Clínicas Veterinárias, Instituto de Ciências Biomédicas de Abel Salazar (ICBAS), Universidade do Porto (UP), Rua de Jorge Viterbo Ferreira, Porto, Portugal

3 REQUIMTE/LAQV – U. Porto –Porto/Portugal, Departamento de Engenharia Metalúrgica e Materiais, Faculdade de Engenharia, Universidade do Porto, Porto, Portugal

4 Departamento de Ciências Veterinárias, Centro de Ciência Animal e Veterinária (CECAV), Universidade de Trás-os-Montes e Alto Douro, UTAD, Quinta de Prados, Vila Real, Portugal

References

[1] Wojtkiewicz DM, Saunders J, Domeshek L, Novak CB, Kaskutas V, Mackinnon SE. Social impact of peripheral nerve injuries. Hand. 2015;**10**(2):161-167. DOI: 10.1007/s11552-014-9692-0

[2] Antoniadis G, Kretschmer T, Pedro MT, König RW, Heinen CP, Richter H-P. Iatrogenic nerve injuries. Deutsches Aerzteblatt International. 2014;**111**(16):273-279. DOI: 10.3238/arztebl.2014.0273

[3] Navarro X. Functional evaluation of peripheral nerve regeneration and target reinnervation in animal models: A critical overview. European Journal of Neuroscience. 2016;**43**(3):271-286. DOI: 10.1111/ejn.13033

[4] Lutz AB, Barres BA. Contrasting the glial response to axon injury in the central and peripheral nervous systems. Developmental Cell. 2014;**28**(1):7-17. DOI: 10.1016/j.devcel.2013.12.002

[5] Faroni A, Mobasseri SA, Kingham PJ, Reid AJ. Peripheral nerve regeneration: Experimental strategies and future perspectives. Advanced Drug Delivery Reviews. 2015;**82**:160-167. DOI: 10.1016/j.addr.2014.11.010

[6] Sulaiman W, Gordon T. Neurobiology of peripheral nerve injury, regeneration, and functional recovery: From bench top research to bedside application. The Ochsner Journal. 2013;**13**(1):100-108. DOI: 10.4172/2155-6105.1000292

[7] Fitzgerald M, McKelvey R. Nerve injury and neuropathic pain—A question of age. Experimental Neurology. 2016;**275**:296-302. DOI: 10.1016/j.expneurol.2015.07.013

[8] Houschyar K, Momeni A, Pyles M, Cha J, Maan Z, Duscher D, et al. The role of current techniques and concepts in peripheral nerve repair. Plastic Surgery International. 2016;**2016:**Article ID 4175293. DOI: 10.1155/2016/4175293

[9] Kemp SW, Cederna PS, Midha R. Comparative outcome measures in peripheral regeneration studies. Experimental Neurology. 2016. DOI: 10.1016/j.expneurol.2016.04.011

[10] Jonsson S, Wiberg R, McGrath AM, Novikov LN, Wiberg M, Novikova LN, et al. Effect of delayed peripheral nerve repair on nerve regeneration, Schwann cell function and target muscle recovery. PloS One. 2013;**8**(2):e56484. DOI: 10.1371/journal.pone.0056484

[11] Simon NG, Spinner RJ, Kline DG, Kliot M. Advances in the neurological and neurosurgical management of peripheral nerve trauma. Journal of Neurology, Neurosurgery & Psychiatry. 2015;**87**(2):198-208. DOI: 10.1136/jnnp-2014-310175

[12] Bhangra KS, Busuttil F, Phillips JB, Rahim AA. Using stem cells to grow artificial tissue for peripheral nerve repair. Stem Cells International. 2016;**2016:**Article ID 7502178. DOI: 10.1155/2016/7502178

[13] Ge L, Jiang M, Duan D, Wang Z, Qi L, Teng X, et al. Secretome of olfactory mucosa mesenchymal stem cell, a multiple potential stem cell. Stem Cells International. 2016;**2016:**Article ID 1243659. DOI: 10.1155/2016/1243659

[14] Grinsell D, Keating C. Peripheral nerve reconstruction after injury: A review of clinical and experimental therapies. BioMed Research International. 2014;**2014:**Article ID 698256. DOI: 10.1155/2014/698256

[15] Biazar E, Khorasani M, Montazeri N, Pourshamsian K, Daliri M, Rezaei M, et al. Types of neural guides and using nanotechnology for peripheral nerve reconstruction. International Journal of Nanomedicine. 2010;**5**:839-852. DOI: 10.2147/IJN.S11883

[16] Bove GM. Epi-perineurial anatomy, innervation, and axonal nociceptive mechanisms. Journal of Bodywork and Movement Therapies. 2008;**12**(3):185-190. DOI: 10.1016/j.jbmt.2008.03.004

[17] Chhabra A, Ahlawat S, Belzberg A, Andreseik G. Peripheral nerve injury grading simplified on MR neurography: As referenced to Seddon and Sunderland classifications. The Indian Journal of Radiology & Imaging. 2014;**24**(3):217. DOI: 10.4103/0971-3026.137025

[18] Seddighi A, Nikouei A, Seddighi AS, Zali AR, Tabatabaei SM, Sheykhi AR, et al. Peripheral nerve injury: A review article. International Clinical Neuroscience Journal. 2016;**3**(1):1-6. DOI: 10.1155/2014/698256

[19] Brull R, Hadzic A, Reina MA, Barrington MJ. Pathophysiology and etiology of nerve injury following peripheral nerve blockade. Regional Anesthesia and Pain Medicine. 2015;**40**(5):479-490. DOI: 10.1097/AAP.0000000000000125

[20] Hogan QH. Pathophysiology of peripheral nerve injury during regional anesthesia. Regional Anesthesia and Pain Medicine. 2008;**33**(5):435-441. DOI: 10.1016/j.rapm.2008.03.002

[21] Burnett MG, Zager EL. Pathophysiology of peripheral nerve injury: A brief review. Neurosurgical Focus. 2004;**16**(5):1-7. DOI: 10.3171/foc.2004.16.5.2

[22] Sekiguchi KJ, Shekhtmeyster P, Merten K, Arena A, Cook D, Hoffman E, et al. Imaging large-scale cellular activity in spinal cord of freely behaving mice. Nature Communications. 2016;**7**:Article number 11450. DOI: 10.1038/ncomms11450

[23] Hainline BW. Peripheral nerve injury in sports. Continuum: Lifelong Learning in Neurology. 2014;**20**(6, Sports Neurology):1605-1628. DOI: 10.1212/01.CON.0000458971.86389.9c

[24] Uzun N, Tanriverdi T, Savrun FK, Kiziltan ME, Sahin R, Hanimoglu H, et al. Traumatic peripheral nerve injuries: Demographic and electrophysiologic findings of 802 patients from a developing country. Journal of Clinical Neuromuscular Disease. 2006;**7**(3):97-103. DOI: 10.1097/01.cnd.0000203641.38887.63

[25] Lim TK, Shi XQ, Johnson JM, Rone MB, Antel JP, David S, et al. Peripheral nerve injury induces persistent vascular dysfunction and endoneurial hypoxia, contributing to the genesis of neuropathic pain. The Journal of Neuroscience. 2015;**35**(8):3346-3359. DOI: 10.1523/JNEUROSCI.4040-14.2015

[26] Kim DD, Asif A, Kataria S. Presentation of neurolytic effect of 10% lidocaine after perineural ultrasound guided injection of a canine sciatic nerve: A pilot study. The Korean Journal of Pain. 2016;**29**(3):158-163. DOI: 10.3344/kjp.2016.29.3.158

[27] Farber SJ, Saheb-Al-Zamani M, Zieske L, et al. Peripheral nerve injury after local anesthetic injection. Anesthesia & Analgesia. 2013;**117**:731-739. DOI: 10.1213/ANE.0b013e3182a00767

[28] Cheng K-I, Wang H-C, Wu Y-C, Tseng K-Y, Chuang Y-T, Chou C-W, et al. Sciatic nerve intrafascicular lidocaine Injection-induced peripheral neuropathic pain: Alleviation by systemic minocycline administration. The Clinical Journal of Pain. 2016;**32**(6):513-521. DOI: 10.1097/AJP.0000000000000293

[29] Menorca RM, Fussell TS, Elfar JC. Nerve physiology: Mechanisms of injury and recovery. Hand Clinics. 2013;**29**(3):317-330. DOI: 10.1016/j.hcl.2013.04.002

[30] Hall S. The response to injury in the peripheral nervous system. Bone & Joint Journal. 2005;**87**(10):1309-1319. DOI: 10.1302/0301-620X.87B10.16700

[31] Dubový P, Klusáková I, Hradilová Svíženská I. Inflammatory profiling of Schwann cells in contact with growing axons distal to nerve injury. BioMed Research International. 2014;**2014**:7. DOI: 10.1155/2014/691041

[32] Deumens R, Bozkurt A, Meek MF, Marcus MA, Joosten EA, Weis J, et al. Repairing injured peripheral nerves: Bridging the gap. Progress in Beurobiology. 2010;**92**(3):245-276. DOI: 10.1016/j.pneurobio.2010.10.002

[33] Li N, Downey JE, Bar-Shir A, Gilad AA, Walczak P, Kim H, et al. Optogenetic-guided cortical plasticity after nerve injury. Proceedings of the National Academy of Sciences. 2011;**108**(21):8838-8843. DOI: 10.1073/pnas.1100815108

[34] Au NPB, Kumar G, Asthana P, Tin C, Mak YL, Chan LL, et al. Ciguatoxin reduces regenerative capacity of axotomized peripheral neurons and delays functional recovery in pre-exposed mice after peripheral nerve injury. Scientific Reports. 2016;**6**:26809. DOI: 10.1038/srep26809

[35] Seddon H. Three types of nerve injury. Brain. 1943;**66**(4):237-288. DOI: 10.1093/brain/66.4.237

[36] Choi EJ, Choi YM, Jang EJ, Kim JY, Kim TK, Kim KH. Neural ablation and regeneration in pain practice. The Korean Journal of Pain. 2016;**29**(1):3-11. DOI: 10.3344/kjp.2016.29.1.3

[37] Campbell WW. Evaluation and management of peripheral nerve injury. Clinical Neurophysiology. 2008;**119**(9):1951-1965. DOI: 10.1016/j.clinph.2008.03.018

[38] Sunderland S. A classification of peripheral nerve injuries producing loss of function. Brain. 1951;**74**(4):491-516. DOI: 10.1093/brain/74.4.491

[39] Mackinnon S, Dellon A. Diagnosis of Nerve Injury. Surgery of the Peripheral Nerve. New York: Thieme; 1988. pp. 74-79.

[40] Dahlin L. Techniques of peripheral nerve repair. Scandinavian Journal of Surgery. 2008;**97**(4):310-316. DOI: http://sjs.sagepub.com/content/97/4/310.full.pdf

[41] Lundborg G. A 25-year perspective of peripheral nerve surgery: Evolving neuroscientific concepts and clinical significance. The Journal of Hand Surgery. 2000;**25**(3):391-414. DOI: 10.1053/jhsu.2000.4165

[42] Millesi H. Bridging Defects: Autologous Nerve Grafts. Berlin Heidelberg: Springer; 2007. DOI: 10.1007/978-3-211-72958-8_8

[43] Grand AG, Myckatyn TM, Mackinnon SE, Hunter DA. Axonal regeneration after cold preservation of nerve allografts and immunosuppression with tacrolimus in mice. Journal of Neurosurgery. 2002;**96**(5):924-932. DOI: 10.3171/jns.2002.96.5.0924

[44] Kehoe S, Zhang X, Boyd D. FDA approved guidance conduits and wraps for peripheral nerve injury: A review of materials and efficacy. Injury. 2012;**43**(5):553-572. DOI: 10.1016/j.injury.2010.12.030

[45] Safa B, Buncke G. Autograft substitutes: Conduits and processed nerve allografts. Hand Clinics. 2016;**32**(2):127-140. DOI: 10.1016/j.hcl.2015.12.012

[46] Scatena M, Eaton KV, Jackson MF, Lund SA, Giachelli CM. Macrophages: The bad, the ugly, and the good in the inflammatory response to biomaterials. In: Corradetti, B, editor. The Immune Response to Implanted Materials and Devices. Switzerland: Springer; 2017. pp. 37-62. DOI: 10.1007/978-3-319-45433-7_3

[47] Basu B. Corrosion and degradation of implantable biomaterials. In: Biomaterials for Musculoskeletal Regeneration. Singapore: Springer; 2017. pp. 253-289. 10.1007/978-981-10-3059-8_8

[48] Belanger K, Dinis TM, Taourirt S, Vidal G, Kaplan DL, Egles C. Recent strategies in tissue engineering for guided peripheral nerve regeneration. Macromolecular Bioscience. 2016;**16**:472-481. DOI:10.1002/mabi.201500367

[49] Peng S-W, Li C-W, Chiu M, Wang G-J. Nerve guidance conduit with a hybrid structure of a PLGA microfibrous bundle wrapped in a micro/nanostructured membrane. International Journal of Nanomedicine. 2017;**12**:421. DOI: 10.2147/IJN.S122017

[50] Meek MF, Coert JH. US Food and Drug Administration/Conformit Europe-approved absorbable nerve conduits for clinical repair of peripheral and cranial nerves. Annals of Plastic Surgery. 2008;**60**(1):110-116. DOI: 10.1097/SAP.0b013e31804d441c

[51] Kaur G, editor. Biomaterials influencing human lives. In: Bioactive Glasses. Cham, Switzerland: Springer; 2017. pp. 1-20. 10.1007/978-3-319-45716-1_1

[52] Kokai LE, Lin Y-C, Oyster NM, Marra KG. Diffusion of soluble factors through degradable polymer nerve guides: Controlling manufacturing parameters. Acta Biomaterialia. 2009;**5**(7):2540-2550. 10.1016/j.actbio.2009.03.009

[53] Stammen JA, Williams S, Ku DN, Guldberg RE. Mechanical properties of a novel PVA hydrogel in shear and unconfined compression. Biomaterials. 2001;**22**(8):799-806. DOI: 10.1016/S0142-9612(00)00242-8

[54] Hardin-Young J, Carr RM, Downing GJ, Condon KD, Termin PL. Modification of native collagen reduces antigenicity but preserves cell compatibility. Biotechnology and Bioengineering. 1996;**49**(6):675-682. DOI: 10.1002/(SICI)1097-0290(19960320)49:6<675::AID-BIT9>3.0.CO;2-L

[55] Stang F, Fansa H, Wolf G, Keilhoff G. Collagen nerve conduits–assessment of biocompatibility and axonal regeneration. Bio-medical Materials and Engineering. 2005;**15**(1, 2):3-12.

[56] Tyner TR, Parks N, Faria S, Simons M, Stapp B, Curtis B, et al. Effects of collagen nerve guide on neuroma formation and neuropathic pain in a rat model. The American Journal of Surgery. 2007;**193**(1):e1-e6. DOI: 10.1016/j.amjsurg.2006.08.026

[57] Hudson TW, Liu SY, Schmidt CE. Engineering an improved acellular nerve graft via optimized chemical processing. Tissue Engineering. 2004;**10**(9-10):1346-1358. DOI: 10.1089/ten.2004.10.1346

[58] Evans GR. Peripheral nerve injury: A review and approach to tissue engineered constructs. The Anatomical Record. 2001;**263**(4):396-404. DOI: 10.1002/ar.1120

[59] Tabesh H, Amoabediny G, Nik NS, Heydari M, Yosefifard M, Siadat SR, et al. The role of biodegradable engineered scaffolds seeded with Schwann cells for spinal cord regeneration. Neurochemistry International. 2009;**54**(2):73-83. DOI: 10.1016/j.neuint.2008.11.002

[60] Pereira T, Gärtner A, Amorim I, Almeida A, Caseiro A, Armada-da-Silva PA, et al. Promoting nerve regeneration in a neurotmesis rat model using poly(DL-lactide-caprolactone) membranes and mesenchymal stem cells from the Wharton's jelly: In vitro and in vivo analysis. BioMed Research International. 2014;**2014**:302659. DOI: 10.1155/2014/302659

[61] Meek MF, Jansen K. Two years after in vivo implantation of poly (DL-lactide-ε-caprolactone) nerve guides: Has the material finally resorbed? Journal of Biomedical Materials Research Part A. 2009;**89**(3):734-738. DOI: 10.1002/jbm.a.32024

[62] Meek MF, Den Dunnen WF. Porosity of the wall of a Neurolac® nerve conduit hampers nerve regeneration. Microsurgery. 2009;**29**(6):473-478. DOI: 10.1002/micr.20642

[63] Gaudin R, Knipfer C, Henningsen A, Smeets R, Heiland M, Hadlock T. Approaches to peripheral nerve repair: Generations of biomaterial conduits yielding to replacing autologous nerve grafts in craniomaxillofacial surgery. BioMed Research International. 2016;**2016**:18 p. DOI: 10.1155/2016/3856262

[64] Yi J-S, Lee H-J, Lee H-J, Lee I-W, Yang J-H. Rat peripheral nerve regeneration using nerve guidance channel by porcine small intestinal submucosa. Journal of Korean Neurosurgical Society. 2013;**53**(2):65-71. DOI: 10.3340/jkns.2013.53.2.65

[65] Madduri S, Feldman K, Tervoort T, Papaloïzos M, Gander B. Collagen nerve conduits releasing the neurotrophic factors GDNF and NGF. Journal of Controlled Release. 2010;**143**(2):168-174. DOI: 10.1016/j.jconrel.2009.12.017

[66] Sariguney Y, Yavuzer R, Elmas C, Yenicesu I, Bolay H, Atabay K. Effect of platelet-rich plasma on peripheral nerve regeneration. Journal of reconstructive microsurgery. 2008;**24**(03):159-167. DOI: 10.3944/AOTT.2014.13.0029

[67] Casañas J, de la Torre J, Soler F, García F, Rodellar C, Pumarola M, et al. Peripheral nerve regeneration after experimental section in ovine radial and tibial nerves using synthetic nerve grafts, including expanded bone marrow mesenchymal cells: Morphological and neurophysiological results. Injury. 2014;**45**:S2-S6. DOI: 10.1016/S0020-1383(14) 70003-8

[68] Kobolak J, Dinnyes A, Memic A, Khademhosseini A, Mobasheri A. Mesenchymal stem cells: Identification, phenotypic characterization, biological properties and potential for regenerative medicine through biomaterial micro-engineering of their niche. Methods. 2016;**99**:62-68. DOI: 10.1016/j.ymeth.2015.09.016

[69] Caceres ZB, Peress L, Rameshwar P, Fernandez-Moure JS. Not all Stem Cells are Created Equal. 2016. Available from: http://www.avidscience.com/wp-content/uploads/2016/07/SCR-16-01.pdf

[70] Friedenstein A, Chailakhyan R, Gerasimov U. Bone marrow osteogenic stem cells: In vitro cultivation and transplantation in diffusion chambers. Cell Proliferation. 1987;**20**(3):263-272. DOI: 10.1111/j.1365-2184.1987.tb01309.x

[71] Dominici M, Le Blanc K, Mueller I, Slaper-Cortenbach I, Marini F, Krause D, et al. Minimal criteria for defining multipotent mesenchymal stromal cells. The International Society for Cellular Therapy position statement. Cytotherapy. 2006;**8**(4):315-317. DOI: 10.1080/14653240600855905

[72] Kingham PJ, Kalbermatten DF, Mahay D, Armstrong SJ, Wiberg M, Terenghi G. Adipose-derived stem cells differentiate into a Schwann cell phenotype and promote neurite outgrowth in vitro. Experimental Neurology. 2007;**207**(2):267-274. DOI: 10.1016/j.expneurol.2007.06.029

[73] Duan D, Lu M. Olfactory mucosa: A rich source of cell therapy for central nervous system repair. Reviews in the Neurosciences. 2015;**26**(3):281-293. DOI: 0.1515/revneuro-2014-0065

[74] Goldstein BJ, Goss GM, Choi R, Saur D, Seidler B, Hare JM, et al. Olfactory basal stem cells: Contribution of Polycomb group proteins to renewal in a novel c-Kit+ culture model and in vivo. Development. 2016:dev. 142653. DOI: 10.1242/dev.142653

[75] Goldstein BJ, Hare J, Lieberman S, Casiano R. Adult human nasal mesenchymal stem cells have an unexpected broad anatomic distribution. International Forum of Allergy & Rhinology. 2013;**3**:550-555. DOI: 10.1002/alr.21153

[76] Tomé M, Lindsay SL, Riddell JS, Barnett SC. Identification of nonepithelial multipotent cells in the embryonic olfactory mucosa. Stem Cells. 2009;**27**:2196-2208. DOI: 10.1002/stem.130

[77] Lindsay SL, Johnstone SA, Mountford JC, Sheikh S, Allan DB, Clark L, et al. Human mesenchymal stem cells isolated from olfactory biopsies but not bone enhance CNS myelination in vitro. Glia. 2013;**61**(3):368-382. DOI: 10.1002/glia.22440

[78] Shafiee A, Kabiri M, Ahmadbeigi N, Yazdani SO, Mojtahed M, Amanpour S, et al. Nasal septum-derived multipotent progenitors: A potent source for stem cell-based regenerative medicine. Stem Cells and Development. 2011;**20**(12):2077-2091. DOI: 10.1089/scd.2010.0420

[79] Marshall CT, Guo Z, Lu C, Klueber KM, Khalyfa A, Cooper NG, et al. Human adult olfactory neuroepithelial derived progenitors retain telomerase activity and lack apoptotic activity. Brain Research. 2005;**1045**(1):45-56. DOI: 10.1016/j.brainres.2005.03.041

[80] Huang YS, Li I, Chueh SH, Hueng DY, Tai MC, Liang CM, et al. Mesenchymal stem cells from rat olfactory bulbs can differentiate into cells with cardiomyocyte characteristics. Journal of Tissue Engineering and Regenerative Medicine. 2015;**9**(12):E191-E201. DOI: 10.1002/term.1684

[81] Delorme B, Nivet E, Gaillard J, Häupl T, Ringe J, Devèze A, et al. The human nose harbors a niche of olfactory ectomesenchymal stem cells displaying neurogenic and osteogenic properties. Stem Cells and Development. 2009;**19**(6):853-866. DOI: 10.1089/scd.2009.0267

[82] Johnstone SA, Liley M, Dalby MJ, Barnett SC. Comparison of human olfactory and skeletal MSCs using osteogenic nanotopography to demonstrate bone-specific bioactivity of the surfaces. Acta Biomaterialia. 2015;**13**:266-276. DOI: 10.1016/j.actbio.2014.11.027

[83] Hauser S, Widera D, Qunneis F, Müller J, Zander C, Greiner J, et al. Isolation of novel multipotent neural crest-derived stem cells from adult human inferior turbinate. Stem Cells and Development. 2011;**21**(5):742-756. DOI: 10.1089/scd.2011.0419

[84] Stamegna J-C, Girard SD, Veron A, Sicard G, Khrestchatisky M, Feron F, et al. A unique method for the isolation of nasal olfactory stem cells in living rats. Stem Cell Research. 2014;**12**(3):673-679. DOI: 10.1016/j.scr.2014.02.010

[85] Girard SD, Devéze A, Nivet E, Gepner B, Roman FS, Féron F. Isolating nasal olfactory stem cells from rodents or humans. JoVE (Journal of Visualized Experiments). 2011;**54**:e2762. DOI: 10.3791/2762

[86] Ercolin ACM, Roballo KCS, Casals JB, Pieri NCG, Souza AF, Barreto RdSN, et al. Rabbit olfactory stem cells. Isolation protocol and characterization. Acta Cirurgica Brasileira. 2016;**31**(1):59-66. DOI: 10.1590/S0102-865020160010000009

[87] ALTUNBAŞ K, Yaprakci MV, Celik S. Isolation and characterization of olfactory stem cells from canine olfactory mucosa. Kafkas Universitesi Veteriner Fakultesi Dergisi. 2016;**22**(2). DOI: 10.9775/kvfd.2015.14277

[88] Rui K, Zhang Z, Tian J, Lin X, Wang X, Ma J, et al. Olfactory ecto-mesenchymal stem cells possess immunoregulatory function and suppress autoimmune arthritis. Cellular & Molecular Immunology. 2015. DOI: 10.1038/cmi.2015.82

[89] Féron F, Perry C, Girard SD, Mackay-Sim A. Isolation of adult stem cells from the human olfactory mucosa. Neural Progenitor Cells: Methods and Protocols. 2013;**1059**:107-114. DOI: 10.1007/978-1-62703-574-3_10

[90] Diaz-Solano D, Wittig O, Ayala-Grosso C, Pieruzzini R, Cardier JE. Human olfactory mucosa multipotent mesenchymal stromal cells promote survival, proliferation, and differentiation of human hematopoietic cells. Stem Cells and Development. 2012;**21**(17):3187-3196. DOI: 10.1089/scd.2012.0084

[91] Salgado AJ, Reis RL, Sousa NJ, Gimble JM. Adipose tissue derived stem cells secretome: Soluble factors and their roles in regenerative medicine. Current Stem Cell Research & Therapy. 2010;**5**(2):103-110. DOI: 10.2174/157488810791268564

[92] Yin F, Guo L, Meng C-Y, Liu Y-J, Lu R-F, Li P, et al. Transplantation of mesenchymal stem cells exerts anti-apoptotic effects in adult rats after spinal cord ischemia-reperfusion injury. Brain Research. 2014;**1561**:1-10. DOI: 10.1016/j.brainres.2014.02.047

[93] Sze SK, de Kleijn DP, Lai RC, Tan EKW, Zhao H, Yeo KS, et al. Elucidating the secretion proteome of human embryonic stem cell-derived mesenchymal stem cells. Molecular & Cellular Proteomics. 2007;**6**(10):1680-1689. DOI: 10.1074/mcp

[94] Frischknecht R, Chang K-J, Rasband MN, Seidenbecher CI. Neural ECM molecules in axonal and synaptic homeostatic plasticity. Progress in Brain Research. 2013;**214**:81-100. DOI: 10.1016/B978-0-444-63486-3.00004-9

[95] Porter D, Weremowicz S, Chin K, Seth P, Keshaviah A, Lahti-Domenici J, et al. A neural survival factor is a candidate oncogene in breast cancer. Proceedings of the National Academy of Sciences. 2003;**100**(19):10931-10936. DOI: 10.1073/pnas.1932980100

[96] Tahir R, Kennedy A, Elsea SH, Dickinson AJ. Retinoic acid induced-1 (Rai1) regulates craniofacial and brain development in Xenopus. Mechanisms of Development. 2014;133:91-104. DOI: 10.1016/j.mod.2014.05.004

[97] Patel SD, Chen CP, Bahna F, Honig B, Shapiro L. Cadherin-mediated cell–cell adhesion: Sticking together as a family. Current Opinion in Structural Biology. 2003;13(6):690-698. DOI: 10.1016/j.sbi.2003.10.007

[98] Lindsay SL, Barnett SC. Are nestin-positive mesenchymal stromal cells a better source of cells for CNS repair? Neurochemistry International. 2016;1-7. DOI: 10.1016/j.neuint. 2016.08.001

[99] Zilkha-Falb R, Kaushansky N, Kawakami N, Ben-Nun A. Post-CNS-inflammation expression of CXCL12 promotes the endogenous myelin/neuronal repair capacity following spontaneous recovery from multiple sclerosis-like disease. Journal of Neuroinflammation. 2016;13(1):1. DOI: 10.1186/s12974-015-0468-4

[100] Lima C, Escada P, Pratas-Vital J, Branco C, Arcangeli CA, Lazzeri G, Maia C, Capucho C, Hasse-Ferreira A, Peduzzi JD. Olfactory mucosal autografts and rehabilitation for chronic traumatic spinal cord injury. Neurorehabilitation and Neural Repair. 2009;**24**(1):10-22. DOI: 10.1177/1545968309347685

[101] Bas E, Van De Water TR, Lumbreras V, Rajguru S, Goss G, Hare JM, et al. Adult human nasal mesenchymal-like stem cells restore cochlear spiral ganglion neurons after experimental lesion. Stem Cells and Development. 2013;**23**(5):502-514. DOI: 10.1089/scd.2013.0274

[102] Nivet E, Vignes M, Girard SD, Pierrisnard C, Baril N, Devèze A, et al. Engraftment of human nasal olfactory stem cells restores neuroplasticity in mice with hippocampal lesions. The Journal of Clinical Investigation. 2011;**121**(7):2808-2820. DOI: 10.1172/JCI44489

[103] Murrell W, Wetzig A, Donnellan M, Feron F, Burne T, Meedeniya A, et al. Olfactory mucosa is a potential source for autologous stem cell therapy for Parkinson's disease. Stem Cells. 2008;**26**(8):2183-2192. DOI: 10.1634/stemcells.2008-0074

[104] Chen L, Qiu R, Xu Q. Mesenchymal stem cell therapy for neurodegenerative diseases. Journal of Nanoscience and Nanotechnology. 2014;**14**(1):969-975. DOI: 10.1166/jnn.2014.9126

[105] Di Trapani M, Bassi G, Ricciardi M, Fontana E, Bifari F, Pacelli L, et al. Comparative study of immune regulatory properties of stem cells derived from different tissues. Stem Cells and Development. 2013;**22**(22):2990-3002. DOI: 10.1089/scd.2013.0204

[106] McDonald C, Mackay-Sim A, Crane D, Murrell W. Could cells from your nose fix your heart? Transplantation of olfactory stem cells in a rat model of cardiac infarction. The Scientific World Journal. 2010;**10**:422-433. DOI: 10.1100/tsw.2010.40

[107] Ge L, Liu K, Liu Z, Lu M. Co-transplantation of autologous OM-MSCs and OM-OECs: A novel approach for spinal cord injury. Reviews in the Neurosciences. 2016;**27**(3):259-270

Section 2

Perspectives and Clinical Applications

Chapter 6

Mesenchymal Stem Cells: A Future Option for Intervening Disease Management

Harish C. Chandramoorthy,
Vishnu Balaji Radhakrishnan and
Narasimman Gurusamy

Additional information is available at the end of the chapter

http://dx.doi.org/10.5772/intechopen.68515

Abstract

Regeneration, revitalizing and reversal (RRR) are the primordial functions of the stem cells in the field of regenerative medicine. Though there are several cases of successful stem cell transplantation the reversal of metabolic diseases and the acquired secondary complications like chronic renal failure, neuropathy, stroke or vascular diseases are not well studied. The transplanted cells in many cases failed to home or graft in the host with no reason to attribute for such failures. Therefore, it becomes necessary to address these secondary complications with cellular therapy. The oxidative stress of the cells and tissues are attributed to the hostile microenvironment, not suitable for the survival of newly recruited cells. From our few animal studies and published literatures sources elsewhere, we foresee a huge potential for using mesenchymal stem cells (MSCs) to initially combat the secondary cardiovascular and neuronal complications in the management of the metabolic diseases. However, not all the stem cells have been tested in these lines, and further we do not know, whether all the progenitor cells from various sources and origin will behave like MSCs, which needs to be studied extensively.

Keywords: mesenchymal stem cells (MSCs), secondary complications, metabolic diseases, microenvironment

1. Introduction

In the past 4 decades of cell therapy, many hematological diseases, both malignant and non-malignant origin, have been treated with wide success, prominently with hematopoietic stem cells (HSCs) [1]. With the growth of regenerative medicine and stem cell research, various other sources of the progenitor stem cells have been identified at different niches of the organs, and few of them are well characterized and tested for its ability to be better performing than HSCs in general. Mesenchymal stem cells (MSCs) are one such progenitor population identified and well characterized for their ability to differentiate in a rigid stress environment like oxidative stress or reperfusion injury, which would usually kill the cells or tissues [2]. There has not been enough investigation on the response of stem cells or progenitor cells in general to the stimuli of biological or mechanical origin in-vivo. Some of our experiences and literature evidences [3] have shown $[Ca^{2+}]_i$ playing a major role in the death or survival of the stem cells through oxidative stress observed at the site of pathological manifestations [4, 5]. Recent studies have shown the involvement of the mitochondria by its Ca^{2+}-buffering homeostatic mechanisms to be largely playing a role in cell sustenance toward survival and differentiation [6]. Much was taught on the stem cells regenerative capacity by grafting, homing and repairing by differentiation of the transplanted stem cells. However, for many years, there was no mechanistic definition for the failure of the stem cells other than physiological parameters like viability of the cells or volume of the cells used in the transplantation [7–9]. The microenvironment which largely supports repair by mobilization of the MSCs or in general the progenitor stem cells required experimental evidence on the survival time, dose, frequency and preconditioning of the repair area. In many cases, the stress is characterized by the irregular Ca^{2+} homeostasis resulting in triggering of destructive signals like oxidants and transcription factors responsible for the eventual cell death [10]. Further physiologically normal Ca^{2+} signaling is an essential part of the cell growth and differentiation, and when the homeostasis is challenged, the Ca^{2+} acts as a trigger of self-destruction in the matured cells [11, 12]. The role of Ca^{2+} in the progenitor cells may induce signals of survival as observed in the tumor microenvironment, which might result in the destruction by the host cells. **Table 1** gives Ca^{2+} channels associated with the MSCs. It can be noted that MSCs offer a good threshold to these cellular factors resulting in the sustained survival. However, these Ca^{2+} thresholds are broken when the disturbance of the cellular Ca^{2+} is transferred to mitochondria, resulting in the loss of the mitochondrial membrane potential ($\Delta\psi_m$) and leading cellular ROS (cROS) mediated to mitochondrial ROS (mROS) and thereby apoptotic signals skewing the cells toward death phenotype [3, 5]. Cellular mechanisms like survival, death or differentiation require a clear understanding on the normal calcium homeostasis, thereby equilibrium between $[Ca^{2+}]_c$ and $[Ca^{2+}]_m$ existing within the cells [13]. Cells of different tissue origins and physiological functions differ in their ability to respond to these stress signals while general speculation is that progenitor cells, either resident at the niche or mobilized to the site of damage, usually have higher threshold which makes its activity of regeneration successful [14]. There are studies which indicate the dose dependency of the MSCs for successful regeneration, and we speculate that the ability of the MSCs to tolerate the stress at the pathological site is the mechanism behind the dose dependency [15, 16]. However, another dimension of MSC's potential is in the therapeutic modulation of the given disease conditions or at least in animal models, through release of inducible factors without direct involvement of the MSCs by division or differentiation [17, 18]. In such cases, the tissue revival post MSC treatment shows no trace of the transplanted cells by the common tracking methods like 5(6)-Carboxyfluorescein diacetate N-succinimidyl ester (CFSE) chase or Green Fluorescent Protein (GFP). Additionally, MSCs are known for their immunomodulation capability and stromal character in the regeneration of the organs and

Channel/receptor	Type of MSCs	Species	Differentiation	Functional expression
Voltage-gated Ca^{2+} channels (VGCC)				
VGCC:LT	AMSCs	Human	Undifferentiated	No
VGCC:LT	AMSCs	Human	Undifferentiated neuronal	No
VGCC:LP/QN	AMSCs	Rat	Undifferentiated neuronal	No/yes
VGCC:LT	BMSCs	Human	Undifferentiated	Yes
VGCC:LT	BMSCs	Human	Undifferentiated	Yes
VGCC:LP/QTNR	BMSCs	Human	Undifferentiated	Yes
VGCC:LP/QN	BMSCs	Murine	Neuronal	Yes
VGCC:LT	BMSCs	Rat	Undifferentiated	Yes
VGCC:L	BMSCs	Rat	Osteogenic	Yes
VGCC:LP/QN	BMSCs	Rat	Undifferentiated neuronal	No/yes
Intracellular Ca^{2+} stores				
InsP3 R RyR	AMSCs	Human	Undifferentiated	Yes
InsP3	AMSCs	Human	Adipocyte	Yes
InsP3	AMSCs	Human	Adipocyte	Yes
InsP3 R1-3 RyR 1-3	BMSCs	Murine	Neuronal	Yes
InsP3RyR	BMSCs	Human	Undifferentiated	Yes
InsP3	BMSCs	Human	Adipocytes	Yes
P2 purinergic receptors				
P2X, P2Y1	AMSCs	Human	Adipogenic osteogenic	Yes
P2XP2Y	AMSCs	Rat	Undifferentiated neuronal	Yes
P2Y2	BMSCs	Rat	Undifferentiated	Yes
P2XP2Y	BMSCs	Rat	Undifferentiated neuronal	Yes
P2Y1P2X	BMSCs	Human	Undifferentiated	Yes
P2Y1	BMSCs	Human	Adipogenic	Yes
Oxytocin (OT) and vasopressin (AVP) receptors				
AVP V1a, AVP V1b AVP V2	AMSCs	Human	Adipogenic	Yes
OT R	AMSCs	Mouse	Neuronal	–
OT R AVP-V1a AVP V2	BMSCs	Rat	Undifferentiated	Yes

Channel/receptor	Type of MSCs	Species	Differentiation	Functional expression
OT R AVP-V1	BMSCs	Rat	Undifferentiated neuronal	Yes
OT R	AMSCsBMSCs	Human	Adipogenic osteogenic	Yes

Abbreviations: AMSCs, adipose tissue derived mesenchymal stromal cells; AVP, vasopressin; BMSCs, bone marrow mesenchymal stromal cells; $InsP_3$, inositol 1,4,5-trisphosphate receptor; LVA, low voltage activated Ca^{2+} channels; OT, oxytocin; OT R, oxytocin receptor; and RyR, ryanodine receptor. **Table 1** is modified and reproduced from the original **Table 1** with written permission from Forostyak et al. [27].

Table 1. Expression of the Ca^{2+} channels in the MSCs.

structural elements in the organ or tissue [19]. There are few reports and studies on the scope and wide use of MSCs in cellular therapy either individually or combined with HSCs, creating a microenvironment for better homing, grafting and differentiation for HSCs [20, 21]. From the above observations, it is clear that alterations in the microenvironment are crucial for MSC's behavior toward differentiation or other modulation properties. Not only changes associated with Ca^{2+} but also changes in the oxygen concentration can alter the MSCs behavior drastically. There is no clear-cut explanation on what makes MSCs unique though it has been well studied for its in-vitro and in-vivo differentiation as well as therapeutic ability without integration or differentiation at the site of transplantation [22]. Few observations like loss of differentiation capacity at higher passages can be dubbed to senescence observed in vivo or in many failure models of MSC cell therapy [23]. Thus, the cellular senescence can be attributed as the MSCs respond to prolonged or higher oxidative stress encountered at the affected tissues [24–26]. But still the promising aspect of MSCs is from their anatomical locations like Bone Marrow (BM), where these cells are at a constant interaction with the immune cells. We do not know whether this aspect of the BM MSCs is responsible of enhanced expression for the cytokine receptors or its functional expression of the inducible soluble factors or its immunomodulatory properties.

However, the scope of the current topic is to check how far the MSCs without any subsect difference are useful as a promising allogeneic source for the functional restoration of the organ or tissues. Addressing the issue of higher threshold for the MSCs to counteract the oxidative stress, It is well known that MSC's immunomodulate the host immune responses and secrete factors for therapeutic amelioration of the disease complications. We do have substantial data to directly relate the ability of secretome for the therapeutic activities with controlled release ex-vivo in regulated bioreactors. In all these aspects, the reactivity of the MSCs in the microenvironment toward various signals decides the survival, differentiation, modulation or the reactivity toward the repair signals.

2. Mesenchymal stem cells react differently to stress pathology

Cellular stress is mostly mechanosensitive or chemosensitive in nature. Many studies have shown that intracellular Ca^{2+} signaling is closely interconnected with mechanical properties of a cell. The flow of calcium from the extracellular matrix (ECM) through mechanosensitive calcium channels like transient receptor potential (TRP) or Stromal Interaction Molecule (STIM), Ca2+ release-activated Ca2+ (CRAC) channels is closely interconnected to the spatiotemporal intracellular Ca^{2+} signaling (**Figure 1**). Adult differentiated cells exhibit varied calcium dynamics depending on their anatomical location, tissue origins and physiological functions [27]. Cells of cardiac and vascular tissues, for example, withstand more stress, and their Ca^{2+} buffering ability is higher than other cells [28, 29]. There are many studies in various matured cells on the patterns of the Ca^{2+} oscillations regulated by signaling proteomes [30].

Largely, current understanding of mitochondrial Ca^{2+} homeostasis and regulation by the mitochondrial uniporter (MCU), a Ca^{2+} transmembrane protein identified in recent years, have made it more easier in understanding the cell reactivity to the external stress [5, 13]. When the threshold of the cells to withstand the Ca^{2+} oscillations is exceeding the buffering limits, the cells are skewed to death phenotype by oxidative mechanisms [31]. The threshold of the progenitor cells like MSCs makes it unique in understanding the Ca^{2+} homeostasis, for example, human MSCs (hMSCs) exhibit spontaneous Ca^{2+} oscillations, a phenomenon not routine in other matured cells and progenitors with a few exceptions [32] though like other cell types in MSCs Ca^{2+} oscillations are triggered by influx of extracellular Ca^{2+} and release from endoplasmic reticulum (ER) via inositol 1,4,5-trisphosphate receptors (IP3Rs) and ryanodine receptors by calcium-induced calcium release [27]. There are studies that suggest mesenchymal stem cells respond to the extracellular Ca^{2+} levels sensed by calcium sensing receptor (CaSR) in the cell membrane for its proliferation and differentiation [33]. Though low Ca^{2+} levels are favorable for all cells in general, higher Ca^{2+} levels beyond the threshold are detrimental to MSCs. In general, the physiological role of the Ca^{2+} homeostasis largely regulates differentiation, proliferation and cell survival at the site of repair [30]. Studies have revealed the Ca^{2+} handling properties of the precursors are similar to the adult differentiated cells as observed in the neuronal precursors compared with differentiated neuronal cells. There are reports of enhanced Ca^{2+} accumulation in the precursors or embryonic cell types; however, the success of the differentiation largely depends on the microenvironment of the tissue where the progenitors are deployed [34–37]. Further the intracellular compartmentalization and capacity of the various organelles response to heavy $[Ca^{2+}]_i$ is another factor, which might be a factor for sustained survival of the transplanted MSCs. The apparent Ca^{2+} threshold of cells $[Ca^{2+}]_i$ per say basal or resting is ~50–100 nM. These physiological levels of the $[Ca^{2+}]_i$ can rapidly rise to ~1–10 μM on stimulation with mechano or chemosensitive factors [5]. The regulation and balance of Ca^{2+} homeostasis do not stop here when these signals can activate the ER to release the stored intracellular Ca^{2+} which thereby promotes the stress inside cell. The role of the mitochondria and its ability to buffer $[Ca^{2+}]_i$ are several folds higher than the cytoplasmic threshold, and thereby the role of mitochondria cannot be undermined in the survival of the progenitor cells, especially stromal origin cells like MSCs [38]. Hence, the pathological fate of the transplanted or mobilized MSCs does not only depend on the homeostasis of $[Ca^{2+}]_i$ but also on the $[Ca^{2+}]_m$ in evading the stress phenotype for better differentiation and repair [39, 40]. Many studies on the isolated mitochondria suggest that the Ca^{2+} buffering capacity of mitochondria is 100-fold higher than the basal or resting $[Ca^{2+}]_i$ or intracellular release on the stimulation of Ca^{2+} release in the cytoplasm [41–43]. This phenomenon is observed during the physiological stress arising due to ischemia or reperfusion and can be experimentally induced with a known Ca^{2+} agonist like histamine or thapsigargin.

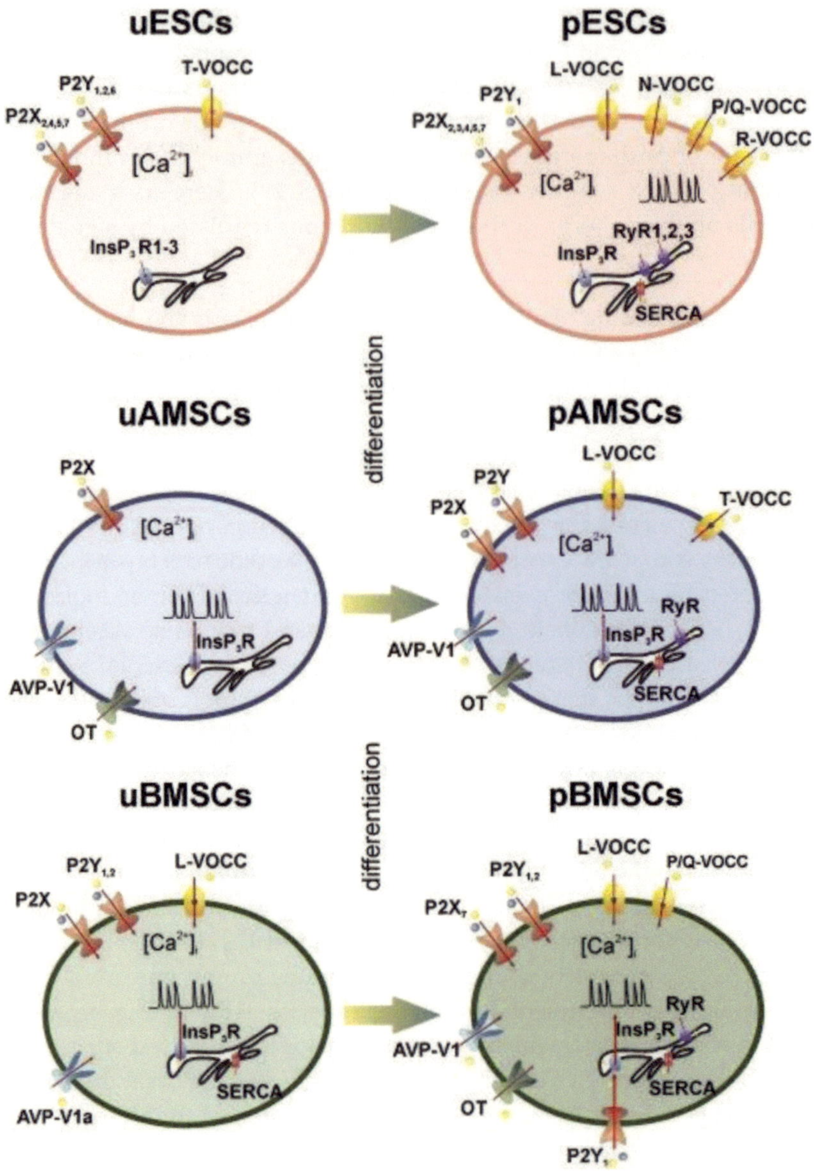

Figure 1. A schematic drawing showing the functional expression of Ca^{2+}-sensitive channels and receptors in ESCs, AMSCs and BMSCs. In particular, VGCC, InsP3, inositol trisphosphate receptors (InsP3R), RyR, P2 purinergic, vasopressin and oxytocin receptors, as well as spontaneous Ca^{2+} oscillations and sarcoendoplasmic reticulum Ca^{2+} ATPase (SERCA pump), are shown. Reproduced with written permission from Forostyak et al. [27].

3. Mechanism behind the mesenchymal stem cell repair

Traditional understanding on therapeutic properties of the MSCs or any type of progenitor stem cells is by direct homing, differentiation and repopulation with the normal phenotype tissues at the site of pathology [1]. However, in some cases, the transplantation is not successful and does not have a clear-cut reason for such failure in spite of all favorable pre- and pro-clinical parameters [44]. The current understanding on the mechanism of the MSC therapy, when supplied exogenously, is homing at the sites of injury and differentiates into adult cell type. In few cases, though we do not know the fate of the MSCs post transplantation or do not follow the traditional understanding, however, the lesions are healed. These phenomena opened a new area of insightful research on what actually the MSCs do at the niche apart from proliferation and differentiation at the site of tissue damage. The term microenvironment simply implies on suitable or favorable conditions promoted by recruited progenitor MSCs at the site of pathology [45, 46]. Transplanted MSCs release soluble factors like cytokines, chemokines, interleukins, secondary messenger molecules and insoluble or physical factors like biomechanical forces, ions and so on for the cell survival. The released factors not only modulate cell death but also induce pro-survival mechanisms. These factors further enrich the tissue repair mechanisms reversing the pathology [18, 47]. The question of the resident stem cells and their failure to resolve the pathology is another important area which is unclear. In case, if the microenvironment is unfavorable for the resident or mobilized progenitor cells, how far can the transplanted cells create a conducive environment to sustain the hostile tissue for repair? There are few well-documented studies, which show the micro-physiology of the microenvironment, like changes associated with oxygen concentration and physiological stress, which can strongly affect the behavior of the MSCs [48–50]. The other factors, which affect the microenvironment, are the immune cells and other soluble and insoluble factors. These altogether affect the desired outcome of the transplanted stem cells. The local immune response to the MSCs results in the induction of the inflammatory mediators, which are not favorable for the MSCs to divide or differentiate [51, 52]. Therefore, microenvironment plays a crucial role in the success of therapeutic MSCs.

Basically, for any cell to act or react, stimuli or cell-to-cell interactions are required [53]. There are many modes by which these interactions or signal transmissions can take place. Further, these signal transmissions are different with the normal or pathological scenario. One explanation is that the release of the cytokines like IL-6 and Vascular Endothelial Growth Factor (VEGF) can induce pro-survival and oppose apoptosis as observed in the tumor microenvironment [54]. The best-explained mechanism is inter- and intra-cellular transmission of the mechanical stimuli, which affect the gene expression of the pro-survival factors [55]. It is unclear how the mechanical forces are tuned into biological signals of life and destruction. Further, these mechanostatic forces are responsible for large number of transcriptional gene regulations affecting the progression or repair of the tissue pathology. Many studies have explained the link between the mechanical stimuli and the Ca^{2+} homeostasis [56, 57]. Mechanical stimuli activate the Ca^{2+} from the ER within the cells or potentiate the entry of extracellular Ca^{2+} which further triggers the transcriptional regulation of the pro-survival cellular factors [58].

The repairing capabilities of MSCs have been reported in various tissues, including the brain, heart, kidney, pancreas and skin [59–62]. The mechanism through which the MSC-mediated tissue regeneration may vary from type of injury or tissues involved. For an instance, the increased expression of stromal cell-derived factor 1 (SDF-1) at the site of injury can attract the MSCs to the injured tissue [63, 64]. The expression of C-X-C chemokine receptor type 4 (CXCR4) by MSCs regulates the adhesion of MSCs to endothelial cells. This has been shown to be a critical step for MSCs to migrate to the target tissue. Thus, the expression of the CD184

(Fusin) is important to expedite the interaction between SDF-1 and CXCR4 system, which play an important role in the survival and migration of bone marrow stromal cells after transplantation into mice cerebral infarct [64]. MSCs can enhance the angiogenesis at the injured tissue, where the level of TGF-β1 is dramatically increased. TGF-β1 is known to stimulate the synthesis of VEGF in MSCs in order to promote the angiogenesis [65] which may augment the endothelial progenitor functions. Formation of new tissues and organs in the embryo requires the transitions from mesenchyme into epithelium that is the mesenchymal-epithelial transition [66]. We cannot speculate whether such a property of the mesenchymal-epithelial transition is observed at late progenitor stages of the MSCs. Further, such activity needs to be clearly elucidated.

Capabilities of MSCs to differentiate into hepatocytes, insulin-producing cells, neural cells, osteoblasts, chondrocytes, adipocytes, fibroblasts and so on are well documented and reproducible by many studies [67]. These properties are not only observed in in-vitro conditions but also in some in-vivo small animal studies, which have revealed the transformation (differentiation) of the MSCs into adult lineages [68]. These are further explained with the presence of the tracker-like GFP [69], indicating the newly formed cells with the presence of the GFP. In human studies, many types of the MSCs expressing pancreatic duodenal homeobox 1 (*Pdx1*) gene have been shown to differentiate into insulin-producing cells or functions of pancreatic β cells [70]. There are many studies showing successful regeneration of skeletal tissues such as bone, cartilage, tendon and intervertebral discs from various sources of MSCs, including MSCs from the foreskin and dental pulp [71–74]. In some preclinical studies, a set of MSCs expressing exogenous glial cell-derived neurotrophic factor (GDNF) and brain-derived neurotrophic factor (BDNF) have shown to reduce stroke-induced lesion volume and further improve neovascularization [75, 76].

There are undoubtedly many in-vitro and in-vivo studies addressing the direct repair potential and the uses of providing conducive environment for the repair by the MSCs. What needs to be addressed here is whether all the subsets of the MSCs located at various anatomical niches are capable of performing the repair irrespective of their small deviations in the surface marker expression. Looking at the other functions of the MSCs, such as immune suppressive or modulatory effects, the therapeutic infusions of MSCs in experimental models of autoimmune encephalomyelitis showed reduced infiltration of T cells and macrophages followed by a reduction of demyelination in the brain and in the spinal cord [77–80]. Repeated administration of MSCs derived from a patient's mother completely cured a young patient suffering from severe grade IV graft-versus-host disease (GVHD); this is another observation, which clearly showed modulation of the properties of the infused MSCs paving a way for another dimension of the MSCs repairing property [21].

4. Mesenchymal stem cells: A tailor-made therapeutic approach in the future of medicine

Today, there is a growing need for novel technologies to restore, maintain and enhance organ function. Since the 1990s, stem cells have originated as a novel therapeutic option for regenerative medicine. Human embryonic stem (ES) cells, mesenchymal stem cells (MSCs) and induced pluripotent stem cells (iPSC) have appeared as potential sources for therapeutic intervention for future.

There has long been a need for unique approaches to challenge the world of diseases and disorders. The skeletal tissue damage is one such clinical condition which requires restoration, maintenance and enhanced organ function. The use of skeletal-derived stromal cells (MSCs) is a better option and an attractive choice. Though human ES cells, MSCs and iPSCs are regarded as potential sources for regenerative medicine and tissue engineering applications [81], they remained predominantly in the realm of laboratory-based in vitro investigation and in vivo animal modeling; however, more recently, a number of research centers have bridged the translational gap, from bench to clinic with few successes. Although the potential of MSCs to regenerate various tissues is known, it is increasingly renowned that the MSCs can exert immune and inflammation modulatory effects [82] through a large number of secreted bioactive factors including anti-scarring, angiogenic, anti-apoptotic as well as factors enhancing tissue remodeling [83, 84]. This mechanism may elucidate the interesting observation of the presence of therapeutically relevant outcomes of MSCs after systemic or local transplantation in a number of tissue injury models, for example, ischemic brain injury and myocardial infarction in the presence of low tissue engraftment of MSCs [85]. Though we do not know the success of these cases in humans, it is still promising unless trials are initiated in these areas of translational research.

The number of the clinical trials using MSCs till 2017 is furnished in **Figure 2**. Interestingly, both autologous and allogeneic MSCs have been employed in these studies as they are believed to be less immunogenic. According to National Institutes of Health (NIH) clinical trials database, predominantly bone and cartilage regeneration (23%), neuronal (21%), cardiovascular (16%) and autoimmune disorders [9] have been highly focused among other therapeutic approaches using MSCs.

The source and environmental niches are playing the critical role on MSCs; they have to be considered while studying their biological activity and clinical applications. Furthermore, the continuous search for novel and potent sources that might be suitable for specific regenerative applications is needed. Recently, we compared the MSC-like cell populations obtained from alternative sources: the human adipose tissue, adult skin and newborn foreskin, with the standard phenotype of human bone marrow MSC. Our whole genome analysis has revealed a common MSC molecular signature composed of 33 CD markers including known MSC markers and several novel markers, for example CD165, CD276 and CD82. MSCs obtained from different sources exhibit significant differences in their proliferation, multipotency and molecular phenotype, which should be considered before applications in the clinical protocols [86]. The skin-derived stromal cells have shown the endothelial lineage differentiation in-vitro, and the angiogenic role with potential contribution to blood vessel formation in ex-vivo Chorioallantoic Membrane Assay (CAM) model is an excellent start for the pre-clinical considerations for the skin-derived MSCs. Therefore, human skin stromal cells are valuable resources that might be useful in applications requiring enhanced angiogenesis or in areas such as ischemic diseases [87–89]. Furthermore, these cells could be employed in tissue engineering and cell-based therapy in which vascularization is an essential component.

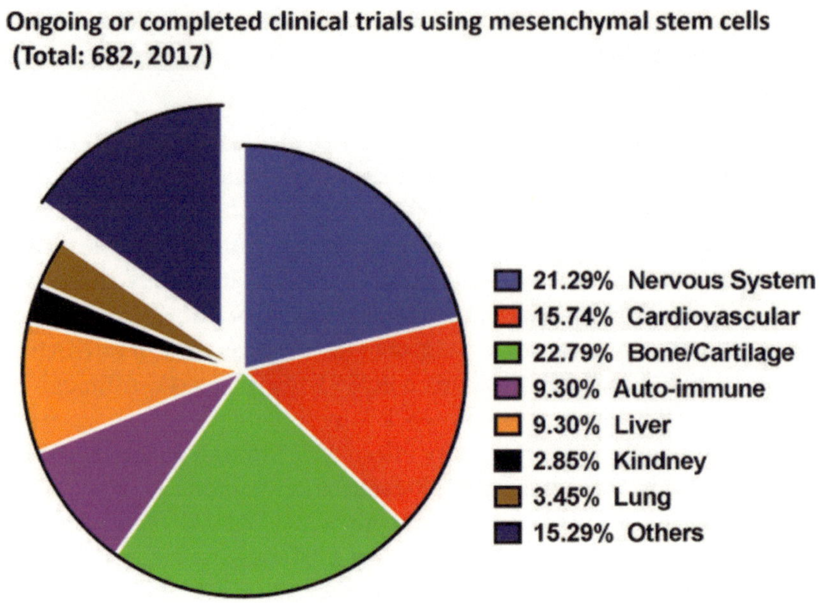

Figure 2. A pie chart showing the ongoing and/or completed clinical trials with MSCs (total of 682, 2017), adapted from http://clinicaltrials.gov.

Currently, several MSC-based therapeutic protocols are being tested in an increasing number of clinical conditions in phase I/II and III clinical trials. At the website of the National Institutes of Health, the USA (http://clinicaltrials.gov), overall, till 2017, the status pertaining to hMSCs-based clinical trials shows 682 studies and in that 438 were closed (including completed, 168, and withdrawn 12), 134 are unknown and finally 244 are under recruiting conditions. Results from these clinical trials are expected to have major impact on the treatment of several disease conditions.

Much progress has been made over the last decade in stem cell technology, and a steady stream of clinical applications and trials have followed on these advances. However, the approaches outlined above provide only limited evidence of current status [90, 91]. To date, there remains a paucity of randomized controlled trials to demonstrate the efficacy of many of these tissue-engineered/stem cell approaches. Thus, to date, it is difficult to recommend any of these strategies as standard therapy. Nevertheless, advances in basic research as well as from clinical trials of MSC-based therapy are expected to provide options for therapeutic interventions for tissue regeneration in multiple organs that will address the current unmet needs of an increasing number of patients suffering from age-related degenerative diseases.

5. Conclusion

Though MSCs have shown some promising therapeutic and transplantation potential, its use in regenerative medicine is primitive. In many aspects of the therapeutic approach, the results of the MSC applications are varied as well as affirmative, suggesting that more research needs to be carried out. The critical feature of the MSCs is their activation in the microenvironment or modulation of or by the host immune system, which makes it much more difficult to understand and study the mechanism of regeneration. There are various opinions on the route of administration of MSCs like in vivo, or local transplantation on site of the organ on the tissue repair is still a subject of debate. Many studies cited in these chapters are individual observations at various centers and still need translation to bedside from the bench. The few clinical trials listed are at different phases and collectively may require more time for MSCs successful clinical applications.

Acknowledgements

We thank Dr. Ahmed AL Hakami, Chairman, Center for Stem Cell Research, College of Medicine, King Khalid University, Saudi Arabia, for his constant encouragement and support.

Author details

Harish C. Chandramoorthy[1*], Vishnu Balaji Radhakrishnan[2] and Narasimman Gurusamy[3]

*Address all correspondence to: ccharishjabali@gmail.com

1 Center for Stem Cell Research Unit, College of Medicine, King Khalid University, Abha, Saudi Arabia

2 Stem Cell Unit, Department of Anatomy, College of Medicine, King Khalid University Hospital, King Saud University, Riyadh, Saudi Arabia

3 Department of Pharmacology, College of Pharmacy, King Khalid University, Abha, Saudi Arabia

References

[1] Squillaro T, Peluso G, Galderisi U. Clinical trials with mesenchymal stem cells: An update. Cell Transplantation. 2016;**25**(5):829-848. [Epub: October 2, 2015]

[2] Fernandez Vallone VB, Romaniuk MA, Choi H, Labovsky V, Otaegui J, Chasseing NA. Mesenchymal stem cells and their use in therapy: What has been achieved? Differentiation; Research in Biological Diversity. 2013;**85**(1-2):1-10. [Epub: January 15, 2013]

[3] Doonan PJ, Chandramoorthy HC, Hoffman NE, Zhang X, Cardenas C, Shanmughapriya S, et al. LETM1-dependent mitochondrial Ca^{2+} flux modulates cellular bioenergetics and proliferation. FASEB Journal: Official Publication of the Federation of American Societies for Experimental Biology. 2014;**28**(11):4936-4949. [Epub: August 1, 2014]

[4] Haidara MA, Assiri AS, Youssef MA, Mahmoud MM, Eajaz A, Al-Hakami A, et al. Differentiated mesenchymal stem cells ameliorate cardiovascular complications in diabetic rats. Cell and Tissue Research. 2015;**359**(2):565-575. [Epub: November 22, 2014]

[5] Mallilankaraman K, Doonan P, Cardenas C, Chandramoorthy HC, Muller M, Miller R, et al. MICU1 is an essential gatekeeper for MCU-mediated mitochondrial Ca(2+) uptake that regulates cell survival. Cell. 2012;**151**(3):630-644. [Epub: October 30, 2012]

[6] Rharass T, Lemcke H, Lantow M, Kuznetsov SA, Weiss DG, Panakova D. Ca^{2+}-mediated mitochondrial reactive oxygen species metabolism augments Wnt/beta-catenin pathway activation to facilitate cell differentiation. The Journal of Biological Chemistry. 2014;**289**(40):27937-27951. [Epub: August 16, 2014]

[7] Lin HT, Otsu M, Nakauchi H. Stem cell therapy: An exercise in patience and prudence. Philosophical Transactions of the Royal Society of London Series B, Biological Sciences. 2013;**368**(1609):20110334. [Epub: November 21, 2012]

[8] Galderisi U, Giordano A. The gap between the physiological and therapeutic roles of mesenchymal stem cells. Medicinal Research Reviews. 2014;**34**(5):1100-1126. [Epub: May 29, 2014]

[9] Dalle JH, Peffault de Latour R. Allogeneic hematopoietic stem cell transplantation for inherited bone marrow failure syndromes. International Journal of Hematology. 2016;**103**(4):373-379. [Epub: February 14, 2016]

[10] Becquart P, Cruel M, Hoc T, Sudre L, Pernelle K, Bizios R, et al. Human mesenchymal stem cell responses to hydrostatic pressure and shear stress. European Cells & Materials. 2016;**31**:160-173. [Epub: February 20, 2016]

[11] Barradas AM, Fernandes HA, Groen N, Chai YC, Schrooten J, van de Peppel J, et al. A calcium-induced signaling cascade leading to osteogenic differentiation of human bone marrow-derived mesenchymal stromal cells. Biomaterials. 2012;**33**(11):3205-3215. [Epub: January 31, 2012]

[12] Lan F, Lee AS, Liang P, Sanchez-Freire V, Nguyen PK, Wang L, et al. Abnormal calcium handling properties underlie familial hypertrophic cardiomyopathy pathology in patient-specific induced pluripotent stem cells. Cell Stem Cell. 2013;**12**(1):101-113. [Epub: January 8, 2013]

[13] Mallilankaraman K, Cardenas C, Doonan PJ, Chandramoorthy HC, Irrinki KM, Golenar T, et al. MCUR1 is an essential component of mitochondrial Ca^{2+} uptake that regulates cellular metabolism. Nature Cell Biology. 2012;**14**(12):1336-1343. [Epub: November 28, 2012]

[14] Louhivuori LM, Louhivuori V, Wigren HK, Hakala E, Jansson LC, Nordstrom T, et al. Role of low voltage activated calcium channels in neuritogenesis and active migration of embryonic neural progenitor cells. Stem Cells and Development. 2013;**22**(8):1206-1219. [Epub: December 14, 2012]

[15] Sister ML, Brown F, Johnson C. An employer follow-up study of De Paul Hospital School of Nursing classes of 1967, 1968, 1969, and 1970. Virginia Nurse Quarterly. 1974;**42**(4):53-65. [Epub: January 1, 1974]

[16] Winkler T, von Roth P, Matziolis G, Mehta M, Perka C, Duda GN. Dose-response relationship of mesenchymal stem cell transplantation and functional regeneration after severe skeletal muscle injury in rats. Tissue Engineering Part A. 2009;**15**(3):487-492. [Epub: August 5, 2008]

[17] Chevillotte-Brivet P, Salou G, Meunier-Lemesle D. Missense exonic mitochondrial mutation in cytochrome b gene of Saccharomyces cerevisiae, resulting in core protein deficiency in complex III of the respiratory chain. Current Genetics. 1987;**12**(2):111-117. [Epub: January 1, 1987]

[18] Kawai T, Katagiri W, Osugi M, Sugimura Y, Hibi H, Ueda M. Secretomes from bone marrow-derived mesenchymal stromal cells enhance periodontal tissue regeneration. Cytotherapy. 2015;**17**(4):369-381. [Epub: January 18, 2015]

[19] Smulewicz JJ. Lead lines at the iliac crest and "early" diagnosis of lead poisoning. The American Journal of the Medical Sciences. 1974;**267**(1):49-51. [Epub: January 1, 1974]

[20] Battiwalla M, Hematti P. Mesenchymal stem cells in hematopoietic stem cell transplantation. Cytotherapy. 2009;**11**(5):503-515. [Epub: September 4, 2009]

[21] Zhao K, Liu Q. The clinical application of mesenchymal stromal cells in hematopoietic stem cell transplantation. Journal of Hematology & Oncology. 2016;**9**(1):46. [Epub: May 20, 2016]

[22] Kariminekoo S, Movassaghpour A, Rahimzadeh A, Talebi M, Shamsasenjan K, Akbarzadeh A. Implications of mesenchymal stem cells in regenerative medicine. Artificial Cells, Nanomedicine, and Biotechnology. 2016;**44**(3):749-757. [Epub: January 14, 2016]

[23] Estrada JC, Torres Y, Benguria A, Dopazo A, Roche E, Carrera-Quintanar L, et al. Human mesenchymal stem cell-replicative senescence and oxidative stress are closely linked to aneuploidy. Cell Death & Disease. 2013;**4**:e691. [Epub: June 29, 2013]

[24] Bonab MM, Alimoghaddam K, Talebian F, Ghaffari SH, Ghavamzadeh A, Nikbin B. Aging of mesenchymal stem cell in vitro. BMC Cell Biology. 2006;**7**:14. [Epub: March 15, 2006]

[25] Wagner W, Horn P, Castoldi M, Diehlmann A, Bork S, Saffrich R, et al. Replicative senescence of mesenchymal stem cells: A continuous and organized process. PloS One. 2008;**3**(5):e2213. [Epub: May 22, 2008]

[26] Turinetto V, Vitale E, Giachino C. Senescence in human mesenchymal stem cells: Functional changes and implications in stem cell-based therapy. International Journal of Molecular Sciences. 2016;**17**(7):1164.

[27] Forostyak O, Forostyak S, Kortus S, Sykova E, Verkhratsky A, Dayanithi G. Physiology of Ca(2+) signalling in stem cells of different origins and differentiation stages. Cell Calcium. 2016;**59**(2-3):57-66. [Epub: February 26, 2016]

[28] Kalogeris T, Baines CP, Krenz M, Korthuis RJ. Cell biology of ischemia/reperfusion injury. International Review of Cell and Molecular Biology. 2012;**298**:229-317. [Epub: August 11, 2012]

[29] Soffe R, Baratchi S, Tang SY, Nasabi M, McIntyre P, Mitchell A, et al. Analysing calcium signalling of cells under high shear flows using discontinuous dielectrophoresis. Scientific Reports. 2015;**5**:11973. [Epub: July 24, 2015]

[30] Tonelli FM, Santos AK, Gomes DA, da Silva SL, Gomes KN, Ladeira LO, et al. Stem cells and calcium signaling. Advances in Experimental Medicine and Biology. 2012;**740**:891-916. [Epub: March 29, 2012]

[31] Gandhirajan RK, Meng S, Chandramoorthy HC, Mallilankaraman K, Mancarella S, Gao H, et al. Blockade of NOX2 and STIM1 signaling limits lipopolysaccharide-induced vascular inflammation. The Journal of Clinical Investigation. 2013;**123**(2):887-902. [Epub: January 26, 2013]

[32] Sun S, Liu Y, Lipsky S, Cho M. Physical manipulation of calcium oscillations facilitates osteodifferentiation of human mesenchymal stem cells. FASEB Journal: Official Publication of the Federation of American Societies for Experimental Biology. 2007;**21**(7):1472-1480. [Epub: February 1, 2007]

[33] Xu Z, Yan L, Ge Y, Zhang Q, Yang N, Zhang M, et al. Effect of the calcium sensing receptor on rat bone marrow-derived mesenchymal stem cell proliferation through the ERK1/2 pathway. Molecular Biology Reports. 2012;**39**(7):7271-7279. [Epub: February 9, 2012]

[34] Janowski E, Cleemann L, Sasse P, Morad M. Diversity of Ca^{2+} signaling in developing cardiac cells. Annals of the New York Academy of Sciences. 2006;**1080**:154-164. [Epub: November 30, 2006]

[35] Itzhaki I, Rapoport S, Huber I, Mizrahi I, Zwi-Dantsis L, Arbel G, et al. Calcium handling in human induced pluripotent stem cell derived cardiomyocytes. PloS One. 2011;**6**(4):e18037. [Epub: April 13, 2011]

[36] Lee P, Klos M, Bollensdorff C, Hou L, Ewart P, Kamp TJ, et al. Simultaneous voltage and calcium mapping of genetically purified human induced pluripotent stem cell-derived cardiac myocyte monolayers. Circulation Research. 2012;**110**(12):1556-1563. [Epub: May 10, 2012]

[37] Herron TJ. Calcium and voltage mapping in hiPSC-CM monolayers. Cell Calcium. 2016;**59**(2-3):84-90. [Epub: February 29, 2016]

[38] Pinto MC, Kihara AH, Goulart VA, Tonelli FM, Gomes KN, Ulrich H, et al. Calcium signaling and cell proliferation. Cellular Signalling. 2015;**27**(11):2139-2149. [Epub: August 16, 2015]

[39] Rehman J. Empowering self-renewal and differentiation: The role of mitochondria in stem cells. Journal of Molecular Medicine (Berlin). 2010;**88**(10):981-986. [Epub: September 3, 2010]

[40] Pinto MC, Tonelli FM, Vieira AL, Kihara AH, Ulrich H, Resende RR. Studying complex system: Calcium oscillations as attractor of cell differentiation. Integrative Biology: Quantitative Biosciences from Nano to Macro. 2016;**8**(2):130-148. [Epub: January 15, 2016]

[41] Wanet A, Arnould T, Najimi M, Renard P. Connecting mitochondria, metabolism, and stem cell fate. Stem Cells and Development. 2015;**24**(17):1957-1971. [Epub: July 3, 2015]

[42] Spees JL, Lee RH, Gregory CA. Mechanisms of mesenchymal stem/stromal cell function. Stem Cell Research & Therapy. 2016;**7**(1):125. [Epub: September 2, 2016]

[43] Hsu YC, Wu YT, Yu TH, Wei YH. Mitochondria in mesenchymal stem cell biology and cell therapy: From cellular differentiation to mitochondrial transfer. Seminars in Cell & Developmental Biology. 2016;**52**:119-131. [Epub: February 13, 2016]

[44] Choi JY, Chun SY, Kim BS, Kim HT, Yoo ES, Shon YH, et al. Pre-clinical efficacy and safety evaluation of human amniotic fluid-derived stem cell injection in a mouse model of urinary incontinence. Yonsei Medical Journal. 2015;**56**(3):648-657. [Epub: April 4, 2015]

[45] Kim J, Shapiro L, Flynn A. The clinical application of mesenchymal stem cells and cardiac stem cells as a therapy for cardiovascular disease. Pharmacology & Therapeutics. 2015;**151**:8-15. [Epub: February 25, 2015]

[46] Dixit P, Katare R. Challenges in identifying the best source of stem cells for cardiac regeneration therapy. Stem Cell Research & Therapy. 2015;**6**:26. [Epub: April 18, 2015]

[47] Greco SJ, Rameshwar P. Microenvironmental considerations in the application of human mesenchymal stem cells in regenerative therapies. Biologics: Targets & Therapy. 2008;**2**(4):699-705. [Epub: August 27, 2009]

[48] Pekovic V, Hutchison CJ. Adult stem cell maintenance and tissue regeneration in the ageing context: The role for A-type lamins as intrinsic modulators of ageing in adult stem cells and their niches. Journal of Anatomy. 2008;**213**(1):5-25. [Epub: July 22, 2008]

[49] Tower J. Stress and stem cells. Wiley Interdisciplinary Reviews: Developmental Biology. 2012;**1**(6):789-802. [Epub: June 27, 2013]

[50] Chandel NS, Jasper H, Ho TT, Passegue E. Metabolic regulation of stem cell function in tissue homeostasis and organismal ageing. Nature Cell Biology. 2016;**18**(8):823-832. [Epub: July 19, 2016]

[51] Shi Y, Su J, Roberts AI, Shou P, Rabson AB, Ren G. How mesenchymal stem cells interact with tissue immune responses. Trends in Immunology. 2012;**33**(3):136-143. [Epub: January 10, 2012]

[52] Kovach TK, Dighe AS, Lobo PI, Cui Q. Interactions between MSCs and immune cells: Implications for bone healing. Journal of Immunology Research. 2015;**2015**:752510. [Epub: May 23, 2015]

[53] Spaggiari GM, Moretta L. Cellular and molecular interactions of mesenchymal stem cells in innate immunity. Immunology and Cell Biology. 2013;**91**(1):27-31. [Epub: November 14, 2012]

[54] Vallabhaneni KC, Hassler MY, Abraham A, Whitt J, Mo YY, Atfi A, et al. Mesenchymal stem/stromal cells under stress increase osteosarcoma migration and apoptosis resistance via extracellular vesicle mediated communication. PloS One. 2016;**11**(11):e0166027. [Epub: November 5, 2016]

[55] Kim TJ, Sun J, Lu S, Qi YX, Wang Y. Prolonged mechanical stretch initiates intracellular calcium oscillations in human mesenchymal stem cells. PloS One. 2014;**9**(10):e109378. [Epub: October 21, 2014]

[56] Ye B. Ca^{2+} oscillations and its transporters in mesenchymal stem cells. Physiological Research. 2010;**59**(3):323-329. [Epub: August 18, 2009]

[57] Hao J, Zhang Y, Jing D, Shen Y, Tang G, Huang S, et al. Mechanobiology of mesenchymal stem cells: Perspective into mechanical induction of MSC fate. Acta Biomaterialia. 2015;**20**:1-9. [Epub: April 15, 2015]

[58] Kim TJ, Joo C, Seong J, Vafabakhsh R, Botvinick EL, Berns MW, et al. Distinct mechanisms regulating mechanical force-induced Ca(2)(+) signals at the plasma membrane and the ER in human MSCs. eLife. 2015;**4**:e04876. [Epub: February 11, 2015]

[59] Li Y, Chen J, Chopp M. Adult bone marrow transplantation after stroke in adult rats. Cell Transplantation. 2001;**10**(1):31-40. [Epub: April 11, 2001]

[60] Wu Y, Chen L, Scott PG, Tredget EE. Mesenchymal stem cells enhance wound healing through differentiation and angiogenesis. Stem Cells. 2007;**25**(10):2648-2659. [Epub: July 7, 2007]

[61] Kinnaird T, Stabile E, Burnett MS, Shou M, Lee CW, Barr S, et al. Local delivery of marrow-derived stromal cells augments collateral perfusion through paracrine mechanisms. Circulation. 2004;**109**(12):1543-1549. [Epub: March 17, 2004]

[62] Urban VS, Kiss J, Kovacs J, Gocza E, Vas V, Monostori E, et al. Mesenchymal stem cells cooperate with bone marrow cells in therapy of diabetes. Stem Cells. 2008;**26**(1):244-253. [Epub: October 13, 2007]

[63] Imitola J, Raddassi K, Park KI, Mueller FJ, Nieto M, Teng YD, et al. Directed migration of neural stem cells to sites of CNS injury by the stromal cell-derived factor 1alpha/CXC chemokine receptor 4 pathway. Proceedings of the National Academy of Sciences of the United States of America. 2004;**101**(52):18117-18122. [Epub: December 21, 2004]

[64] Shichinohe H, Kuroda S, Yano S, Hida K, Iwasaki Y. Role of SDF-1/CXCR4 system in survival and migration of bone marrow stromal cells after transplantation into mice cerebral infarct. Brain Research. 2007;**1183**:138-147. [Epub: November 3, 2007]

[65] Wang XJ, Dong Z, Zhong XH, Shi RZ, Huang SH, Lou Y, et al. Transforming growth factor-beta1 enhanced vascular endothelial growth factor synthesis in mesenchymal stem cells. Biochemical and Biophysical Research Communications. 2008;**365**(3):548-554. [Epub: November 21, 2007]

[66] Zipori D. Mesenchymal stem cells: Harnessing cell plasticity to tissue and organ repair. Blood Cells, Molecules & Diseases. 2004;**33**(3):211-215. [Epub: November 6, 2004]

[67] Pittenger MF, Mackay AM, Beck SC, Jaiswal RK, Douglas R, Mosca JD, et al. Multilineage potential of adult human mesenchymal stem cells. Science. 1999;**284**(5411):143-147. [Epub: April 2, 1999]

[68] Li W, Ye B, Cai XY, Lin JH, Gao WQ. Differentiation of human umbilical cord mesenchymal stem cells into prostate-like epithelial cells in vivo. PloS One. 2014;**9**(7):e102657. [Epub: July 24, 2014]

[69] Yu J, Su X, Zhu C, Pan Q, Yang J, Ma J, et al. GFP labeling and hepatic differentiation potential of human placenta-derived mesenchymal stem cells. Cellular Physiology and Biochemistry: International Journal of Experimental Cellular Physiology, Biochemistry, and Pharmacology. 2015;**35**(6):2299-2308. [Epub: April 22, 2015]

[70] Karnieli O, Izhar-Prato Y, Bulvik S, Efrat S. Generation of insulin-producing cells from human bone marrow mesenchymal stem cells by genetic manipulation. Stem Cells. 2007;**25**(11):2837-2844. [Epub: July 7, 2007]

[71] Zaidi N, Nixon AJ. Stem cell therapy in bone repair and regeneration. Annals of the New York Academy of Sciences. 2007;**1117**:62-72. [Epub: December 7, 2007]

[72] Mackay AM, Beck SC, Murphy JM, Barry FP, Chichester CO, Pittenger MF. Chondrogenic differentiation of cultured human mesenchymal stem cells from marrow. Tissue Engineering. 1998;**4**(4):415-428. [Epub: January 23, 1999]

[73] Ju YJ, Muneta T, Yoshimura H, Koga H, Sekiya I. Synovial mesenchymal stem cells accelerate early remodeling of tendon-bone healing. Cell and Tissue Research. 2008;**332**(3):469-478. [Epub: April 18, 2008]

[74] Nesti LJ, Li WJ, Shanti RM, Jiang YJ, Jackson W, Freedman BA, et al. Intervertebral disc tissue engineering using a novel hyaluronic acid-nanofibrous scaffold (HANFS) amalgam. Tissue Engineering Part A. 2008;**14**(9):1527-1537. [Epub: August 19, 2008]

[75] Kurozumi K, Nakamura K, Tamiya T, Kawano Y, Kobune M, Hirai S, et al. BDNF gene-modified mesenchymal stem cells promote functional recovery and reduce infarct size in the rat middle cerebral artery occlusion model. Molecular Therapy: The Journal of the American Society of Gene Therapy. 2004;**9**(2):189-197. [Epub: February 5, 2004]

[76] Kurozumi K, Nakamura K, Tamiya T, Kawano Y, Ishii K, Kobune M, et al. Mesenchymal stem cells that produce neurotrophic factors reduce ischemic damage in the rat middle cerebral artery occlusion model. Molecular Therapy: The Journal of the American Society of Gene Therapy. 2005;**11**(1):96-104. [Epub: December 9, 2004]

[77] Le Blanc K, Ringden O. Immunomodulation by mesenchymal stem cells and clinical experience. Journal of Internal Medicine. 2007;**262**(5):509-525. [Epub: October 24, 2007]

[78] Nasef A, Ashammakhi N, Fouillard L. Immunomodulatory effect of mesenchymal stromal cells: Possible mechanisms. Regenerative Medicine. 2008;**3**(4):531-546. [Epub: July 1, 2008]

[79] Zappia E, Casazza S, Pedemonte E, Benvenuto F, Bonanni I, Gerdoni E, et al. Mesenchymal stem cells ameliorate experimental autoimmune encephalomyelitis inducing T-cell anergy. Blood. 2005;**106**(5):1755-1761. [Epub: May 21, 2005]

[80] Le Blanc K, Rasmusson I, Sundberg B, Gotherstrom C, Hassan M, Uzunel M, et al. Treatment of severe acute graft-versus-host disease with third party haploidentical mesenchymal stem cells. Lancet. 2004;**363**(9419):1439-1441. [Epub: May 4, 2004]

[81] Badylak SF, Nerem RM. Progress in tissue engineering and regenerative medicine. Proceedings of the National Academy of Sciences of the United States of America. 2010;**107**(8):3285-3286. [Epub: February 26, 2010]

[82] Aggarwal S, Pittenger MF. Human mesenchymal stem cells modulate allogeneic immune cell responses. Blood. 2005;**105**(4):1815-1822. [Epub: October 21, 2004]

[83] Ortiz LA, Dutreil M, Fattman C, Pandey AC, Torres G, Go K, et al. Interleukin 1 receptor antagonist mediates the antiinflammatory and antifibrotic effect of mesenchymal stem cells during lung injury. Proceedings of the National Academy of Sciences of the United States of America. 2007;**104**(26):11002-11007. [Epub: June 16, 2007]

[84] Togel F, Weiss K, Yang Y, Hu Z, Zhang P, Westenfelder C. Vasculotropic, paracrine actions of infused mesenchymal stem cells are important to the recovery from acute kidney injury. American Journal of Physiology. Renal Physiology. 2007;**292**(5):F1626–F1635. [Epub: January 11, 2007]

[85] Rose RA, Jiang H, Wang X, Helke S, Tsoporis JN, Gong N, et al. Bone marrow-derived mesenchymal stromal cells express cardiac-specific markers, retain the stromal phenotype, and do not become functional cardiomyocytes in vitro. Stem Cells. 2008;**26**(11):2884-2892. [Epub: August 9, 2008]

[86] Al-Nbaheen M, Vishnubalaji R, Ali D, Bouslimi A, Al-Jassir F, Megges M, et al. Human stromal (mesenchymal) stem cells from bone marrow, adipose tissue and skin exhibit differences in molecular phenotype and differentiation potential. Stem Cell Reviews. 2013;**9**(1):32-43. [Epub: April 25, 2012]

[87] Vishnubalaji R, Al-Nbaheen M, Kadalmani B, Aldahmash A, Ramesh T. Comparative investigation of the differentiation capability of bone-marrow- and adipose-derived mesenchymal stem cells by qualitative and quantitative analysis. Cell and Tissue Research. 2012;**347**(2):419-427. [Epub: January 31, 2012]

[88] Vishnubalaji R, Atteya M, Al-Nbaheen M, Oreffo RO, Aldahmash A, Alajez NM. Angiogenic potential of human neonatal foreskin stromal cells in the chick embryo chorioallantoic membrane model. Stem Cells International. 2015;**2015**:257019. [Epub: July 30, 2015]

[89] Vishnubalaji R, Manikandan M, Al-Nbaheen M, Kadalmani B, Aldahmash A, Alajez NM. In vitro differentiation of human skin-derived multipotent stromal cells into putative endothelial-like cells. BMC Developmental Biology. 2012;**12**:7. [Epub: January 28, 2012]

[90] Nadig RR. Stem cell therapy—hype or hope? A review. Journal of Conservative Dentistry: JCD. 2009;**12**(4):131-138. [Epub: June 15, 2010]

[91] Wei X, Yang X, Han ZP, Qu FF, Shao L, Shi YF. Mesenchymal stem cells: A new trend for cell therapy. Acta Pharmacologica Sinica. 2013;**34**(6):747-754. [Epub: June 6, 2013]

Chapter 7

The Proangiogenic Potential of Mesenchymal Stem Cells and Their Therapeutic Applications

Nadeeka Bandara, Shiang Lim, Haiying Chen,
Shuangfeng Chen, Le-Xin Wang and
Padraig Strappe

Additional information is available at the end of the chapter

http://dx.doi.org/10.5772/intechopen.68516

Abstract

Mesenchymal stem cells (MSCs) can be isolated from many tissue types and following *in vitro* culture expansion, large numbers of patient-specific or allogenic cells can be produced for clinical applications. MSCs exhibit anti-inflammatory and immunomodulatory properties and are identified as lacking major histocompatibility complex (MHC) class II molecules. Cellular-based approaches using MSCs to enhance new blood vessel formation have shown promise in preclinical models and preliminary clinical trials. Transplantation of MSCs *in vivo* has significantly enhanced the formation of new blood vessels and promoted the healing of chronic wounds. The proangiogenic potential of MSCs can be further enhanced through gene delivery such as vascular endothelial growth factor (VEGF) or endothelial nitric oxide synthase (eNOS) providing long-term therapeutic expression. In this chapter, we review recent advances on the isolation and characterization of MSCs and *in vivo* applications for promoting angiogenesis. Enhancement of angiogenesis is also required for improved healing in myocardial infarction and cerebral ischemia, and the use of MSCs in these areas will also be reviewed. Furthermore, the combination of MSCs with biomaterials has greatly improved their survival and potency with improved vascularization of tissue-engineered constructs and integration within the host. In summary, this chapter provides an overview of both the basic science supporting the proangiogenic properties of MSCs and their translational use.

Keywords: mesenchymal stem cell, angiogenesis, clinical trials, myocardial infarction, wound healing

1. Stem cells

Stem cells can be broadly described as a group of undifferentiated cells capable of self-renewal (cell division without differentiation) and can subsequently differentiate into specialized cell types [1, 2]. Stem cell division can be described as symmetric and asymmetric [1]. Symmetric cell division yields two daughter cells showing the same characteristics of the parent stem cell and has the potential to differentiate into other lineage cell types [1]. On the other hand, asymmetric cell division yields differentiated cells through the development of lineage-specific intermediate progenitor cells [3]. Progenitor cells are generated as an intermediate state before the stem cell is converted into the fully differentiated cell type [2] and are regarded as being committed to differentiating along a particular cellular developmental pathway. There are two types of resident stem cells, which are categorized as embryonic stem cells (ESCs) or somatic/adult stem cells. ESCs are referred to as pluripotent, an ability to differentiate into all the cell types in the body, whereas adult stem cells are multipotent and demonstrate a restricted ability to differentiate into multiple lineages.

1.1. Embryonic stem cells

Embryonic stem cells (ESC) are a class of unspecialized cells derived from the inner cell mass of a blastocyst, which is an early stage of the embryo containing 200–250 cells [4, 5]. ESCs are pluripotent stem cells, which can differentiate into any cell type represented within three germ layers (mesoderm, ectoderm, and endoderm) [6]. In response to various stimuli during development, ESCs can be differentiated into specialized cells, which have specific roles in the body [7, 8]. There are two key features, which characterize ESCs, pluripotency (the ability to differentiate into all three germ layers, ectoderm, endoderm, and mesoderm) and self-renewal (the ability to go through numerous cycles of cell division while maintaining the undifferentiated state) whereby they are maintained as pure populations of undifferentiated cells in culture for extended periods of time, retaining a normal karyotype unlike tumor cell lines [6]. Over the past two decades, ESCs have been used as a model system for studying the basic processes in mammalian development and cellular differentiation events [9]. ESCs have also provided a valuable platform for regenerative medicine and tissue engineering for the development of future treatments of human diseases. Furthermore, ESCs have been also used as a reference *in vitro* model for understanding key molecular mechanisms, which control cell fate and organogenesis [10].

2. Induced pluripotent stem cells (iPS cells)

To find an alternative pluripotent cell type to ESCs, in 2006, the Japanese scientists Shinya Yamanaka and Kazutoshi Takahashi demonstrated the groundbreaking discovery of induced pluripotent stem cells (iPSCs). iPS cells are artificially created embryonic-like stem cells generated by over expressing four transcription factors in somatic cells such as fibroblasts [11]. These iPS cells exhibited similar features to ESCs. Since iPS cells are artificially created cells, they do not have ethical and immunological problems associated with ESCs. Therefore, iPS cells show potential in cell biological research, including their application in cell therapy, drug screening, and disease modeling.

2.1. Generation of iPS cells

Differentiated cells can be reprogrammed into a pluripotent state by the transfer of nuclear contents into oocytes [12], and the fusion of somatic cells with embryonic stem (ES) cells can also result in reprogramming to a pluripotent state [13]. These studies revealed that oocytes and ES cells contain factors, which may be responsible for the conversion of somatic cells to a pluripotency state. In 2006, Yamanaka and Takahashi demonstrated that, mouse embryonic or adult fibroblasts can be reprogrammed back to an embryonic-like state by the overexpression of four transcription factors, OCT4, SOX2, KLF4, and cMYC [10, 11]. They named these ES-like reprogrammed cells as induced pluripotent stem cells (iPSCs). In 2007, the same investigators demonstrated the generation of iPSCs from human fibroblasts [14]. Yu and colleagues have also reported the generation of human iPSCs from fibroblasts with a slightly different combination of transcription factors, in which KLF4 and cMYC were replaced with NANOG and LIN28 [15]. Both of these iPS cells exhibited similar features to ES cells including morphology, proliferation, ESC-specific gene expression profiles, and teratoma formation. This method of cellular reprogramming has been shown to be universal and can be applied to a variety of cell types such as B-cells [16], liver cells [17], and umbilical cord blood mononuclear cells [18]. Moreover, iPS cells have been generated from different species such as monkey [19], rat [20], and horse [21].

2.2. Limitations of iPS cells in clinical applications

Even though iPS cells have provided a solution for many of the obstacles raised with ESCs, iPS cells also have inherent disadvantages in terms of clinical applications, which include teratoma formation [22] and the use of oncogene cMYC as a reprogramming factor, which can lead to tumorigenesis [23].

A second issue associated with the therapeutic application of iPS cells is their immunogenicity. Transplanted iPS cells have been considered to be immune tolerant by the recipient. However, induction of T-cell-dependent immune response in recipients has been demonstrated [24].

3. Adult stem cells

Adult stem cells or somatic stem cells are multipotent stem cells, which can be found in specific cellular niches of organs and tissues. Adult stem cells are essential for maintaining the health of organs throughout a life time [25]. Somatic stem cells were first identified about 40 years ago with the discovery of hematopoietic stem cells and bone marrow stem cells (mesenchymal stem cells) [26]. Adult stem cells can be found in many tissues such as brain [27], liver [28], heart [29], lung [30], and adipose [31]. Adult stem cells are multipotent; they can self-renew and differentiate to all the cell types in their tissue environment and as well as other lineages such as cardiomyocytes [32], neurons [33], and endothelial cells [34].

The use of adult stem cells in cell therapy applications is currently limited due to several factors:

1. Limited differentiation potential [35].
2. The results obtained in animal models may not be directly translated to humans [35].
3. Loss of proliferative capacity under standard culture conditions as well as the method for the delivery of adult stem cells to the patient may impact on their ability to survive post-transplantation [35, 36].

3.1. Mesenchymal stem cells

Mesenchymal stem cells (MSCs) or mesenchymal stromal cells were discovered by Friedenstein et al. in 1968 [26]. He observed many different types of cells in bone marrow cultures some of which were adherent to tissue culture plastic, showed fibroblastic morphology and formed colonies. These cells were named as colony forming unit fibroblasts (CFU-F) [37] and were found to differentiate into bone, adipose, cartilage, and muscle tissue. Caplan coined the term "mesenchymal stem cells" (MSCs) [38] and MSCs obtained from human bone marrow aspirates were characterized [39]. The BM aspirate was first separated by density gradient separation and plated on tissue culture plastic and the attached cells were counted based on their colony formation ability. Approximately, 0.001–0.01% cells of total cells were found to be MSCs and expressed CD29, CD90, CD71, and CD106 surface markers and were negative for CD45, CD14, and CD34. Importantly, they found that these cells could undergo 40 population doublings *in vitro* over 10 weeks.

MSCs have now been isolated from many other tissue sources such as adipose tissue, umbilical cord blood, placenta, and even from dental pulp. Increasing research interest is in finding stem cells from different organs and focusing on strategies to repair the same organs with autologous stem cells. Interestingly, cell isolated from a variety of different tissues have shown different CD marker expression profile, cellular phenotype, and population doublings. While no definitive single surface marker for MSCs had been described so far, an internationally accepted set of criteria has been established by the International Society for Cellular Therapy (ISCT) [40]. Accordingly, ISCT outlined that MSCs should be positive for CD73, CD90, and CD105, negative for CD19, CD34, CD45, CD11b, and HLA-DR. In addition, they should attach to the plastic tissue culture plates and demonstrate an ability to differentiate to adipocytes, chondrocytes, and osteoblasts *in vitro*.

For cell therapy applications, MSCs are remarkable since they show antiapoptotic and immunomodulatory features providing them with nonimmunogenic properties. MSCs release a variety of cytokines. Therefore, when MSCs were injected to the damaged tissues (kidney) in animal models, they could reduce the apoptotic rate of the surrounding cells, which was mediated by the secretion of several growth factors like VEGF, FGF2, and TGF-β from MSCs cultured in hypoxic conditions [41–43]. More specifically, these studies have shown that infusion of MSCs is important for revascularization, which contributes to the recovery from acute kidney injury in mouse models through the secretion of growth factors. Furthermore, proliferation of T-cells was inhibited when co-cultured with MSCs *in vitro* [44–46]. In addition to T-cells, the activity of other immune responsive cells such as natural killer cells, B-cells, and immature dendritic cells have also been modulated by MSCs [47–49]. In particular, MSCs can inhibit B-cell proliferation by inhibiting the G0/G1 phase through the release of paracrine factors that affect B-cell differentiation and IgM, IgG, and IgA production. The immunomodulatory effects of MSCs on T-cells and NK cells have also been shown to be driven by cytokines such as TGF-β, PGE2, and IL10 [49–53].

Thus, MSCs are important candidates for cellular-based therapies as they feature the following characteristics.

1. A repertoire of defined surface markers and an ability to produce relatively homogenous cultures.

2. Ease of *in vitro* expansion resulting in high cell concentration without significant loss of properties.

3. Ability to differentiate into a variety of different cell types.
4. Possess immunosuppressive attributes, which contribute to their possible use in allogeneic grafting [47–49].

3.2. Clinical trials using mesenchymal stem cells

In the literature, there are many studies investigating the regenerative capacities of MSCs in different disease models generated by employing different nonhuman animal species. Cardiac regeneration, liver regeneration, kidney regeneration, autoimmune diseases, graft versus host disease (GvHD), neurological diseases, pulmonary diseases, osteogenic diseases, and cartilage repair are the most widely studied conditions. Moreover, MSCs are also being investigated extensively by clinical trials, mostly in United States, Europe, and East Asia, with trials investigating MSC use in neurological, liver, bone, heart diseases, GvHD, and some autoimmune diseases such as diabetes. In the following section, the clinical application of MSCs will be discussed and a particular attention will be given to their role in heart disease.

4. Mesenchymal stem cells in vascular repair

The formation of new vessels is the cornerstone of successful cardiac repair. There are three mechanisms of postnatal neovascularization: (1) angiogenesis, (2) arteriogenesis, and (3) postnatal vasculogenesis [54] with progenitor cells migrating from the bone marrow and to site of sites of tissue damage resulting in the generation of new capillaries. Whether the formation of new capillary networks and vessel integration into neighboring tissue is associated with direct differentiation of MSCs to endothelial cells is still unknown or the importance of secreted factors [55, 56]. MSCs have been shown to exist in perivascular niches with similarities to pericytes, which may account for their ability to promote vascularization [57]. Expression of MSC markers has also been detected on the surface perivascular cells without *in vitro* culture, which may point to a very localized depot of progenitors in vessels [58]. *In vitro*, MSCs express α-smooth muscle actin and β-actin filaments [59], whereas *in vivo* studies have shown that MSCs express an endothelial phenotype that can enhance microvascular density [60]. However, contrary evidence has shown that the number of vessels harboring progenitor/adult stem cells is low and that the secretion of proangiogenic factors may be the dominant mechanism associated with vasculogenesis [61] and neoangiogenesis [62]. Interesting work by Chen and colleagues have shown significant increases in the levels of VEGF and basic fibroblast growth factor (bFGF) in MSC-treated rats, which resulted in angiogenesis following intravenous injection 24 h after middle cerebral artery occlusion (MCAO). They further showed significant increases in newly formed capillaries at the boundary of the ischemic lesion in rats treated with MSCs compared with rats treated with phosphate buffered saline (PBS) [62]. Further evidence of MSC-supported neovessel formation, comes from Markel and colleagues [63] who showed that MSCs under-expressing VEGF have significantly less cardio reparative capabilities. In this work, female adult rats were subjected to ischemia-reperfusion injury and following injury, VEGF knockout MSCs or normal MSCs were infused into the coronary circulation. Following MSC treatment, it was observed that VEGF knockout MSCs significantly impaired myocardial function while normal MSCs showed improvement highlighting the importance of VEGF as a paracrine factor associated with MSCs.

4.1. Neovessel formation

The process of neovessel formation is an important event during embryonic development and also in adult tissues following injury such as ischemic infarction. Neovessels from the neighboring normal tissues are needed to form the vessel network and restore blood supply to the damaged tissues. Both ECs and SMCs are essential for the formation of blood vessels; however, the detailed mechanism of SMC migration and differentiation is not fully understood.

Until recently, it was accepted that vessels in adult ischemic tissues could only grow by angiogenic mechanisms, in which the sprouting of mature ECs from pre-existing vessels was likely in response to angiogenic factors. However, recent studies have revealed that endothelial progenitor cells (EPCs) circulate postnatally in peripheral blood. These may be recruited from the bone marrow and incorporate into sites of active neovascularization in ischemic hind limbs, ischemic myocardium, injured corneas, and tumor vasculature [64]. This process is termed postnatal vasculogenesis [65]. EPCs participate in vasculogenesis by the differentiation into endothelial cells (ECs) and thereby promote angiogenesis through the production of angiogenic growth factors [66]. Accumulating evidence has shown that EPCs have a therapeutic potential for vascular repair through promoting the reendothelialization of damaged vessel walls and the neovascularization of ischemic tissues [67, 68].

Bone marrow–derived mesenchymal stem cells (BMSCs) and alternatively named multipotent adult progenitor cells (MAPCs) can be induced to differentiate into endothelial-like cells *in vitro* and subsequently promote neoangiogenesis *in vivo* [69, 70]. The bioactivity of secreted molecules from BMSCs has been shown to increase collateral remodeling and perfusion in ischemic tissues in animal models, again highlighting the importance of paracrine mechanisms following local delivery [71, 72].

Recently, it has been shown that adult BMSCs, under appropriate *in vitro* environmental cues, can be induced to undergo vasculogenic differentiation culminating in microvessel morphogenesis. When rat BMSCs were seeded onto a three-dimensional (3D) tubular scaffold, the maturation and co-differentiation into endothelial and SMC lineages, which led to successful microvessel formation was observed [73]. A separate study showed that locally delivered, activated cardiac progenitor cells (CPCs) could generate coronary vasculature by dividing and differentiating into both ECs and SMCs, restoring blood supply to ischemic myocardium [74].

4.2. Mesenchymal stem cells in cardiac repair

Ischemic heart disease is associated with the highest mortality rate among all diseases (http://www.who.int). There is an urgent need for alternative cell-based therapies to treat cardiovascular diseases. Broadly, ischemic heart diseases are characterized by a shortage of blood supply to different regions of the heart, resulting in these regions undergoing necrosis and apoptosis. With a limited endogenous regeneration available to the mammalian heart, heart transplantation is often the only therapeutic option currently available.

Cell therapy to regenerate damaged cardiac tissue is an exciting alternative to heart transplantation. In 1995, Wakitani et al. reported the generation of cardiomyocytes *in vitro* from rat bone marrow-derived MSCs (rBMSCs) [75]. Following this, several studies reported the successful differentiation of MSCs into cardiomyocytes [76, 77]. Both of these studies demonstrated the *in vitro* generation of beating cardiomyocytes from rat bone marrow MSCs.

Many *in vivo* studies have since been performed to investigate the efficacy of MSCs in cardiovascular regeneration. In 2002, Shake et al. demonstrated that swine bone marrow–derived MSCs could be differentiated into functional cardiomyocytes when injected into the infarcted swine myocardium [78]. On the other hand, when MSCs were injected intracardially in a canine model, the MSCs were differentiated into smooth muscle cells and endothelial cells [79] and further studies showed that, when MSCs were injected into a rat myocardial infarct, there was a significant reduction in the damaged area [80]. Moreover, genetic modification of MSCs to overexpress Akt, exerted a beneficial effect [81], suggesting that genetic modification of MSCs would provide a better platform for cardiovascular repair. It is also possible that Akt may activate mammalian target of rapamycin complex 1 (mTORC1) and forkhead box o3 (Foxo3a), which are acted downstream of Akt to promote cardiomyocyte reprogramming [82].

4.3. Direct MSC stimulation of endogenous repair

MSC transplantation to the heart has been shown by multiple groups to stimulate proliferation and differentiation of endogenous cardiac stem cells [83–85]. Neomyogenesis can be promoted by two related mechanisms through the stimulation of endogenous cardiac stem cells (c-kit + and other lineages such as cardiac fibroblasts) and enhancement of myocyte cell cycling [83]. To demonstrate this, GFP+ allogeneic MSCs were injected into infarcted swine hearts and allowed to form chimeric clusters of immature MSCs and endogenous c-kit+ cardiac stem cells. These clusters exhibited cell-cell interactions mediated by connexin-43 gap junction formation and N-cadherin mechanical connections. Importantly, the endogenous c-kit+ cell population was increased by 20-fold in MSC-treated animals relative to controls; furthermore, the c-kit+ cells showed a high capacity for myocyte lineage commitment [83]. It has been demonstrated that, when MSCs were co-cultured with rat ventricular myocytes, MSCs became actin-positive and formed gap junctions with the native myocytes [86]. Furthermore, an improvement in myocardial wall thickening in pigs with hibernating myocardium, which is a pathology when some segments of the myocardium exhibit abnormalities of contractile function, was induced upon MSC injection [85] compared with controls. This same study also found a fourfold increase in c-kit+ and CD133+ populations that co-expressed Gata4 and Nkx2.5 at 3 days through to 2 weeks in animals receiving MSCs. In a preclinical study, the combination of human MSCs and c-kit+ cardiac stem cells showed enhanced cardiac regeneration [87].

5. Preclinical trials of MSCs for cardiac repair in animal models

Toma et al. showed that human MSCs were differentiated to a cardiac fate when injected into murine hearts [88]. In this study, MSCs labeled with lacZ were injected into the left ventricle of the adult mice, and after 1 week post injection, the lacZ-labeled MSCs morphologically resembled the surrounding host cardiomyocytes; furthermore, they expressed cardiac-specific genes such as α-actinin and cardiac troponin T. It has been shown that MSCs can modulate host immune responses when allogeneic porcine MSCs were injected (2×10^8 cells) intramyocardially into 3-day-old immune-competent porcine-infarcted hearts, this resulted in long-term engraftment and a significant decrease in scar tissue without an inflammatory response [89]. MSCs have also been tested in numerous cardiovascular settings. In a separate study, where porcine MSCs were injected endomyocardially of one of three MSC doses (2.4×10^7, 2.4×10^8, 4.4×10^8 cells) into the porcine heart 5 days after infarction, an improvement in ejection fraction (EF) and a reduction in scar formation were seen in MSC-treated animals [90]. The effect of MSC dosage was examined in ovine models of MI where different doses of ovine BM-derived MSCs (2.5×10^7, 3.75×10^7, 5×10^7 cells) were directly injected into sheep hearts 1 h post MI [91], and improvements in end-diastolic volume were only seen in animals receiving the two lower doses, although the EF increased regardless of the cell dosage [91] suggesting that there may be a therapeutic threshold relating to the total number of cells that can be injected and a beneficial therapeutic outcome. In a study with a different species (canine), chronic myocardial ischemia was produced by the implantation of an ameroid constrictor in the proximal left anterior descending coronary artery (LAD) and diagonal branch ligation, followed by the injection of allogeneic canine MSCs (1×10^8 cells) into the heart resulted in increased EF, vascular density, and a decrease in scar tissue [79]. Furthermore, it has been reported that the region specific administration of allogeneic porcine MSCs (2×10^8 cells) to the border and to infarct zones of porcine myocardium 3 days after MI also reduced scar size by 50% [89] with improvements in EF, left ventricular end-diastolic pressure, relaxation time, and systolic compliance in the treated animals. Furthermore, in a model of acute myocarditis in rats myocardial inflammation was attenuated when autologous rat MSCs (3×10^6 cells) were injected into 10-weeks-old animals [92], together with the increased capillary density in MSC-treated animals.

6. Human clinical trials of MSC-based therapies for cardiac repair

6.1. Acute myocardial infarction

In a phase I randomized study, 53 patients received different doses of allogeneic human MSCs (0.5, 1.6, and 5.0×10^6 hMSCs/kg) 7–10 days post MI [93]. The MSCs were injected intravenously. Six months after infusion, clinical data showed fewer arrhythmic events, and an improved EF. Following the success of this pilot study, a phase II trial was established to investigate whether allogeneic MSCs were as safe and effective as autologous MSCs in patients with left ventricular (LV) dysfunction due to ischemic cardiomyopathy [94]. Upon intravenous infusion of allogeneic MSCs (2×10^7 cells) within 7 days of an acute MI, resulted in reduced cardiac hypertrophy, stress-induced ventricular arrhythmia, heart failure, LV end-diastolic volumes, and increased EF. Interestingly, allogeneic MSCs did not stimulate significant donor-specific alloimmune reactions. In a separate study, Chen and colleagues have injected autologous MSCs (1×10^{11} cells) intracoronarily in patients with subacute MI and observed decreased perfusion defects, improved left ventricular ejection fraction, and left ventricular remodeling 3 months after therapy [95]. Other clinical benefits attributed to MSCs include decreased perfusion defects and improved left ventricular ejection fraction and left ventricular remodeling when MSCs were administered to patients with subacute MI [96]. In addition to bone marrow MSCs, adipose-derived MSCs have also been used to treat acute MI. A trial with 14 patients, which tested the safety of intracoronary injection of freshly isolated adipose-derived MSCs after myocardial infarction [97] demonstrated improved cardiac function, accompanied with a significant improvement in perfusion defect and a 50% reduction in myocardial scar formation.

There are a reported 41 clinical trials in which MSCs have been applied in relation to cardiac injury and repair between 2010 and 2015 [98]. There is also an ongoing clinical trial using adipose-derived MSCs, in patients with chronic myocardial ischemia [71] where they used culture-expanded adipose tissue-derived MSCs. This study has been designed to investigate the safety and efficacy of intramyocardial delivery of VEGF-A165-stimulated autologous adipose tissue-derived MSCs to improve myocardial perfusion and exercise capacity [99]. **Table 1** summarizes completed and ongoing clinical trials.

6.2. Phase III clinical trials

There are six ongoing phase III clinical trials using MSCs. Of note, one of these studies [100] applied autologous MSCs treated *ex-vivo* with cardiogenic growth factors (TGF-β, BMP4, FGF2, cardiotrophin, and α thrombin) to enhance their commitment to the cardiopoietic lineage and investigators reported significant improvements in EF and end-systolic volume compared with controls. Other phase III studies are currently underway, in which one in United States is treating 600 patients with chronic heart failure (https://clinicaltrials.gov/ct2/show/NCT02032004) all the phase III clinical trials currently undergoing are also listed in **Table 1**.

Clinical trial ID	Disease	Phase	No. of patients/status	MSC source	Country
NCT01076920	Chronic ischemic cardiomyopathy	I, II	10/completed	Autologous	France
NCT01219452	Idiopathic dilated cardiomyopathy	Phase I, II	Unknown	Unknown	China
NCT01392105	Acute myocardial infarction	Phase II, III	80/completed	Autologous	South Korea
NCT01394432	Acute myocardial infarction	Phase III	50/recruiting	Autologous	Russia
NCT01392625	Dilated cardiomyopathy	Phase I, II	36/active, not recruiting	Autologous and allogenic	United States
NCT01449032	Myocardial ischemia (MyStromalCell Trial)	Phase II	60/completed	Unknown	Denmark
NCT01291329	Acute myocardial infarction (AMI)	Phase II	160/completed	Autologous	China
NCT01753440	Coronary artery disease and ischemic cardiomyopathy	Phase II, III	30/unknown status	Allogenic	Greece
NCT01759212	end-stage heart failure undergoing left ventricular assist device implantation	Phase II, III	5/unknown status	Allogenic	Greece

Clinical trial ID	Disease	Phase	No. of patients/ status	MSC source	Country
NCT01739777	Cardiopathy in dilated stage, of different etiology	Phase I, II	30/completed	Allogenic	Chile
NCT01720888	Ischemic dilated cardiomyopathy	Phase II	80/active, not recruiting	Autologous	Malaysia
NCT01957826	Idiopathic dilated cardiomyopathy	Phase I, II	70/recruiting	Autologous	Spain
NCT01709279	Ischemic heart failure	Unknown	6/recruiting by invitation	Autologous	Japan
NCT01557543	Revascularization for coronary artery disease with depressed left ventricular function	Phase I	24/active, not recruiting	Autologous	United States
NCT01652209	Acute myocardial infarction	Phase III	135/recruiting	Autologous	South Korea
NCT01610440	Duchenne muscular dystrophy	Phase I, II	15/unknown status	Unknown	China
NCT01946048	Ischemic cardiomyopathy	Phase I	10/unknown status	Allogenic	China
NCT02013674	Chronic ischemic left ventricular dysfunction secondary to myocardial infarction	Phase II	30/active, not recruiting	Allogenic	United States
NCT01913886	Ischemic cardiomyopathy	Phase I, II	10/unknown status	Autologous	Brazil
NCT01781390	Myocardial infarction	Phase II	105/active, not recruiting	Allogenic	Australia, Belgium Denmark,New Zealand
NCT01770613	Myocardial infarction	Phase II	50/active, not recruiting	Allogenic	United States
NCT02398604	Hypoplastic left heart syndrome	Phase I	30/recruiting	Allogenic	United States
NCT02097641	Acute respiratory distress syndrome	Phase II	60/recruiting	Allogenic	United States
NCT02323477	Myocardial infarction	Phase I, II	79/recruiting	Allogenic	Turkey
NCT02387723	Ischemic heart disease and heart failure	Phase I	10/completed	Allogenic	Denmark
NCT02032004	Chronic heart failure due to left ventricular systolic dysfunction of either ischemic or nonischemic etiology	Phase III	600/recruiting	Allogenic	United States Canada
NCT02501811	Ischemic heart failure	Phase II	144/recruiting	Autologous	United States

Clinical trial ID	Disease	Phase	No. of patients/ status	MSC source	Country
NCT02472002	Coronary graft disease in heart transplant patients	Phase I, II	14/recruiting	Autologous	France
NCT02439541	Ischemic cardiomyopathy	Phase I, II	40/recruiting	Unknown	China
NCT02408432	Recent onset anthracycline-associated cardiomyopathy	Phase I	45/recruiting	Allogenic	United States
NCT02509156	Anthracycline-induced cardiomyopathy	Phase I	36/recruiting	Allogenic	United States
NCT02460770	A pilot study to investigate bone marrow-derived mesenchymal stem cells (MSC) administration from left ventricular assist device	Phase I	4/recruiting	Autologous	France
NCT02467387	Nonischemic heart failure	Phase II	23/active, not recruiting	Allogenic	United States
NCT02503280	Chronic ischemic left ventricular dysfunction and heart failure secondary to myocardial infarction	Phase I, II	55/Active, not recruiting	Allogenic	United States
NCT02568956	Ischemic heart disease	Phase I, II	64/active, not recruiting	Autologous	Unknown
NCT02368587	Ischemic cardiomyopathy	Phase II	160/active, not recruiting	Unknown	Unknown
NCT02462330	Chronic ischemic cardiomyopathy and left ventricular dysfunction	Phase II	90/recruiting	Autologous	France
NCT02635464	Chronic ischemic cardiomyopathy	Phase I, II	45/recruiting	Allogenic	China
NCT02504437	Ischemic heart disease	Phase I, II	200/active, not recruiting	Allogenic	Unknown

Table 1. Clinical trials of MSCs for cardiovascular repair.

7. Direct reprogramming of adult stem cells

With increasing use of MSCs in clinical trials, improving the ability of MSCs to become cells of interest has been the main focus of reprogramming. Genetic modification is one approach to convert an adult cell from one developmental lineage to another and this is mainly achieved by overexpression of lineage-restricted transcription factors and various gene transfer methods have been used.

7.1. Gene delivery of reprogramming factors via retroviral vectors

Retroviral vectors are commonly used as gene delivery systems since they are well characterized and they have a high transduction efficiency. For gene delivery approaches, replication of defective viral vectors is used. In these vectors, coding regions for the genes necessary for additional rounds of virion replication and packaging are deleted. Viruses generated from replication-defective vectors can infect their target cells and deliver genes of interest, but avoid triggering the lytic pathway, which would result in cell lysis and death. Replication-defective retroviral vectors can usually package inserts of up to 10 kb. The major disadvantage of the retrovirus-mediated gene delivery approach is the requirement for cells to be actively dividing to allow transduction by the viral vectors. Thus, slowly dividing or nondividing cells such as neurons are difficult to transduce efficiently with retroviruses. Stable integration of retroviral DNA into the host genome provides a platform for the persistent expression of transgenes; however, this may lead to insertional mutagenesis. Proviral integration could occur within a transcriptional active region of the host genome, which could result in dysregulation of gene expression. In a landmark study by Idea et al., using a Moloney murine leukemia virus (MMLV) retrovirus-mediated gene delivery approach, demonstrated that mouse cardiac and dermal fibroblasts could be reprogrammed into cardiac muscle cells using three cardiac-specific transcription factors, Gata4, Mef2c, and Tbx5 [101]. In this expression vector, expressions of the transgenes were driven by the 5′MMLV long terminal repeat (LTR) promoter, which can be silenced by methylation [11]. This method has been used by several groups and the efficiency of reprogramming has been enhanced by using alternative transcription factors or small molecules (summarized in **Tables 2** and **3**). The reprogramming efficiency of the retrovirus-mediated gene delivery approach is partially dependent on the stoichiometry of the delivered transcription factors [102]. It was reported that a higher reprogramming efficiency than that achieved in the original GMT experiment when the stoichiometry of the transcription factors is altered [102]. In this investigation, six different polycistronic lentiviral vectors were constructed to cover all possible combinations of G, M, T with identical internal 2A sequences. Using this approach, it was shown that the splicing order of G, M, T resulted in distinct G, M, and T protein expression levels, when using a polycistronic vector that resulted in higher protein level of Mef2c with lower levels of Gata4 and Tbx5 (MGT vector), which significantly enhanced reprogramming efficiency compared to separate G, M, T transduction as evident by cardiac-specific gene expression such as cTnT. In addition, the MGT vector resulted in more than a 10-fold increase in the number of mature beating cardiomyocytes. On the other hand, addition of an extra transcription factor Hand2 has also resulted in enhanced reprogramming efficiency [103]. In addition, combinations of small molecules such as SB431542, CHIR99021, 6-bromoindirubin-3′-oxime (BIO), and lithium chloride (LiCl) to replace transcription factors have also been reported to induce cardiac reprogramming [104]. Of note, CHIR99021 is a GSK3 inhibitor, which can up-regulate canonical Wnt signaling increased cardiac reprogramming efficiency.

Species	Cell types	Reprogramming factors	Delivery method	References
Mouse	Embryonic fibroblasts	Gata4, Mef2c, Tbx5, Hand2, Nkx2.5	Lentivirus	[109]
Mouse	embryonic fibroblasts	Gata4, Mef2c, Tbx5, miR133	Retrovirus/lentivirus miRNA transfection	[110]
Mouse	Embryonic fibroblasts	Mef2c, Tbx5, Myocd	Lentivirus	[111]
Mouse	Embryonic fibroblasts	Gata4, Tbx5, Mef2c, Myocd, Srf, Mesp1, Smarcd3	Lentivirus	[112]
Mouse	Embryonic and dermal tail tip fibroblast	Gata4, Mef2c, Tbx5, Hand2, Nkx2.5, TGFB inhibitor	Lentivirus	[113]
Mouse	Embryonic cardiac and dermal tail tip fibroblasts	Oct4, Sox2, Klf4	Retrovirus	[114]
Mouse	Neonatal cardiac fibroblasts	miR1, miR133, miR208, miR499, JAK inhibitor I	Plasmid	[115]
Mouse	Postnatal cardiac and dermal tail tip Fibroblast Fibroblasts	Gata4, Mef2c, Tbx5	Retrovirus	[101]
Mouse	Adult cardiac and dermal tail tip fibroblasts	Gata4, Mef2c, Tbx5, Hand2	Retrovirus	[103]
Mouse	$Sca1^{+}$ side population CSCs	Gata4, Mef2c, Tbx5, Myocd	Lentivirus	[116]
Human	Neonatal derma, cardiac and ESC derived fibroblasts	GATA4, MEF2C, TBX5, ESSRG, MESP1	Retrovirus	[117]
Human	Adult dermal and cardiac and fibroblasts	GATA4, MEF2C, TBX5, HAND2, miR1, miR133	Retrovirus	[118]
Human	ADSCs	Gata4, Tbx5, Baf60c	Lentivirus	[32]
Human	ADSCs	GATA4, MEF2C, TBX5, ESRRG, MESP1, MYOCD, ZFPM2	Retrovirus	[119]
Human	ADSCs	Gata4, Mef2c, Tbx5	PEI method	[120]
Human	Fibroblasts	Small molecules	Supplemented with medium	[104]
Mouse	Embryonic fibroblast	MyoD transactivation domain fused Mef2c, Gata4, Tbx5, Hand2	Retrovirus	[121]
Mouse	Dermal tail tip, embryonic and cardiac fibroblasts	Akt1, Gata4, Mef2c, Tbx5, Hand2	Retrovirus	[82]
Mouse	Embryonic fibroblasts	ROCK inhibitor, TGF-β inhibitor, Gata4, Hand2, Mef2c, Tbx5	Retrovirus, and AAV	[122]

Table 2. *In vitro* cardiac reprogramming.

Species	Reprogramming factors	Vector	Delivery method	References
Mouse	Gata4, Mef2c, Tbx5	Retrovirus	Intramyocardial injection	[123]
Mouse	Gata4, Mef2c, Tbx5, Hand2	Retrovirus	Intramyocardial injection	[103]
Mouse	Gata4, Mef2c, Tbx5, Thymosin β4	Retrovirus	Intramyocardial injection	[124]
Mouse	miR1, miR133, miR208, miR499	Lentivirus	Intramyocardial injection	[125]
Mouse	Gata4, Mef2c, Tbx5	Retrovirus	Intramyocardial injection	[126]
Rat	Gata4, Mef2c, Tbx5, Vegf (121, 165, 189)	Lentivirus/adenovirus	Intramyocardial injection	[127]

Table 3. *In vivo* cardiac reprogramming.

8. Angiogenic properties of MSCs combined with biomaterials

Application of MSCs together with a biomaterial to improve vascularization of damaged tissue as in the case of myocardial infarction or to enhance wound healing is an attractive approach to maintain cell viability and localization. MSCs have been incorporated into a wide range of biomaterials including collagen-based hydrogels and cell sheet techniques. Angiogenesis is one component of successful wound healing, which includes wound closure, reducing inflammation, skin regeneration, and remodeling of the extracellular matrix (ECM). The proangiogenic properties within a wound environment include possible direct differentiation into endothelial cells and secretion of proangiogenic molecules. Murine MSCs seeded in a pullulan-collagen hydrogel enhanced healing time in a mouse excisional wound together with increased survival of transplanted cells and secretion VEGF [105]. In an alternative wound model, rats subjected to severe burns and treated with human umbilical cord MSCs showed increased healing accompanied with reduced expression of proinflammatory cytokines IL-1, IL-6, and TNF-alpha [106]. Efficient cell seeding of biomaterial constructs is important for clinical translation and it has been demonstrated that capillary-based uptake of adipose-derived stem cells into a pullulan-collagen hydrogel could be performed rapidly and these cell laden gels demonstrated increased *in vivo* wound healing and secretion of proangiogenic factors. Comparison of the *in vitro* angiogenic capabilities of a range of adult stem cells has shown that bone marrow–derived MSCs were superior to adipose-derived MSCs in terms of tubule formation and VEGF secretion and interestingly placental chorionic villi-derived MSCs also showed promise. The ability of MSCs to show *in vitro* endothelial-like characteristics is strongly dependent on culture conditions and underlying substrate, and the majority of studies only show endothelial-like trans-differentiation in the presence of low serum endothelial media and the use of a matrigel-based extracellular matrix [107]. While nitric oxide has been shown to be an important modulator of the vasculogenic potential of MSCs [61] and nonviral ectopic expression of eNOS promotes endothelial transdifferentiation [108], eNOS, or NO does not appear to be expressed or produced in nondifferentiated MSCs. Adipose-derived MSCs engineered to express eNOS and seeded onto a decellularised human saphenous vein and implanted as an aortal bridge showed viability for up to 2 months in a rabbit model.

9. Conclusions

In conclusion, MSCs have been well documented to have both proangiogenic and myogenic properties and a significant number of clinical trials represent the current efforts to translate this therapeutic potential. Gene modification of MSCs represents a promising strategy with both viral and nonviral vectors to reprogram cells toward endothelial and cardiac lineages and improve the capability of transplanted MSCs to promote neovessel formation and repair of damaged myocardium.

Abbreviations

CFU	colony forming unit
EPCs	endothelial progenitor cells
ESCs	embryonic stem cells
hPL	human plasma lysate
iPSC	induced pluripotent stem cells
MSCs	mesenchymal stem cells

Author details

Nadeeka Bandara[1,2], Shiang Lim[2,4], Haiying Chen[3], Shuangfeng Chen[3], Le-Xin Wang[1,3] and Padraig Strappe[5]*

*Address all correspondence to: p.strappe@cqu.edu.au

1 School of Biomedical Sciences, Charles Sturt University, Wagga Wagga, NSW, Australia

2 O'Brien Institute Department, St. Vincent's Institute of Medical Research, Fitzroy, VIC, Australia

3 Central Laboratory, Liaocheng Peoples Hospital, Liaocheng, Shandong Province, China

4 Department of Surgery, St. Vincent's Hospital, University of Melbourne, Melbourne, VIC, Australia

5 School of Health, Medical and Applied Sciences, Central Queensland University, Rockhampton, Qld, Australia

References

[1] Morrison SJ, Kimble J. Review article—Asymmetric and symmetric stem-cell divisions in development and cancer. Nature. 2006;**441**:1068-1074

[2] Weissman IL, Anderson DJ, Gage F. Stem and progenitor cells: Origins, phenotypes, lineage commitments, and transdifferentiations. Annual Review of Cell and Developmental Biology. 2001;**17**:387-403

[3] Knoblich JA. Mechanisms of asymmetric stem cell division. Cell. 2008;**132**(4):583-597

[4] Thomson JA, Itskovitz-Eldor J, Shapiro SS, Waknitz MA, et al. Embryonic stem cell lines derived from human blastocysts. Science. 1998;**282**:1145-1147

[5] Reubinoff BE, Itsykson P, Turetsky T, Pera MF, et al. Neural progenitors from human embryonic stem cells. Nature Biotechnology. 2001;**19**:1134-1140

[6] Keller G. Embryonic stem cell differentiation: Emergence of a new era in biology and medicine. Genes & Development. 2005;**19**(10):1129-1155

[7] Takasato M, Er PX, Becroft M, Vanslambrouck JM, et al. Directing human embryonic stem cell differentiation towards a renal lineage generates a self-organizing kidney. Nature Cell Biology. 2014;**16**(1):118-126

[8] Soh BS, Ng SY, Wu H, Buac K, et al. Endothelin-1 supports clonal derivation and expansion of cardiovascular progenitors derived from human embryonic stem cells. Nature Communications. 2016;**7**:10774

[9] Huch M, Koo BK. Modeling mouse and human development using organoid cultures. Development. 2015;**142**:3113-3125

[10] Davidson KC, Adams AM, Goodson JM, McDonald CE, et al. Wnt/β-catenin signaling promotes differentiation, not self-renewal, of human embryonic stem cells and is repressed by Oct4. Proceedings of the National Academy of Sciences of the United States of America. 2012;**109**(12):4485-4490

[11] Takahashi K, Yamanaka S. Induction of pluripotent stem cells from mouse embryonic and adult fibroblast cultures by defined factors. Cell. 2006;**126**(4):663-676

[12] Wilmut I, Schnieke AE, McWhir J, Kind AJ, et al. Viable offspring derived from fetal and adult mammalian cells. Nature. 1997;**385**(6619):810-813

[13] Cowan CA, Atienza J, Melton DA, Eggan K. Nuclear reprogramming of somatic cells after fusion with human embryonic stem cells. Science. 2005;**309**(5739):1369-1373

[14] Takahashi K, Tanabe K, Ohnuki M, Narita M, et al. Induction of pluripotent stem cells from adult human fibroblasts by defined factors. Cell. 2007;**131**(5):861-872

[15] Yu J, Vodyanik MA, Smuga-Otto K, Antosiewicz-Bourget J, et al. Induced pluripotent stem cell lines derived from human somatic cells. Science. 2007;**318**(5858):1917-1920

[16] Wada H, Kojo S, Kusama C, Okamoto N, et al. Successful differentiation to T cells, but unsuccessful B-cell generation, from B-cell-derived induced pluripotent stem cells. International Immunology. 2011;**23**(1):65-74

[17] Aoi T, Yae K, Nakagawa M, Ichisaka T, et al. Generation of pluripotent stem cells from adult mouse liver and stomach cells. Science. 2008;**321**(5889):699-702

[18] Wang J, Gu Q, Hao J, Bai D, et al. Generation of induced pluripotent stem cells with high efficiency from human umbilical cord blood mononuclear cells. Genomics, Proteomics & Bioinformatics. 2013;**11**(5):304-311

[19] Liu H, Zhu F, Yong J, Zhang P, et al. Generation of induced pluripotent stem cells from adult rhesus monkey fibroblasts. Cell Stem Cell. 2008;**3**(6):587-590

[20] Takenaka-Ninagawa N, Kawabata Y, Watanabe S, Nagata K, et al. Generation of rat-induced pluripotent stem cells from a new model of metabolic syndrome. PLoS One. 2014;**9**(8):e104462

[21] Lepage SI, Nagy K, Sung HK, Kandel RA, et al. Generation, characterization, and multi-lineage potency of Mesenchymal-like progenitors derived from equine induced pluripotent stem cells. Stem Cells and Development. 2016;**25**(1):80-89

[22] Hong H, Takahashi K, Ichisaka T, Aoi T, et al. Suppression of induced pluripotent stem cell generation by the p53-p21 pathway. Nature. 2009;**460**(7259):1132-1135

[23] Okita K, Ichisaka T, Yamanaka S. Generation of germline-competent induced pluripotent stem cells. Nature. 2007;**448**(7151):313-317

[24] Zhao T, Zhang ZN, Rong Z, Xu Y. Immunogenicity of induced pluripotent stem cells. Nature. 2011;**474**(7350):212-215

[25] Yin T, Li L. The stem cell niches in bone. The Journal of Clinical Investigation. 2006;**116**(5):1195-1201

[26] Friedenstein AJ, Petrakova KV, Kurolesova AI, Frolova GP. Heterotopic of bone marrow. Analysis of precursor cells for osteogenic and hematopoietic tissues. Transplantation. 1968;**6**:230-247

[27] Gonzalez-Perez O. Neural stem cells in the adult human brain. Biological and Biomedical Reports. 2012;**2**(1):59-69

[28] Fiegel HC, Lange C, Kneser U, Lambrecht W, et al. Fetal and adult liver stem cells for liver regeneration and tissue engineering. Journal of Cellular and Molecular Medicine. 2006;**10**(3):577-587

[29] Zhang Y, Sivakumaran P, Newcomb AE, Hernandez D, et al. Cardiac repair with a novel population of mesenchymal stem cells resident in the human heart. Stem Cells. 2015;**33**(10):3100-3113

[30] Schilders KA, Eenjes E, van Riet S, Poot AA, et al. Regeneration of the lung: Lung stem cells and the development of lung mimicking devices. Respiratory Research. 2016;**23**: 17-44

[31] Bunnell BA, Flaat M, Gagliardi C, Patel B, et al. Adipose-derived stem cells: Isolation, expansion and differentiation. Methods. 2008;**45**(2):115-120

[32] Li Q, Guo ZK, Chang YQ, Yu X, et al. Gata4, Tbx5 and Baf60c induce differentiation of adipose tissue-derived mesenchymal stem cells into beating cardiomyocytes. The International Journal of Biochemistry & Cell Biology. 2015;**66**:30-36

[33] Petersen GF, Hilbert BJ, Trope GD, Kalle WH, et al. Direct conversion of equine adipose-derived stem cells into induced neuronal cells is enhanced in three-dimensional culture. Cellular Reprogramming. 2015;**17**(6):419-426

[34] Zhang R, Wang N, Zhang LN, Huang N, et al. Knockdown of DNMT1 and DNMT3a promotes the angiogenesis of human mesenchymal stem cells leading to arterial specific differentiation. Stem Cells. 2016;**34**(5):1273-1283

[35] Bang OY, Kim EH, Cha JM, Moon GJ. Adult stem cell therapy for stroke: Challenges and progress. Journal of Stroke. 2016;**18**(3):256-266

[36] Li L, Chen X, Wang WE, Zeng C. How to improve the survival of transplanted mesenchymal stem cell in ischemic heart? Stem Cells International. 2016;**2016**:9682757

[37] Lanotte M, Allen TD, Dexter TM. Histochemical and ultrastructural characteristics of a cell line from human bone-marrow stroma. Journal of Cell Science. 1981;**50**:281-297

[38] Caplan AI. Mesenchymal stem cells. Journal of Orthopaedic Research. 1991;**9**:641-650

[39] Pittenger MF, Mackay AM, Beck SC, Jaiswal RK, et al. Multilineage potential of adult human mesenchymal stem cells. Science. 1999;**284**:143-147

[40] Dominici M, Le Blanc K, Mueller I, Slaper-Cortenbach I, et al. Minimal criteria for defining multipotent mesenchymal stromal cells. The International Society for Cellular Therapy position statement. Cytotherapy. 2006;**8**:315-317

[41] Togel F, Weiss K, Yang Y, Hu Z, et al. Vasculotropic, paracrine actions of infused mesenchymal stem cells are important to the recovery from acute kidney injury. American Journal of Physiology. Renal Physiology. 2007;**292**:F1626-F1635

[42] Parekkadan B, van Poll D, Suganuma K, Carter EA, et al. Mesenchymal stem cell-derived molecules reverse fulminant hepatic failure. PLoS One. 2007;**2**(9):e941

[43] Block GJ, Ohkouchi S, Fung F, Frenkel J, et al. Multipotent stromal cells are activated to reduce apoptosis in part by upregulation and secretion of stanniocalcin-1. Stem Cells. 2009;**27**:670-681

[44] Di Nicola M, Carlo-Stella C, Magni M, Milanesi M, et al. Human bone marrow stromal cells suppress T-lymphocyte proliferation induced by cellular or nonspecific mitogenic stimuli. Blood. 2002;**99**(10):3838-3843

[45] Krampera M, Glennie S, Dyson J, Scott D, et al. Bone marrow mesenchymal stem cells inhibit the response of naive and memory antigen-specific T cells to their cognate peptide. Blood. 2003;**101**:3722-3729

[46] Le Blanc K, Tammik C, Rosendahl K, Zetterberg E, et al. HLA expression and immunologic properties of differentiated and undifferentiated mesenchymal stem cells. Experimental Hematology. 2003;**31**(10):890-896

[47] Corcione A, Benvenuto F, Ferretti E, Giunti D, et al. Human mesenchymal stem cells modulate B-cell functions. Blood. 2006;**107**:367-372

[48] Sotiropoulou PA, Perez SA, Gritzapis AD, Baxevanis CN, et al. Interactions between human mesenchymal stem cells and natural killer cells. Stem Cells. 2006;**24**:74-85

[49] Spaggiari GM, Capobianco A, Abdelrazik H, Becchetti F, et al. Mesenchymal stem cells inhibit natural killer-cell proliferation, cytotoxicity, and cytokine production: Role of indoleamine 2,3-dioxygenase and prostaglandin E2. Blood. 2008;**111**:1327-1333

[50] Gonzalez-Rey E, Gonzalez MA, Varela N, O' Valle F, et al. Human adipose-derived mesenchymal stem cells reduce inflammatory and T cell responses and induce regulatory T cells in vitro in rheumatoid arthritis. Annals of the Rheumatic Diseases. 2010;**69**(1):241-248

[51] Tasso R, Ilengo C, Quarto R, Cancedda R, et al. Mesenchymal stem cells induce functionally active T-regulatory lymphocytes in a paracrine fashion and ameliorate experimental autoimmune uveitis. Investigative Ophthalmology & Visual Science. 2012;**53**:786-793

[52] Nemeth K, Keane-Myers A, Brown JM, Metcalfe DD, et al. Bone marrow stromal cells use TGF-beta to suppress allergic responses in a mouse model of ragweed-induced asthma. Proceedings of the National Academy of Sciences of the United States of America. 2010; **107**:5652-5657

[53] Qu X, Liu X, Cheng K, Yang R, et al. Mesenchymal stem cells inhibit Th17 cell differentiation by IL-10 secretion. Experimental Hematology. 2012;**40**:761-770

[54] Asahara T, Kawamoto A, Masuda H. Concise review: Circulating endothelial progenitor cells for vascular medicine. Stem Cells. 2011;**29**:1650-1655

[55] Williams AR, Hare JM. Mesenchymal stem cells: Biology, pathophysiology, translational findings, and therapeutic implications for cardiac disease. Circulation Research. 2011;**109**:923-940

[56] Premer C, Blum A, Bellio MA, Schulman IH, et al. Allogeneic mesenchymal stem cells restore endothelial function in heart failure by stimulating endothelial progenitor cells. eBioMedicine. 2015;**2**(5):467-475

[57] Cai X, Lin Y, Friedrich CC, Neville C, et al. Bone marrow derived pluripotent cells are pericytes which contribute to vascularization. Stem Cell Reviews. 2009;**5**:437-445

[58] Crisan M, Yap S, Casteilla L, Chen CW, et al. A perivascular origin for mesenchymal stem cells in multiple human organs. Cell Stem Cell. 2008;**3**:301-313

[59] Davani S, Marandin A, Mersin N, Royer B, et al. Mesenchymal progenitor cells differentiate into an endothelial phenotype, enhance vascular density, and improve heart function in a rat cellular cardiomyoplasty model. Circulation. 2003;**108**(Suppl 1):II253-II258

[60] Psaltis PJ, Zannettino AC, Worthley SG, et al. Concise review: Mesenchymal stromal cells: Potential for cardiovascular repair. Stem Cells. 2008;**26**:2201-2210

[61] Gomes SA, Rangel EB, Premer C, Dulce RA, et al. S-nitrosoglutathione reductase (GSNOR) enhances vasculogenesis by mesenchymal stem cells. Proceedings of the National Academy of Sciences of the United States of America. 2013;**110**:2834-2839

[62] Chen J, Zhang ZG, Li Y, Wang L, et al. Intravenous administration of human bone marrow stromal cells induces angiogenesis in the ischemic boundary zone after stroke in rats. Circulation Research. 2003;**92**(6):692-699

[63] Markel TA, Wang Y, Herrmann JL, Crisostomo PR, et al. VEGF is critical for stem cell-mediated cardioprotection and a crucial paracrine factor for defining the age threshold in adult and neonatal stem cell function. American Journal of Physiology. Heart and Circulatory Physiology. 2008;**295**:H2308-H2314

[64] Asahara T, Masuda H, Takahashi T, Kalka C, et al. Bone marrow origin of endothelial progenitor cells responsible for postnatal vasculogenesis in physiological and pathological neovascularization. Circulation Research. 1999;**85**(3):221-228

[65] Luttun A, Tjwa M, Moons L, Wu Y, et al. Revascularization of ischemic tissues by PlGF treatment, and inhibition of tumor angiogenesis, arthritis and atherosclerosis by anti-Flt1. Nature Medicine. 2002;**8**(8):831-840

[66] Miyamoto Y, Suyama T, Yashita T, Akimaru H, et al. Bone marrow subpopulations contain distinct types of endothelial progenitor cells and angiogenic cytokine-producing cells. Journal of Molecular and Cellular Cardiology. 2007;**43**:627-635

[67] Rafii S, Lyden D. Therapeutic stem and progenitor cell transplantation for organ vascularization and regeneration. Nature Medicine. 2003;**9**:702-712

[68] Kim J, Kim M, Jeong Y, Lee WB, et al. BMP9 induces cord blood-derived endothelial progenitor cell differentiation and ischemic neovascularization via ALK1. Arteriosclerosis, Thrombosis, and Vascular Biology. 2015;**35**(9):2020-2031

[69] Al-Khaldi A, Eliopoulos N, Martineau D, Lejeune L, et al. Postnatal bone marrow stromal cells elicit a potent VEGF-dependent neoangiogenic response in vivo. Gene Therapy. 2003;**10**(8):621-629

[70] Reyes M, Dudek A, Jahagirdar B, Koodie L, et al. Origin of endothelial progenitors in human postnatal bone marrow. The Journal of Clinical Investigation. 2002;**109**(3):337-346

[71] Kinnaird T, Stabile E, Burnett MS, Lee CW, et al. Marrow-derived stromal cells express genes encoding a broad spectrum of arteriogenic cytokines and promote in vitro and in vivo arteriogenesis through paracrine mechanisms. Circulation Research. 2004;**94**(5):678-685

[72] Cai M, Shen R, Song L, Lu M, et al. Bone marrow mesenchymal stem cells (BM-MSCs) improve heart function in swine myocardial infarction model through paracrine effects. Scientific Reports. 2016a;**6**:28250

[73] Valarmathi MT, Davis JM, Yost MJ, Goodwin RL, et al. A three-dimensional model of vasculogenesis. Biomaterials. 2009;**30**:1098-1112

[74] Tillmanns J, Rota M, Hosoda T, Misao Y, et al. Formation of large coronary arteries by cardiac progenitor cells. Proceedings of the National Academy of Sciences of the United States of America. 2008;**105**(5):1668-1673

[75] Wakitani S, Saito T, Caplan AI. Myogenic cells derived from rat bone marrow mesenchymal stem cells exposed to 5-azacytidine. Muscle & Nerve. 1995;**18**(12):1417-1426

[76] Makino S, Fukuda K, Miyoshi S, Konishi F, et al. Cardiomyocytes can be generated from marrow stromal cells in vitro. The Journal of Clinical Investigation. 1999;**103**(5):697-705

[77] Bittira B, Kuang JQ, Al-Khaldi A, Shum-Tim D, et al. In vitro preprogramming of marrow stromal cells for myocardial regeneration. The Annals of Thoracic Surgery. 2002;**74**:1154-1159

[78] Shake JG, Gruber PJ, Baumgartner WA, Senechal G, et al. Mesenchymal stem cell implantation in a swine myocardial infarct model: Engraftment and functional effects. Annals of Thoracic Surgery. 2002;**73**(6):1919-1926

[79] Silva GV, Litovsky S, Assad JAR, Sousa AL, et al. Mesenchymal stem cells differentiate into an endothelial phenotype, enhance vascular density, and improve heart function in a canine chronic ischemia model. Circulation. 2005;**111**(2):150-156

[80] Tang J, Xie Q, Pan G, Wang J, et al. Mesenchymal stem cells participate in angiogenesis and improve heart function in rat model of myocardial ischemia with reperfusion. European Journal of Cardio-Thoracic Surgery. 2006;**30**(2):353-361

[81] Lim SY, Kim YS, Ahn Y, Jeong MH, et al. The effects of mesenchymal stem cells transduced with Akt in a porcine myocardial infarction model. Cardiovascular Research. 2006;**70**:530-542

[82] Zhou H, Dickson ME, Kim MS, Bassel-Duby R, et al. Akt1/protein kinase B enhances transcriptional reprogramming of fibroblasts to functional cardiomyocytes. Proceedings of the National Academy of Sciences of the United States of America. 2015;**112**(38): 11864-11869

[83] Hatzistergos KE, Quevedo H, Oskouei BN, Hu Q, et al. Bone marrow mesenchymal stem cells stimulate cardiac stem cell proliferation and differentiation. Circulation Research. 2010;**107**:913-922

[84] Loffredo FS, Steinhauser ML, Gannon J, Lee RT. Bone marrow-derived cell therapy stimulates endogenous cardiomyocyte progenitors and promotes cardiac repair. Cell Stem Cell. 2011;**8**:389-398

[85] Suzuki G, Iyer V, Lee TC, Canty JM Jr. Autologous mesenchymal stem cells mobilize cKit+ and CD133+ bone marrow progenitor cells and improve regional function in hibernating myocardium. Circulation Research. 2011;**109**:1044-1054

[86] Xu M, Wani M, Dai YS, Wang J, et al. Differentiation of bone marrow stromal cells into the cardiac phenotype requires intercellular communication with myocytes. Circulation. 2004;**110**:2658-2665

[87] Williams AR, Hatzistergos KE, Addicott B, McCall F, et al. Enhanced effect of combining human cardiac stem cells and bone marrow mesenchymal stem cells to reduce infarct size and to restore cardiac function after myocardial infarction. Circulation. 2013;**127**:213-223

[88] Toma C, Pittenger MF, Cahill KS, Byrne BJ, et al. Human mesenchymal stem cells differentiate to a cardiomyocyte phenotype in the adult murine heart. Circulation. 2002;**105**:93-98

[89] Amado LC, Schuleri KH, Saliaris AP, Boyle AJ, et al. Multimodality noninvasive imaging demonstrates in vivo cardiac regeneration after mesenchymal stem cell therapy. Journal of the American College of Cardiology. 2006;**48**:2116-2124

[90] Hashemi SM, Ghods S, Kolodgie FD, Parcham-Azad K, et al. A placebo controlled dose-ranging, safety study of allogenic mesenchymal stem cells injected by endomyocardial delivery after an acute myocardial infarction. European Heart Journal. 2008;**29**:251-259

[91] Houtgraaf JH, de Jong R, Kazemi K, de Groot D, et al. Intracoronary infusion of allogeneic mesenchymal precursor cells directly after experimental acute myocardial infarction reduces infarct size, abrogates adverse remodeling, and improves cardiac function. Circulation Research. 2013;**113**:153-166

[92] Ohnishi S, Yanagawa B, Tanaka K, Miyahara Y, et al. Transplantation of mesenchymal stem cells attenuates myocardial injury and dysfunction in a rat model of acute myocarditis. Journal of Molecular and Cellular Cardiology. 2007;**42**:88-97

[93] Hare JM, Traverse JH, Henry TD, Dib N, et al. A randomized, double-blind, placebo-controlled, dose-escalation study of intravenous adult human mesenchymal stem cells (prochymal) after acute myocardial infarction. Journal of the American College of Cardiology. 2009;**54**:2277-2286

[94] Hare JM, Fishman JE, Gerstenblith G, DiFede Velazquez DL, et al. Comparison of allogeneic vs autologous bone marrow-derived mesenchymal stem cells delivered by transendocardial injection in patients with ischemic cardiomyopathy: The POSEIDON randomized trial. Journal of the American Medical Association. 2012;**308**:2369-2379

[95] Chen SL, Fang WW, Ye F, Liu YH, et al. Effect on left ventricular function of intracoronary transplantation of autologous bone marrow mesenchymal stem cell in patients with acute myocardial infarction. The American Journal of Cardiology. 2004a;**94**:92-95

[96] Houtgraaf JH, den Dekker WK, van Dalen BM, Springeling T, et al. First experience in humans using adipose tissue-derived regenerative cells in the treatment of patients with st-segment elevation myocardial infarction. Journal of the American College of Cardiology. 2012;**59**:539-540

[97] Perin EC, Sanz-Ruiz R, Sanchez PL, et al. Adipose-derived regenerative cells in patients with ischemic cardiomyopathy: The precise trial. American Heart Journal. 2014;**168**:88-95 e82

[98] Singh A, Singh A, Sen D. Mesenchymal stem cells in cardiac regeneration: A detailed progress report of the last 6 years (2010-2015). Stem Cell Research & Therapy. 2016;**7**(1):82

[99] Mushtaq M, DiFede DL, Golpanian S, Khan A, et al. Rationale and design of the percutaneous stem cell injection delivery effects on neomyogenesis in dilated cardiomyopathy (the poseidon-dcm study): A phase i/ii, randomized pilot study of the comparative safety and efficacy of transendocardial injection of autologous mesenchymal stem cell vs. allogeneic mesenchymal stem cells in patients with non-ischemic dilated cardiomyopathy. Journal of Cardiovascular Translational Research. 2014;**7**:769-780

[100] Bartunek J, Behfar A, Dolatabadi D, Vanderheyden M, et al. Cardiopoietic stem cell therapy in heart failure: The c-cure (cardiopoietic stem cell therapy in heart failure) multicenter randomized trial with lineage-specified biologics. Journal of the American College of Cardiology. 2013;**61**:2329-2338

[101] Ieda M, JD F, Delgado-Olguin P, Vedantham V, et al. Direct reprogramming of fibroblasts into functional cardiomyocytes by defined factors. Cell. 2010;**142**(3):375-386

[102] Wang L, Liu Z, Yin C, Asfour H, et al. Stoichiometry of Gata4, Mef2c, and Tbx5 influences the efficiency and quality of induced cardiac myocyte reprogramming. Circulation Research. 2015;**116**:237-244

[103] Song K, Nam YJ, Luo X, Qi X, et al. Heart repair by reprogramming non-myocytes with cardiac transcription factors. Nature. 2012;**485**:599-604

[104] Cao N, Huang Y, Zheng J, Spencer CI, et al. Conversion of human fibroblasts into functional cardiomyocytes by small molecules. Science. 2016;**352**(6290):1216-1220

[105] Rustad KC, Wong VW, Sorkin M, Glotzbach JP. Enhancement of mesenchymal stem cell angiogenic capacity and stemness by a biomimetic hydrogel scaffold. Biomaterials. 2012;**33**(1):80-90

[106] Liu L, Yu Y, Hou Y, Chai J, et al. Human umbilical cord mesenchymal stem cells transplantation promotes cutaneous wound healing of severe burned rats. PLoS One. 2014;**9**(2):e88348

[107] Aguilera V, Briceño L, Contreras H, Lamperti L, et al. Endothelium trans differentiated from Wharton's jelly mesenchymal cells promote tissue regeneration: Potential role of soluble pro-angiogenic factors. PLoS One. 2014;**9**(11):e111025

[108] Bandara N, Gurusinghe S, Lim SY, Chen H, et al. Molecular control of nitric oxide synthesis through eNOS and caveolin-1 interaction regulates osteogenic differentiation of adipose derived stem cells by modulation of Wnt/β-catenin signalling. Stem Cell Research & Therapy. 2016;**7**(1):182

[109] Addis RC, Ifkovits JL, Pinto F, Kellam LD, et al. Optimization of direct fibroblast reprogramming to cardiomyocytes using calcium activity as a functional measure of success. Journal of Molecular and Cellular Cardiology. 2013;**60**:97-106

[110] Muraoka N, Yamakawa H, Miyamoto K, Sadahiro T, et al. MiR-133 promotes cardiac reprogramming by directly repressing Snai1 and silencing fibroblast signatures. The EMBO Journal. 2014;**33**:1565-1581

[111] Protze S, Khattak S, Poulet C, Lindemann D, et al. A new approach to transcription factor screening for reprogramming of fibroblasts to cardiomyocyte-like cells. Journal of Molecular and Cellular Cardiology. 2012;**53**:323-332

[112] Christoforou N, Chellappan M, Adler AF, Kirkton RD, et al. Transcription factors MYOCD, SRF, Mesp1 and SMARCD3 enhance the cardio-inducing effect of GATA4, TBX5, and MEF2C during direct cellular reprogramming. PLoS One. 2013;**8**:e63577

[113] Ifkovits JL, Addis RC, Epstein JA, Gearhart JD. Inhibition of TGFbeta signaling increases direct conversion of fibroblasts to induced cardiomyocytes. PLoS One. 2014;**9**:e89678

[114] Efe JA, Hilcove S, Kim J, Zhou H, et al. Conversion of mouse fibroblasts into cardiomyocytes using a direct reprogramming strategy. Nature Cell Biology. 2011;**13**:215-222

[115] Jayawardena TM, Egemnazarov B, Finch EA, Zhang L, et al. MicroRNA-mediated in vitro and in vivo direct reprogramming of cardiac fibroblasts to cardiomyocytes. Circulation Research. 2012;**110**(11):1465-1473

[116] Belian E, Noseda M, Abreu Paiva MS, Leja T, et al. Forward programming of cardiac stem cells by homogeneous transduction with MYOCD plus TBX5. PLoS One. 2015;**10**:e0125384

[117] JD F, Stone NR, Liu L, Spencer CI, et al. Direct reprogramming of human fibroblasts toward a cardiomyocyte-like state. Stem Cell Reports. 2013;**1**:235-247

[118] Nam YJ, Song K, Luo X, Daniel E, et al. Reprogramming of human fibroblasts toward a cardiac fate. Proceedings of the National Academy of Sciences of the United States of America. 2013;**110**:5588-5593

[119] Wystrychowski W, Patlolla B, Zhuge Y, Neofytou E, et al. Multipotency and cardiomyogenic potential of human adipose-derived stem cells from epicardium, pericardium, and omentum. Stem Cell Research & Therapy. 2016;**7**(1):84

[120] Park E, Takimoto K. A long-lasting cardiomyogenic gene expression by PEI-based transfection induces endogenous cardiac mRNAs in human adipose-derived stem cells. Biochemical and Biophysical Research Communications. 2016;**479**(1):12-16

[121] Hirai H, Katoku-Kikyo N, Keirstead SA, Kikyo N. Accelerated direct reprogramming of fibroblasts into cardiomyocyte-like cells with the MyoD transactivation domain. Cardiovascular Research. 2013;**100**:105-113

[122] Zhao Y, Londono P, Cao Y, Sharpe EJ, et al. High-efficiency reprogramming of fibroblasts into cardiomyocytes requires suppression of pro-fibroticsignalling. Nature Communications. 2015;**6**:8243

[123] Qian L, Huang Y, Spencer CI, Foley A, et al. In vivo reprogramming of murine cardiac fibroblasts into induced cardiomyocytes. Nature. 2012;**485**:593-598

[124] Srivastava D, Ieda M, Fu J, Qian L. Cardiac repair with thymosin beta4 and cardiac reprogramming factors. Annals of the New York Academy of Sciences. 2012;**1270**:66-72

[125] Jayawardena TM, Finch EA, Zhang L, Zhang H, et al. MicroRNA induced cardiac reprogramming in vivo: Evidence for mature cardiac myocytes and improved cardiac function. Circulation Research. 2015;**116**(3):418-424

[126] Inagawa K, Miyamoto K, Yamakawa H, Muraoka N, et al. Induction of cardiomyocyte-like cells in infarct hearts by gene transfer of Gata4, Mef2c, and Tbx5. Circulation Research. 2012;**111**:1147-1156

[127] Mathison M, Gersch RP, Nasser A, Lilo S, et al. In vivo cardiac cellular reprogramming efficacy is enhanced by angiogenic preconditioning of the infarcted myocardium with vascular endothelial growth factor. Journal of the American Heart Association. 2012;**1**:e005652

Chapter 8

Current Applications of Mesenchymal Stem Cells for Cartilage Tissue Engineering

Lizeth Fuentes-Mera, Alberto Camacho,
Nidia K. Moncada-Saucedo and
Víctor Peña-Martínez

Additional information is available at the end of the chapter

http://dx.doi.org/ 10.5772/intechopen.68172

Abstract

Articular cartilage injuries caused by traumatic/mechanical progressive degeneration result in joint pain, swelling, the consequent loss of joint function, and eventually osteoarthritis. Articular tissue possesses a poor ability to regenerate that further complicates the therapeutic approaches. Mesenchymal stem cells (MSCs) have emerged as a promising alternative treatment. Recently, it has been reported that a wide variety of strategies ranging from merely using cells in the injured area to employ biofunctional substitutes in which cells are harmonizing with scaffolding and growth factors to create an engineered cartilage tissue.

This chapter reviews the state-of the-art in cartilage tissue engineering focused on tissue engineering approaches designed to recapitulate the native development of cartilage and its tridimensional structure as an osteochondral unit. Since the production of hypertrophied tissue is one of the most critical challenges to overcome in chondral tissue regeneration, here we show new strategies to minimize hypertrophy in cartilage. Finally, the efficacy and safety of different treatments of cartilage in current clinical trials will be discussed.

While the framework provides new features and benefits concerning the strategies for articular tissue regeneration, this chapter presents a set of tools to improve approaches to orthopedic regenerative medicine based on the use of MSCs.

Keywords: MSCs, MSCs-subpopulations, cartilage regeneration, cartilage tissue engineering, hypertrophy

1. Introduction

The chondrocytes are the only cells found in cartilage. The chondrocytes demonstrate distinctive properties such as being metabolically active in order to maintain the renewal of the extracellular matrix (ECM) by synthesizing collagens, proteoglycans, hyaluronic acid, and glycoproteins. Restoration of the cartilage damage is still challenging for orthopedic medicine due to its poor ability to regenerate [1].

Mesenchymal stem cells (MSCs) have potential applications in tissue engineering, and regenerative medicine represents an attractive option for repairing lesions in cartilage. Stem cell–based therapies that harmonize with tissue-engineering technologies, and biomaterials are vital for the continuous advance of cartilage regenerative medicine [2, 3].

Once the relationship between structure function in normal and damaged tissues is understood and the development of biological substitutes for the repair or regeneration can be reached. To develop a biological substitute, tissue engineering uses scaffolds, cells, and growth factors. Each of these elements alone is able to promote tissue regeneration, but composites fabricated in combination would be more effective [4, 5].

The objective of the present chapter is, therefore, to describe the cellular and molecular framework in which chondrocyte differentiation develops and the articular tissue responds to the injury.

The maintenance of the chondrogenic phenotype during *in vitro* expansion and avoidance of hypertrophy of MSC-derived chondrocytes remains a challenge in cell-based strategies. Since chondrocyte differentiation is regulated by various signalling pathways, including fibroblast growth factor (FGF), transforming growth factor-β (TGF-β), and wingless/Int (WNT)/β-catenin, the role of these growth factors is analyzed. Furthermore, we show some strategies to minimize hypertrophy in cartilage [6–8].

Even though an ideal protocol for cartilage regeneration is yet to be established, approaches involving cells, biomaterials, and technology of tissue engineering will advance firmly toward effective clinical application.

2. Cartilage tissue

Cartilage is a type of connective tissue whose function is to protect the bones of the diarthrodial joints from the frictional forces associated with the load and impact support [1]. Hyaline or articular cartilage is heterologous, with varying density and organization according to the depth of its zones [9]. Articular cartilage is predominantly avascular, aneural, and alymphatic, so the main route for nutrition is through synovial fluid and assisted by mechanical compression forces [10].

The articular cartilage forms a thin layer of tissue with variable thickness depending on the body location. In humans, it ranges from 1 to 4 mm depending on the joint [11]. This tissue has viscoelastic ability, giving it the characteristic of deforming in order to increase the total contact surface with the consequent reduction of stress and increase the resistance to damage caused by applied loads. This function depends on the organization of the macromolecules in the extracellular matrix, particularly the arrangement and orientation of the collagen fibers [12].

Articular cartilage possesses a coefficient of friction between 0.002 and 0.02; the quality of the synovial fluid, the elastic deformation of the cartilage, and the effusion of the liquid from it are the factors able to decrease it. There are also factors that increase the coefficient of friction, such as alteration in the continuity at the surface of the cartilage (fibrillation) [13].

2.1. Chondrocytes

Cartilage consists of a single type of specialized cells called chondrocytes, representing approximately 5–10% of the tissue [14]; chondrocytes are embedded and clustered in the extracellular matrix (ECM) secreted by themselves (**Figure 1A**). Its function is influenced by changes in the ECM itself, as much as by changes in cell membrane pressure, age, and certain growth factors. These cells depolymerize and eliminate the ECM to enlarge their lacunae, a process featured in the endochondral ossification [13]. They are also present in isolation or organized in isogenic groups, depending on their metabolic activity, i.e., the cells that have just divided are active chondrocytes possessing a very large Golgi apparatus and a basophilic cytoplasm, which means that it can be stained with basic stains with net positive charge such as hematoxylin. These characteristics indicate that a protein synthesis is being performed and are initially located in the same lacunae, but as they secrete new intercellular matrix, they are separated; on the contrary, initial chondrocytes with low or no activity have a clear cytoplasm and a small Golgi apparatus [13].

2.2. Extracellular matrix

More than 98% of the articular cartilage corresponds to the ECM. The extracellular matrix is a dynamic network of macromolecules self-assembled. It is composed of water, gases, metabolites, cations and predominantly of collagens, noncollagenous glycoproteins, hyaluronan, and proteoglycans. ECM is able to regulate cell behavior influencing its proliferation and maturation processes. Therefore, it is not only scaffolding for chondrocytes but also functioning as a reservoir for growth factors and cytokines and modulates the cell activation state (**Figure 1A**) [13].

2.2.1. Water

The water occupies between 60 and 80% of the ECM volume; its function is to allow the deformation of the cartilage in response to stress, and it is also important for cartilage nutrition and joint lubrication. Approximately 30% of the water is contained in the intrafibrillar space within the collagen; however, a small percentage fills the intracellular space. The rest is concentrated in the pore space of the matrix. Interestingly, the ability of the articular cartilage to withstand significant loads comes from the frictional resistance to water flow and the pressurization of water within the matrix. When the amount of water increases up to 90%, as in the osteoarthritis (OA), it causes increased permeability, which in turn causes a decrease in resistance and compromises elastic abilities [13].

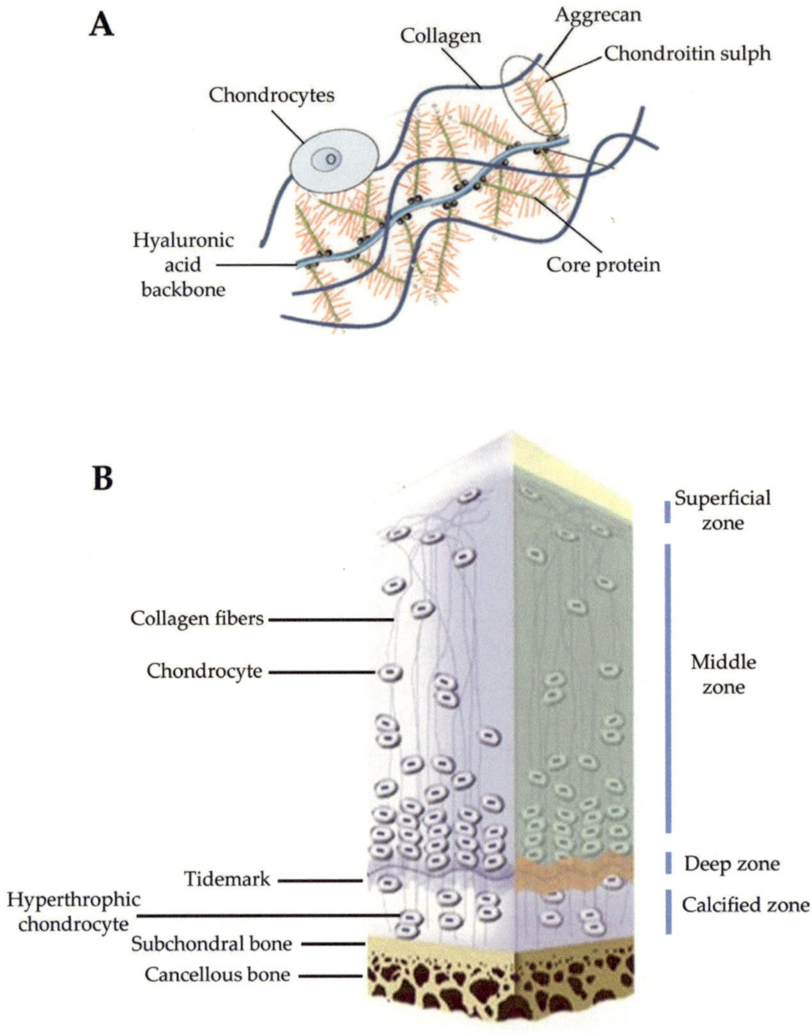

Figure 1. Components of ECM and 3D structure of the articular cartilage. (A) Model of extracellular matrix proteins showing their interaction with each other: collagens (mostly type II collagen); proteoglycan monomers, GAGs as aggrecan, and chondroitin sulphate covalently bound to the core protein. The proteoglycan monomers are assembled onto hyaluronic acid to form aggregates of proteoglycans. (B) The 3D organization of normal articular cartilage: the four areas of the cartilage are highlighted. The superficial zone, where flattened chondrocytes are located; the middle zone containing elongated chondrocytes; the deep zone, where the chondrocytes are arranged in columns and at the bottom, the calcified zone. (A) is modified from Izadifar et al. [15], and (B) from Minas et al. [16].

2.2.2. Collagen

Collagen is the most abundant structural macromolecule in ECM, and it makes up about 60% of the dry weight of cartilage. Collagen types present in cartilage are I, II, IV, V, VI, IX, and XI; however, type II collagen represents 90–95% of the total amount. Type II collagen forms fibrils, and fibers intertwines with proteoglycan, while the minor collagens stabilize the fibril network. Collagen X alone is present in osteochondral ossification phases and is therefore associated with cartilage calcification. Collagen VI is thought to form hexagonal nets in the vicinity of cells where it can bind to collagen II fibrils through matrilin-4, and biglycan nevertheless can be found in OA processes [17].

2.2.3. *Proteoglycans*

Proteoglycans (PGs) represent 10–15% of the ECM and are the main noncollagenous proteins present in cartilage. These macromolecules secreted by chondrocytes are responsible for the compression of cartilage. PGs are composed of one or more linear glycosaminoglycan (GAGs) chains covalently attached. The most frequent GAG is chondroitin sulfate (of which there are two subtypes, chondroitin-4-sulfate and chondroitin-6-sulfate), then keratan sulfate (or keratan sulfate or keratin sulfate), and dermatan sulfate. GAGs bind to a protein core, thus structuring the aggrecan. These aggrecans are bound by hyaluronic acid-binding proteins (a nonsulfated GAG) to form aggregates of PGs. It is important to mention that chondroitin-4-sulfate decreases over the years, whereas chondroitin-6-sulfate remains constant over the years, on the other hand, keratan sulfate and dermatan sulfate tend to increase with age. The half-life of PGs is 3 months, having the ability to retain water being responsible for the porous structure of cartilage [17].

Aggrecan occupies the interfibrillar space of the cartilage ECM and gives cartilage its osmotic properties, a critical feature to its ability to resist compressive loads [17].

The nonaggregating proteoglycans decorin, biglycan, and fibromodulin are also able to interact with collagen. These molecules differ from glycosaminoglycan in composition and function. Decorin and fibromodulin interact with the type II collagen fibrils and have a role in fibrillogenesis and interfibril interactions, whereas biglycan mainly interacts with collagen VI (**Figure 1A**) [18].

2.3. Extracellular glycoproteins

Among these are anchorine CII, fibronectin, laminin, and integrins. Their functions are to connect to the chondrocytes with the ECM, whereas integrins are the most important since they are able to interact with cellular receptors and influence migration, proliferation, and differentiation of chondrocytes [19].

Cartilage usually has protease inhibitors that help in the continuous renewal of the ECM constituent. The composition of ECM varies depending on the cartilage layer and the proximity to the chondrocytes, in this way, PGs rich in keratan-sulfate accumulate in the internal ECM, whereas in the territorial ECM, PGs rich in chondroitin sulfate are abundantly found.

Integrins into focal adhesions contribute to the activation of signaling pathways in the cell, promoting changes in cell survival, proliferation, and gene expression [20]. *In vitro* studies demonstrate that chondrocytes may interact with various proteins of the ECM such as fibronectin, laminin, vitronectin, osteopontin, bone sialoprotein II, and collagen types I, II, and VI through different integrins [21, 22].

2.4. Tridimensional structure of the articular cartilage

Till this point, we have reviewed the cellular and molecular components of the articular tissue, but how are they connected to each other?

The articular cartilage has a complex three-dimensional structure. Chondrocytes, collagen fibers, and ECM allow separate articular cartilage into various zones: the superficial zone, the middle zone, the deep zone, and the calcified zone (**Figure 1B**). The space between these zones allows in turn identifying three regions: the pericellular region, the territorial region, and the interterritorial region.

The superficial zone is thin and protects deeper layers from shear stresses. Mainly composed of types II and IX collagen packed tightly and aligned parallel to the articular surface (**Figure 1B**), contains flattened chondrocytes, which are in contact with synovial fluid. This zone is responsible for the tensile properties of cartilage. Below the superficial zone, the middle (transitional) zone is found, representing a bridge between the superficial and deep zones. The middle zone contains spherical chondrocytes at low density, proteoglycans, and thicker collagen fibrils. The middle zone is responsible for resistance to compressive forces. The deep zone provides the highest resistance to compressive forces. In the deep zone are found the largest diameter collagen fibrils in a radial arrangement and a low quantity of water. The chondrocytes are organized in columnar orientation, parallel to the collagen fibers and perpendicular to the joint line. Finally, the calcified layer of hypertrophic chondrocytes attaches the cartilage to the bones through anchoring the collagen fibrils of the deep zone to subchondral bone. The tidemark discriminate the deep zone from the calcified cartilage.

3. Cartilage injuries and osteoarthritis

Articular cartilage injuries are able to stimulate significant musculoskeletal morbidity in young and in aging patient populations. Restoration of joint damage to date represents a major challenge for medicine since they cannot heal spontaneously, and over time can also lead to the development of osteoarthritis.

The grading of articular cartilage lesions is performed through instrumented palpation of the lesion and via direct observation by arthroscopy [23, 24]. The most complete grading system is established by the International Cartilage Repair Society (ICRS) [25]. The ICRS grading system is based on the depth of the lesion and the degree to which subchondral bone is involved and classified the lesion as follows: Grade 0 for a normal joint; Grade 1 for nearly normal, featured by superficial lesions, soft indentation, and/or superficial fissures and cracks; Grade 2 for abnormal lesions extending down to <50% of cartilage depth; Grade 3 for severely abnormal lesions where cartilage defects are extending down to >50% of cartilage depth as well as down to calcified layer and down to but not through the subchondral bone; and Grade 4 for severely abnormal lesions where blisters are included [25].

Articular cartilage has limited ability for intrinsic repair. The injured chondrocytes (by a superficial or partial-thickness injuries) since early stages have a defective metabolic capacity and unable to maintain the normal PG concentration. This triggers other modifications in the tissue, such as increased tissue hydration and fibrillar disorganization of the collagen [26, 27]. These changes favor an increased transmission force to the subchondral bone. Exceeding the capacity of the subchondral bone makes the impact on the damaged cartilage even greater. In response to the events, chondrocytes proliferate, and thus production of matrix molecules at the site of injury increases; however, the new matrix fails to restore the native surface [26]. When the injury reaches subchondral bone (full-thickness injuries), the inflow of pluripotent marrow elements is observed [28]. The migrating mesenchymal stem cells produce type I collagen fibers to fill the full-thickness defect with fibrocartilage. Fibrocartilage fails to provide the necessary functions needed by the articular cartilage [29].

The strategies for articular cartilage lesions treatment can be classified into palliative such as physiotherapy and systemic pain relief medications; reparative such as debridement, knee joint lavage, arthroscopic abrasion arthroplasty, microfracture, and marrow stimulation techniques; restorative such as high tibial osteotomy, unicompartmental knee arthroplasty, and total knee arthroplasty; and transplantation such as osteochondral transplantation (osteochondral grafting), mosaicplasty, and autologous chondrocyte transplantation [27, 30].

Osteoarthritis (OA) is a chronic joint disorder classified into primary and secondary according to the etiology. OA is characterized by the progressive breakdown of the articular cartilage likewise changes in the subchondral bone, synovium, and muscles [31]. In early-stage OA, remodeling and bone loss of both trabecular and cortical subchondral areas are enhanced, while late-stage OA is featured by remodeling and an increased subchondral plate densification [32].

Mesenchymal stem cells (MSCs) are a promising option for the treatment of OA. MSCs are multipotent progenitor cells with self-renewal abilities, high plasticity, and anti-inflammatory properties; moreover, the capability to differentiate into different lineages including chondrocytes [33].

Despite extensive preclinical research with promising results, because of its technical limitations such as definition on the optimal cell source, processing, assembly with scaffolding, and administration modality, the use in patients is not yet overwhelming, and the design of a systematic procedure is still to be addressed [34, 35, 153, 154].

4. Cartilage tissue engineering

The application of cells into scaffolds, as tissue engineering do, makes cartilage regeneration strategies complex but allows to orchestrate the process efficiently [155]. Tissue engineering (TI) can be defined as the combined use of cells, biomaterials, and chemical factors to repair injured or diseased tissues. At the moment, it combines the contribution of cells that are placed on a scaffold, where the factors that accelerate its proliferation can be added; this composite is then transplanted at the site of the lesion in order to achieve tissue regeneration [36].

TI has the potential to provide long-lasting solutions to tissue damage and tissue loss, and engineering cartilage is not an exception to this approach. In fact, due to its limited ability to self-repair, cartilage is an ideal candidate for tissue engineering.

The critical point of the strategies based on TI is the expansion of cells in culture to be able to generate more cells for the production of tissue *in vitro* or for the implantation of cells in suspension or on scaffolds for regeneration of the tissue *in vivo* [37]. Another challenge to overcome regarding this strategy is to achieve the merging of the composite after implantation. The integration of the implanted tissue with the organ requires remodeling, degradation, and formation of new tissue. The remodeling of the implanted tissue is essential for its functionality [37].

TI makes possible the *in vitro* tissue synthesis, and then, the functional abilities of the composite can be evaluated before implantation. The main disadvantage is the partial absence of physiological and mechanical stimuli during their formation, which does not allow an adequate cellular regulation and spatial development of the tissue, and the decrease of its mechanical quality is observed as a consequence.

In the past decades, the strategies were designed without considering the cartilage as a complex tissue with a functionally that stratified three-dimensional structure. Nowadays, efforts are focused on achieving the landmarks in the process of cartilage formation with the development of a multiphase implant that recapitulates the cartilage as an osteochondral unit [156].

Cartilage tissue engineering combines a cell source, biomaterials, and growth or differentiation factors. Useful cell sources include autologous chondrocytes, minced autologous cartilage, and mesenchymal stem cells (from bone marrow, muscle, synovium, or adipose) [4, 6, 7, 24–28]. Regarding scaffolds, they may be fabricated with natural (e.g., collagen) or synthetic materials and designed as monophasic (chondral phase) or multiphasic (imitating the osteochondral unit) [6–8, 23, 38–41, 157, 158].

Chondroinductive growth factors are essentially members of the transforming growth factor-β (TGF-β) superfamily, some members of fibroblast growth factor (FGF) family and insulin-like growth factor-1 (IGF-1). These growth factors can be added to a culture medium to induce chondrogenic differentiation or through gene delivery, and, more recently, by nanoparticle delivery [4, 27, 29–32].

4.1. Cell source

Cell-based therapy is a biological therapy, involving the use of cells to develop new tissues or repair damage tissues. Therapies have been designed in order to generate a neocartilage in an attempt to offer the patients with chondral injuries an improvement in the quality of life or a long-lasting cure.

Autologous implantation of chondrocytes (ACT), intra-articular injection of meniscus with stem cells, and autologous matrix-induced chondrogenesis represent the current strategies for cartilage repairing by cell-based therapies. Furthermore, approaches using cell therapy with tissue engineering and biomaterials are increasing [42].

The optimal cell source for cartilage tissue engineering is not yet well established. The goal is to select a cell source that can be isolated by simple methods, are able to expand, and capable of being cultured to synthesize cartilage-specific molecules. The sources range from chondrocytes, fibroblasts, and stem cells to genetically modified cells [159–161].

Although chondrocytes have been the most used at first since they are found in native cartilage and have been extensively studied, they currently do not represent the best option.

Likewise, fibroblasts are easily and abundantly obtained and under treatment with lactic acid, they can acquire a chondrogenic phenotype.

Furthermore, stem cells can be expanded through several passages maintaining the differentiation potential. Additionally, all of these cells can be modified genetically to induce or enhance chondrogenesis. Adult MSCs are able to differentiate into osteoblasts, adipocytes, muscles, and chondrocytes and are, therefore, a suitable cellular source for tissue regeneration. Recent evidence indicates that there is great variability in the ability of differentiation between stem cells from specific tissues.

4.1.1. *Chondrocytes*

To perform autologous chondrocytes implantation, a full-thickness sample from a low-weight-bearing region of the joint is taken by biopsy during a first arthroscopic operation to collect a chondrocyte population.

The recovered tissue is enzymatically treated to isolate chondrocytes, which are then *ex vivo* expanded under conditions that preserve cell viability till yielding ~12–48 million cells [43].

During a second operation, the chondrocytes are implanted into the debrided cartilage defect. This technique avoids potential immune complications or viral infections from transplanting allogeneic cells or foreign materials [44]. Nevertheless, two operations are needed, and a long recovery time (6–12 months) is required to ensure neotissue maturation.

Several studies have shown that chondrocytes "dedifferentiate" into fibrochondrocytes in culture [45]. However, according to Martinez et al., they can redifferentiate and express chondrocytic markers after being cultured into a 3D *in vitro* culture system [46].

Moreover, *ex vivo* culturing of the chondrocytes reduced production of type II collagen and PGs upon expansion in monolayer culturing [47]. This process has been known as dedifferentiation, so the analysis of different markers of chondrocytic maturity as BMP-2, FGFR-3, and COL2A1 is necessary to confirm a stable chondrocyte phenotype.

The growth and the expression of type II collagen have been assessed after autologous chondrocytes implantation to substantiate the expansion of chondrocytes. These markers were found ineffective in predicting the capacity of expanded cells to produce stable cartilage tissue [48].

Xenogeneic and allogeneic chondrocyte have been studied as alternative chondrocytes cell sources. However, these cells can be involved in the induction of immune responses and diseases transmission. Thus, more studies are needed to overcome such issues in the field of allogeneic and xenogeneic chondrocytes.

4.1.2. *Mesenchymal stem cells*

MSCs have higher proliferation rates than chondrocytes and possess a vast differentiation potential toward a chondrogenic, also they are easy to collect from several tissues, such as adipose tissue, synovial membrane, and bone marrow. In addition, MSCs produces several extracellular matrix macromolecules involved in cartilage function, including fibronectin, collagen(s), PGs, and glycosylaminoglycans (GAGs), as well as a wide range of cytokines, growth factors, chemokines, and colony stimulating factors [49].

4.1.2.1. Bone marrow–derived MSCs

Bone marrow–derived stem cells (BM-MSCs) are one of the relevant stem cell choices for tissue engineering, and different studies have reported a potential of these cells for cartilage repair and as a treatment of the osteoarthritis. BM-MSCs can be differentiated into chondrocytes in a variety of culture conditions, usually involving induction with TGF-β and in a tridimensional environment (e.g., cell pellets and micromasses). The addition of TGF-β enhances chondrogenesis; however, the degree of chondrogenesis depends on the culture method or scaffolding [50]. In addition to TGF-β, other growth factors as BMP-6 and IGF-1 during *in vitro* culture also affect chondrogenesis as evidenced by enhanced type II collagen and aggrecan expression and accumulation [51].

Coculture system with chondrocytes is another approach used to promote chondrocyte differentiation of MSCs. Cell proliferation and positive expression of type II collagen have been observed, and this is due to growth factor secretions and cell-cell interactions as well as the microenvironment created by the chondrocytes [52].

Limitation on the use of BM-MSCs is the mechanical integrity of the matrix they produce which is poor in GAGs content. Moreover, in BM-MSCs undergoing chondrogenic induction, a high expression of COLX, MMP13, and ALP markers was observed by *in vitro* pellet culture; this profile frames a hypertrophic process. Likewise, stem cells derived from adipose tissue (AD-MSCs) were also associated with the development of hypertrophy, as demonstrated by type X collagen over-expression and up-regulation of ALP activity [53].

4.1.2.2. Adipose tissue–derived MSCs

Adipose tissue–derived MSCs (AD-MSCs) have a mesenchymal lineage as demonstrated by the expression of surface markers, such as CD105, CD73, and CD90, and posses the potential to differentiate into chondrocytes. Differentiation is achieved under a high density as micromass cultures or embedded in scaffolds in the presence of TGF-β, ascorbate, and dexamethasone, especially in combination with a tridimensional culture environment [54].

Lai et al. analyze chondrogenic potential of human adipose–derived stem cells (hAD-MSCs) using three-dimension biomimetic hydrogels. In addition, the effect of TGF-β3 supplementation was also included. They demonstrated that in the presence of TGF-β3, the expression levels of aggrecan and type II collagen expression were significantly up-regulated. However, expression levels of type I- and X- collagen were also significantly enhanced, which indicates a fibrotic repair [55].

To overcome the fibrotic repair, Zhu et al. developed a strategy of programmed application of TGFβ3 and NSC23766 (a Rac1 inhibitor) to commit the hyaline cartilage differentiation of adipose-derived stem cells (AD-MSCs) for joint cartilage repair. The efficacy of AD-MSCs with programmed application of TGFβ3 and Rac1 inhibitor for cartilage regeneration was analyzed in a rat model of osteochondral defects. The results showed that TGFβ3 promoted AD-MSCs chondro-lineage differentiation, and that the administration of NSC23766 after 7 days postindcution prevented AD-MSC-derived chondrocytes from hypertrophy *in vitro* and *in vivo* (**Figure 2**) [56].

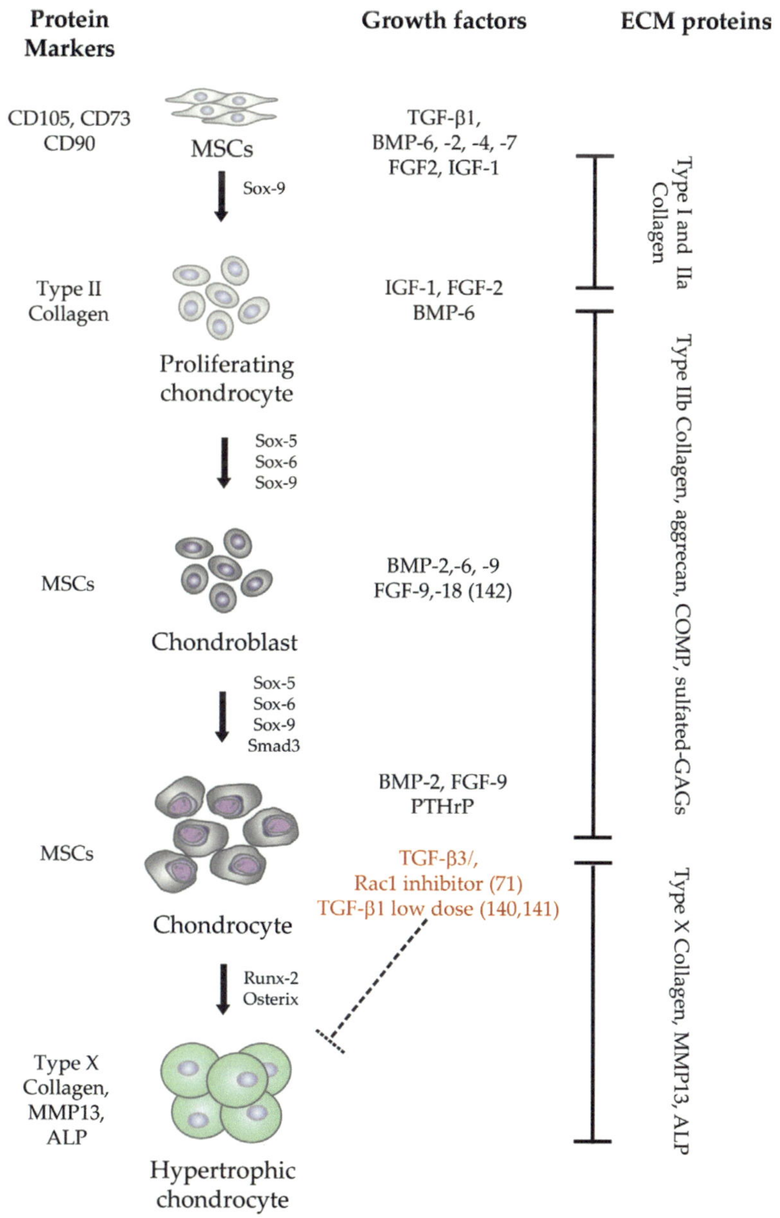

Figure 2. Chondrocyte differentiation from MSCs. Scheme showing the main transcription factors and growth factors involved in the production of extracellular matrix proteins in the articular cartilage.

In vitro culture, using growth factors like FGF-2 and BMP-6, influences the chondrogenesis of AD-MSCs as they are able to generate matrix proteins with accumulation of sulfated-GAGs. The use of FGF-2 increases cell proliferation and chondrogenesis through up-regulation of FGF-R2 and Sox 9 [57]. Besides, BMP-6 enhances expression of aggrecan and type II collagen (**Figure 2**) [58].

4.1.2.3. Synovium-derived MSCs

In the pursuit for a healthy cartilage regeneration through autologous transplantation, it has been discovered that synovial can be a valuable source of SCs for the effective induction of chondrogenesis and the production of high-quality cartilage *in vitro* [59, 60] and *in vivo* [61].

Synovium-derived MSCs (SDSCs) have a more effective chondrogenic potential than stem cells from other sources. Sakaguchi et al. revealed that human SDSCs have greater expansion and chondrogenic ability *in vitro* than MSCs from classical sources as bone marrow, periosteum, and adipose tissue. They also note that SDSCs generated larger pellets and a more structured ECM consistent with toluidine blue cartilage matrix staining, concluding that these are a superior source for chondrogenesis than AD-MSCs [62].

Extracellular matrix deposited by SDSCs overcomes two of the main problems related to the development and maturation of chondrocytes: dedifferentiation and chondrocyte redifferentiation [63]. Both processes are beaten because the tissue engineered cartilage matrix secreted by SDSCs is rich in collagen-II and aggrecan but not collagen-I or collagen-X and is mechanically similar to articular cartilage [64].

MSCs culture in general possesses inherent cell heterogeneity; however, for tissue engineering applications, it is imperative to start with a well-defined cell population, particularly, since it has been demonstrated that MSCs subpopulations are featured by a distinct regeneration potential. In this regard, the isolation of subpopulations from SDSCs cultures has been reported using various surface markers in order to sort by flow cytometry.

Arufe et al. reported the isolation from SDSCs of a $CD271^+$ subpopulation which showed high expression of SOX9, aggrecan, and COL2A1 at day 46 of chondrogenic induction; however, the expression of COL10A1 was observed [35, 65]. Meanwhile, the $CD105^+$ subpopulation reached a homogeneous cellular culture, and it was shown that after a chondrogenic induction, the increase in SOX9 expression was efficiently accompanied by an extracellular matrix rich in type II collagen with no evidence of fibrocartilage [35, 66].

More recently, in 2013, another subpopulation with efficient chondrogenic potential was reported, and the $CD73^+CD39^+$ cell subpopulation showed higher expression levels of SOX9 and a significantly greater chondrogenic potency than the $CD73^+CD39^-$ cell subpopulation and the original SDSCs population [35, 67].

4.2. Growth and transcription factors in chondrocyte

Chondrocytes differentiation from MSCs, chondrocyte morphology maintenance, and cartilage matrix formation are processes driven by differentiation and growth factors. A number of extracellular signalling molecules and growth factors as members of the fibroblast growth factor (FGF), hedgehog, transforming growth factor-β (TGF-β) and bone morphogenic protein (BMP), insulin-like growth factor (IGF), and wingless/Int (Wnt) glycoproteins are all key regulators of chondrocyte cell condensation and chondrogenic differentiation.

Cartilage formation begins by the condensation and then the differentiation of MSCs to prechondrocytes; thus, the cells first express types I and IIA collagen, and begins to synthesize GAGs and adhesion-related proteins such as cadherin [68]. This cascade of events is in response to the effect of factors, such as some members of the TGF-β superfamily (TGF-β1, -β2, and -β3), which are able to induce the synthesis of fibronectin, tenascin, and syndecan [69].

As prechondrocytes, the expression of transcription factors as SOX9, L-SOX5, and SOX6 became relevant until prechondrocytes reach the maturation stage and produce an ECM rich in collagen fibers (collagen types II, IX and XI) and PGs [70]. The main indicator of chondrocytes maturation is represented by type II collagen [71].

4.2.1. Insulin-like growth factor

Insulin-like growth factor (IGF) is a protein capable of regulating cell growth, differentiation, migration, and survival, and there are two types: IGF-1 and IGF-2. The activity of IGF-1 and IGF-2 (ligands) is collectively regulated by IGF-1R and IGF-2R (receptors), IGF-binding proteases, and IGF-binding proteins.

IGF-1 plays an important role in cell proliferation; it is mainly expressed in the liver although also in brain, heart, lung, bone, placenta, and testes and also produced by chondrocytes [72] thus having an autocrine and/or paracrine regulation [73].

IGFs in the early stages of chondrogenic differentiation induce the proliferation of chondrocytes and stimulate the differentiation of MSCs into prechondrocytes. IGFs act through the type I receptor tyrosine kinase (IGF-1R) that triggers mitogenic activity, regulated by extracellular kinase-kinase signals (MEK, ERK, and MAPK) and via the phosphatidylinositol 3-kinase/Akt (PI3K/Akt) pathway. Several investigations in animal models *in vivo* demonstrated that IGF significantly promotes the growth and matrix synthesis in articular cartilage and also improves the synthesis of proteoglycans and type II collagen [74].

4.2.2. The transforming growth factor-β

The transforming growth factor-β (TGF-β) is a superfamily of polypeptides and contains different factors, including TGF-β, inhibins, activins, and bone morphogenetic proteins (BMPs); the most promising for cartilage tissue engineering is TGF-β and bone morphogenetic proteins (BMPs), highlighting BMP-2, GDF-5, and BMP-7 [75].

TGF-β generally induces differentiation of MSCs into chondrocytes, stimulates their proliferation, increases ECM production, and inhibits endochondral ossification.

The three isotypes of TGF-β (β1, β2, and β3) are secreted in their inactive form and are activated only when dissociated from a peptide associated with latency (LAP). TGF-β initiates signaling by binding to the serine/threonine kinase types I and II receptors on the cell surface, which then propagates the signal through the phosphorylation of the R-Smad protein (Smad 2 and 3). TGF-β signaling is also negatively regulated by I-Smad (Smad 6 and 7), which interferes with R-Smad phosphorylation. TGF-β also activates mitogen-activated protein kinases (MAPK) [74].

BMPs are also multifunctional polypeptides having a key role in chondrogenesis by promoting terminal differentiation [76]. During first stages of chondrogenesis, the BMPs induce the expression of the N-cadherin thus promoting cell-cell interaction [77] indispensable for SOX expression. BMP signalling pathway enhances type X collagen promoter activity resulting in chondrocyte hypertrophy [78]. Furthermore, *in vitro* culturing BMP promotes the up-regulation of type II collagen and aggrecan [79].

4.2.3. *Fibroblast growth factors*

Fibroblast growth factors (FGF) belong to a family of polypeptides that are involved in several functions including cell proliferation, differentiation, survival, and motility, essential during cartilage growth, development, and repair. FGFs play a crucial role in the maintenance of stem cells and their activation [80]. FGF binding to fibroblast growth factor receptor (FGFR) leads to the activation of signaling pathways, including PI3K, Src, phospholipase Cγ, MAPKs, ERK, and p38. In particular, two members of the FGF family, basic fibroblast growth factor (bFGF, also known as FGF-2) and fibroblast growth factor-18 (FGF-18), play an important regulatory role in maintaining homeostasis of the cartilage matrix [74].

4.2.4. *Vascular endothelial growth factor*

Vascular endothelial growth factor (VEGF) is a hormone that seems to be important in the cartilage growth plate, where it supports the formation of endochondral bone, apparently by attracting endothelial cells from the bone marrow. It has been speculated that this factor could promote the mitotic cycle of chondrocytes, although little is known about this feature [81].

4.2.5. *Platelet-derived growth factor*

Platelet-derived growth factor (PDGF) is a potent mitogenic and chemotactic factor for all cells of mesenchymal origin, including chondrocytes and mesenchymal stem cells. It is related to increased cell proliferation and proteoglycan production [82].

4.2.6. *Transcription factors*

Many transcription factors are involved during chondrogenesis and endochondral ossification depending on the location of the chondrocytes within the articular cartilage. There are three main transcription factors involved in chondrogenesis: Sox9, L-Sox5, and Sox6.

Sox9 is involved in the condensation phase of MSCs into prechondrocytes when it is expressed in cells that initiate an aggregation that afterward stimulates the expression of cartilage-specific markers and inhibits terminal differentiation of chondrocytes. Sox9 also induces the expression of the L-SOX5 and SOX6 transcription factors, which definitely compromise the MSCs to develop in the chondrogenic lineage [83].

4.3. Scaffolds

4.3.1. Biomaterials

A significant research has been focused toward developing biomaterials that can mimic cartilage matrix, thus restoring function at the defect site. The biomaterials should satisfy three significant criteria: mechanical properties according to those of existing cartilage, integration with adjacent cartilage, and adequate biodegradability.

Natural biomaterials have the advantage as better cellular interaction due to the presence of ligands that can facilitate adhesion, in addition to promoting the activation of various chondrogenic activation pathways. The most common natural biomaterials are collagen, gelatine, lysozyme, Matrigel™, hyaluronic acid, fibrin, chitosan, agarose, and alginate [84].

As chondrocytes are surrounded by a hyaluronan-based pericellular matrix, they have been designed as hydrogels consisting of hyaluronan [85]. Chondrocytes attach firmly to hyaluronan-based matrices, stimulating as a result the matrix production [86]. Despite these advantages, the newly formed matrix is characterized by insufficient mechanical integrity [87].

Alginate is a natural biopolymer derived from brown algae and is composed of homopolymeric blocks of L-glucuronic acid and D-manuronic acid [88] and is widely used in biomedicine due to its biocompatibility and low toxicity [89]. Such materials require the use of divalent gelling cations to form a network structure, which allows the cells used to promote tissue regeneration to be trapped [90].

In contrast to natural polymers, synthetic polymers provide a better control of the structural and mechanical features. Polyglycolic acid (PGA) and polylactic acid (PLA) are the most used since they degrade by hydrolysis at rates depending on the selecting monomers [91]. A major disadvantage of synthetic polymers is that they do not offer specific biological functions [92]. Its functionalization with biological motifs or bioactive molecules facilitates cell adhesion and consequently stimulating matrix production [76].

Fragments derived from the cartilage have been also used for application in joint regeneration; however, it is mandatory to eliminate the cellular component to avoid an immune response when implanted in an *in vivo* model; therefore, they are decellularized by different methods [93, 94]. It has been observed that they support the production of type II collagen and proteoglycans; in addition, they minimize the hypertrophy of the newly formed tissues with the cooperation of growth factors [95].

4.3.2. Biological scaffolds and its use in the treatment of chondral lesions

A better understanding of the molecular structure and functional role of extracellular matrix components in the physiology of the cartilage [96] supported the construction of scaffolds which mimick cartilage microenvironment.

Scaffold-based approaches possess several advantages compared to scaffold-free: such as increased control to fill the cartilage defect according to the features and size of the lesion; no surgical procedures are required to obtain tissue from the patient; increased graft stability that influence recovery time for the patient. Most important, since the chondrocytes are cultured in a 3D environment, dedifferentiation is a minimal concern, thus, the cells are able to produce a more hyaline-like cartilage [97].

The geometry and microarchitecture of scaffolds are important factors that determine cell adhesion and migration, as well as the preservation of the synthesized matrix. [98]. In addition, the pore size is a critical feature; it should be large enough to allow the migration of the cells and thus promote the production of ECM, [99] but small enough to establish a large surface area for cell adhesion. It is recommended to generate scaffolds with a pore size ranging between 300 and 350 nm [100].

Successful cartilage regeneration is closely related to the ability of the scaffold to support the chondrocyte proliferation rate and to the differentiation of MSCs within a tissue-engineered 3D matrix [101]. At the same time, it becomes imperative to characterize the quality of the expanded MSC as well as to avert the development of hypertrophic chondrocytes [49, 102].

5. Minimizing the development of hypertrophic tissue in cell-based therapies

Chondrogenic hypertrophy is characterized by an increase of the cell volume as well as remodeling of ECM [103]. The increase in the volume is the result of intracellular and extracellular osmolarity variations where aggrecans are the main contributors.

The hypertrophic differentiation of chondrocytes is a process that gradually leads to the mineralization of cartilage. The main factors involved are the transcription factor runt-related transcription factor 2 (RUNX2) and myocyte enhancer factor-2C (MEF2C). Both promote the expression of proteins that determine terminal differentiation, including matrix metalloproteinase 13 (MMP13) [104], type X collagen [105], alkaline phosphatase (ALP), and vascular endothelial growth factor (VEGF) [106]; all of them functionally contribute to endochondral ossification. MMP13 is a key modulator in this process as it degrades type II collagen and aggrecan [107]. Furthermore, type X collagen serves as a framework for the calcification through matrix vesicles [108], ALP hydrolyzes pyrophosphate to inorganic phosphate that, in the presence of calcium, forms hydroxyapatite (**Figure 2**) [109].

Among the growth factors that are directly involved in the process are transforming growth factors (TGFs) and bone morphogenetic proteins, which are able to initiate cartilage differentiation but often lead to hypertrophy and calcification, since TGFβ3 is also the active signal pathway during endochondral ossification [110].

Moreover, Woods et al. has reported that Rac1 functions as a positive regulator in governing chondrocyte hypertrophy, maturation, and calcification [111] through up-regulation of type X collagen, MMP13, and ADTAMTS-5 that induce hypertrophy and chondrocyte calcification [112].

Therefore, Rac1 is an important target for controlling the development of hypertrophy has been demonstrated that the inhibition of Rac1 activity overcomes not only chondrocyte hypertrophy and calcification but also alleviates osteoarthritis progression [56, 112–115].

TGF-βs and BMPs bind to specific receptors to recruit type I receptor to trigger signalling of their specific receptor-Smads. The importance of Smad-dependent TGF-β- and BMP-signalling pathways for cartilage and bone formation has been well established whose synergic or antagonistic function depends on the microenvironment [116].

However, the control of the activity of some members of the TGF superfamily as TGF-β1 is important to control the subsequent maturation of chondrocytes and the consequent mineralization [117]. TGFβ1 is a key factor in the maintenance of chondrocyte phenotype. BMP9 and TGFβ1 dose-dependently synergized on Smad2 phosphorylation and showed an additive effect on expression of Smad3-dependent genes. Furthermore, the addition of a low dose of TGFβ1 (1 ng/ml) diminishes expression of early markers of cellular hypertrophy Alpl and type X collagen (**Figure 2**) [118, 119].

Likewise, *in vitro* addition of TGFβ3 in the later stages of differentiation has been shown to control the fibrotic process. Pei et al. found that in synovium-derived stem cells (SDSCs), TGF-β3 enhancing collagen II and sulfated glycosaminoglycan minimize the expression of collagen I in the repair of partial-thickness cartilage defects in porcine SDSC pellets compared with TGF-β1 [120].

Meanwhile, blocking of BMP signalling during chondrogenesis of MSC restricts Type X collagen and MMP13 expression from cartilage at maintained collagen type II and enhanced SOX9 expression [116]. Thus, the manipulation of BMP-signalling (essentially BMP4) is able to shift chondrogenesis of the MSCs toward a nonhypertrophic phenotype. Dexheimer et al. addressed this concern by the inhibition of Smad1/5/9-signalling using dorsomorphin [121]. This seems to be a good strategy to potentiate chondrogenesis and also inhibit hypertrophic differentiation; however, now the efforts should be focused on establishing a timeline on specific cellular models at which point the inhibition of pSmad1/5/9 signaling should be carried out. Inhibition of pSmad1/5/9 signalling apparently stopped chondrogenesis or decelerated MSC differentiation toward hypertrophy depending on the time of treatment initiation.

FGF factors play an important regulatory role in maintaining homeostasis of the cartilage matrix [74, 164]. Correa et al. designed an elegant sequential protocol based on the addition of FGF2, 9, and 18 on bone marrow–derived hMSC. The growth factors are added as follows: increased cell proliferation and priming (FGF2 [d0 to d3]); stimulated early chondrogenic differentiation by shifting the chondrogenic program earlier (TGF-β, FGF9/FGF18 [d4 to d14]); enhanced ECM production (d14 to d21); and delayed terminal hypertrophy (FGF9/FGF18 [d21 to d28]). To highlight, in the proterminally differentiating conditions, both FGF9 and FGF18 were able to reduce Runx2 expression and the activity of the hypertrophy-specific marker alkaline phosphatase (ALP) (**Figure 2**) [122].

As has already been established, TGF-β and bone morphogenetic protein are key factors for chondrogenesis. They are capable of initiating signals in mitogen-activated protein kinase (MAPK) pathway essentially of extracellular signal-regulated kinase ERK-1 and ERK-2 [123, 124].

PD98059 is one of the ERK inhibitors that suppressed hypertrophy in the chondrogenesis from bone marrow–derived hMSCs, inhibiting the cascade signalling upstream of ERK1/2 activation [125, 126]. Lee et al. [123] constructed a PD98059-impregnated poly(lactic-co-glycolic acid)

(PLGA) scaffold and demonstrated that it effectively suppresses the hypertrophy of hBM-MSCs that have been differentiated toward the chondrocytic lineage in basic chondrogenic medium (Dulbecco's modified Eagle's medium/Ham's F-12 nutrient mixture: DMEM/F-12 supplemented with 1% insulin-transferrin-sodium selenite [ITS], 50 μM ascorbate-2-phosphate, 1×10^{-7} M dexamethasone, 1 mM sodium pyruvate, and 50 μM l-proline). The efficiency of the system was challenged in a rabbit model where the main observations were the abundant presence of type-II collagen in ECM with the absence of types I and type X collagens [123].

As described in the above example, scaffolds can additionally be a useful tool for the release of elements that both regulate the differentiation process and control the development of fibrotic tissue. Inhibitors are not the only elements that can be incorporated into scaffolding systems; the addition of growth factors to scaffold systems has also been reported to reduce the development of hypertrophic chondrocytes, with successful results.

Mimicking the native tissue architecture is critical for effective cartilage regeneration. Kim et al. developed a multifunctional system based on TGF-β3 encapsulated PLCL scaffold using human adipose-derived stem cells (ADSCs) to promote chondrogenesis. They demonstrated the release of TGF-β3 molecules for 8 weeks, which remained in the PLCL matrix. Furthermore, this scaffolding system formed a hyaline cartilage-specific lacunae structure and minimized the hypertrophy of differentiated chondrocytes [127].

Fibrocartilage is a dense, fibrous version of a cartilage. It has been shown that a lack of integration of the graft in the lesion area promotes poor functional properties and limits complete regeneration of the defect.

The scaffold design should offer hierarchical structure, desired mechanical properties (stiffness, elasticity, permeability, and diffusion), and ability for adapting into the anatomical shapes. The use of scaffolds that are composed of one type of biomaterial, with homogeneous porosity and architecture, and used a single-cell type limits the integration with the surrounding tissue and encourages a fibrotic process. Instead, the design of stratified or gradient scaffold mimics the structural and mechanical features of a native osteochondral unit. In order to achieve stratification and composition, composite scaffolds are assembled through a multilayered scaffold design; in this way, structural templates for the cartilaginous layer, the tidemark, and calcified cartilage, and the subchondral bone are generated [163].

A biphasic scaffold design was reported, based on a silk-protein scaffold constituting the cartilage phase and a silk-coated strontium-hardystonite-gahnite ceramic scaffold constituting the bone phase, and both phases are cellularized with human mesenchymal stem cells (hMSCs). For the biphasic scaffold, there were noticeable to significant increases in Sox-9, collagen type II, and aggrecan in addition to low type X collagen expression levels compared with the chondral single-phase version of the implant [39].

Ho et al. designed a biphasic implant comprising of a polycaprolactone (PCL) cartilage scaffold and a PCL-tricalcium phosphate as bone scaffold; it was seeded with mesenchymal stem cells (MSCs), and the cartilage and bone constructs were maintained in the MSC expansion media prior to implantation into critically sized osteochondral defects in a pig model. After 6 months, the cartilaginous repair was observed with a low occurrence of fibrocartilage. Furthermore, the functional cartilage restoration was demonstrated by a high Young's modulus [40].

The chondral phase was derived from bovine decellularized articular cartilage ECM, while the osseous phase was composed of a PLGA/β-TCP wrapped with type I collagen. The biphasic scaffolds was cellularized by BMSCs induced with chondrogenic and osteogenic medium and implanted into osteochondral defects in a rabbit knee model. By a histological evaluation, the presence of an uniform neocartilage surface, a clear fusion of neocartilage, a regenerated subchondral bone with a well-defined tidemark, and no evidence of fibrocartilage was demonstrated [41].

6. The efficacy and safety of different treatments of cartilage in current clinical trials

Current surgical treatments for symptomatic cartilage lesions include reparative and reconstructive treatments. The former employ techniques that stimulate tissue cells to form hyaline cartilage, such as the microfracture technique, perforation, abrasion arthroplasty, and biological procedures involving cell culture, such as autologous chondrocyte implantation (ACI), of the latter derives the development of biocompatible three-dimensional scaffolds, where chondrocytes, stem cells, or chondrocyte-like cells can be grown. The latter involves mosaicplasty, such as the transplantation of autografts or allografts, which are composed of mature hyaline cartilage attached to the subchondral bone [128]. The type of treatment to be used will depend on the size of the lesion, the location, and the type of activity of the patient.

ACI has been the treatment most often used to treat large knee cartilage injuries. Previous studies have compared ACI with mosaicplasty, microfracture, and matrix-guided microfracture; where the ACI obtains better results (88%) than mosaicplasty (69%), presenting hyaline cartilage in half of the biopsies; however, statistically significant results are restricted to the medial condylar area [129]. ACI has been compared to microfracture and mosaicplasty but has never been compared with simple arthroscopic debridement and rehabilitation alone. At present, the first study that will detail, with a high level of evidence, the results of comparing ACI with simple debridement and physiotherapy in symptomatic lesions of full thickness of the knee is being carried out. This study aims to increase the clinical and economic knowledge between these techniques in the short and long term [130].

Recently, surgical treatments have been complemented with the use of autologous biological materials such as PRP and mesenchymal stem cells [162]. Recently, the use of autologous PRP coupled with the microfracture technique has been associated with better short-term clinical and functional outcomes, especially in pain [131–133]; in addition, arthroscopic implantation of AD-MSC combined with the microfracture technique has also been associated with a decrease in pain [134]. Similarly, the use of AD-MSC [135, 136] or BM-MSC [137, 138] has been reported as a safe therapeutic alternative via intraarticular injection in patients with osteoarthritis, which report a significant improvement in pain levels. Another complementary therapy is the use of hyaluronic acid, which improves articular cartilage repair in combination with autologous peripheral blood stem cells via intraarticular postoperative injection, complementing the technique of arthroscopic subchondral perforation [139].

Gene therapy techniques represent an alternative strategy for gene transfer for therapeutic purposes. Through gene therapy, proteins are synthesized at the site of the defect or injury where they directly influence the natural microenvironment. Growth factors most frequently synthesized by engineered cartilage cells to regenerate damaged cartilage include BMPs, IGF-1, and TGF-βs 1, 2, and 3.

The application of genetically engineered cartilage in clinical trials begins to generate results. Mont et al. have reported on a series of clinical trials in which the effect of injecting genetically engineered chondrocytes virally transduced with TGF-β1 (GEC-TGF-β1) into the knees of 54 patients with osteoarthritis has been evaluated. After 12 or 24 weeks of treatment, patients reported a degenerative process delayed compared to the placebo. It should be noted that the studies demonstrating cartilage regeneration are missing [140].

Moreover, they also evaluate the efficacy to safely regenerate cartilage by a phase II clinical trial in 102 (ClinicalTrials.gov NCT01221441) [141] and 27 patients (ClinicalTrials.gov NCT02341378) [142], respectively. Patients expressed decreased pain and improvement in function and physical capacity; thus, treatment injection of GEC-TGF-β1 seems to improve symptoms and pain due to knee osteoarthritis.

Although gene therapy is very promising for cartilage repair, much remains to be understood about the clinical results obtained. It is crucial to control several aspects such as the duration of transgenic expression, as well as the identification and selection of a therapeutic factor that is clinically useful for cartilage regeneration.

With the boom in tissue engineering in recent years and innovation in the area of biomaterials, doctors have new options for treating chondral lesions. To date, the clinical use of these materials is limited; very few polymers have been used for clinical trials in cartilage tissue engineering. The list includes collagen, polymers based on hyaluronan and fibrin, because they have a biomimetic structure similar to the native articular cartilage. Chondral or osteochondral grafts consist in the surgical transfer of mature tissue to a cartilage defect. The graft could be an autologous tissue transfer from a nonload-bearing zone and cultivated on a porated scaffold.

Within this narrow list is highlighted the collagen, which is the main component of ECM. Zheng et al. have analyzed the efficacy of an implant constructed with chondrocytes cultured onto collagen type I/III scaffolds in 56 patients with OA. The results mainly evidenced the maintenance of a chondrocytic phenotype, as well as a good integration of the implant in the injured area and the production of aggrecan and type II collagen. These data demonstrated the regeneration of hyaline cartilage tissue 6 months after treatment [143]. Furthermore, based on the information contained in the website, *clinicaltrials.gov*, clinical trials using stem cells cultivated mainly on collagen scaffolds are ongoing.

Assor et al. conduct a clinical trial to demonstrate the efficacy and safety of BM-MSCs cultured on a collagen type I scaffold dotted with hydroxyapatite to regenerate articular cartilage defects of the knee (**Table 1** NCT01159899). Otherwise, Giannini et al. studied the critical points of the regenerative treatment with BM-MSCs embedded in equine collagen type I scaffold (**Table 1** NCT02005861). Both trials are still in process and have not shown final results.

Clinical trials.gov identifier	Phase	Brief description	Status	No. patients	Principal investigator
NCT00850187	Phase 1	Autologous transplantation of mesenchymal stem cells (MSCs) cultured on a collagen I scaffold in full-thickness articular cartilage	Completed	6	Leila Taghiyar,
NCT01159899	Early Phase 1	Autologous BM-MSCs, stimulated with a protein matrix cultured in a collagen hydroxyapatite scaffold	Unknown	50	Michel Assor
NCT02005861	Not provided	Bone marrow aspirate concentrate cultivated onto an equine collagen type 1 scaffold (IOR-G1, Novagenit, Mezzolombardo, TN, Italy)	Recruiting	140	Sandro Giannini
NCT02659215	Not provided	Bone marrow aspirate concentrate mixed with a hyaluronan-based scaffold (Hyalofast®)	Active, not recruiting	200	Alberto Gobbi
NCT01282034	Phase 4	Bioceramic, multilayered scaffold: Type I equine collagen and Magnesium enriched-Hydroxyapatite (MaioRegenv®)	Completed	145	Maurilio Marcacci
NCT01471236	Phase 4	A cell-free system based on a biphasic, porous resorbable scaffold. Cartilage phase consist in a modified aragonite and HA, while bone phase in aragonite (Agili-C implant).	Recruiting	65	Elizaveta Kon

Source: https://clinicaltrials.gov.

Table 1. Current clinical trials based on cartilage tissue engineered approaches for the treatment of chondral and osteochondral diseases.

Hyaluronate is also reported in the clinical trials; Gobbi leads a study to evaluate the efficiency of Hyalofast®, together with bone marrow aspirate containing the MSCs without the need for processes for cell isolation and *ex vivo* expansion. This methodology has the disadvantage that the number of MSCs destined for the repair of the chondral tissue is not controlled nor the cellularity of the aspirate (**Table 1** NCT02659215).

New studies evaluate the performance, stability, safety, and viability of the surgical procedure using biomimetic osteochondral implants. One of the most studied and relevant procedure that is currently marketed as MaioRegen® (**Table 1** NCT01471236); this is a three-phase implant that considers the biomechanics of the chondral phase, the tidemark, and subchondral bone. Its composition is based on a collagen type-I hydroxyapatite gradient. MaioRegen® has demonstrated throughout its clinical studies that its technique is safe and that the clinical results in the short- and medium-term follow-up are effective in a large population of patients [144–148], even attributed better results than other commercial scaffolds ChondroMimetic™ [149] and TruFit CB™ [11, 150–152]. Despite the encouraging *in vitro* results of a wide variety of osteochondral scaffolds, most are in the early stages of development.

In general, clinical trials present current techniques for treating chondral lesions as safe. However, in order to determine the best surgical option in the treatment of symptomatic chondral defects, a rigorous clinical trial should be developed, where there is prospective control, randomization control, even feeding control, and rehabilitation control so that on the basis of comparative results, the surgeon provides an effective treatment.

7. Conclusion

This chapter presents an overview of the advances in the design of new cell-based therapies in conjunction with the tools that the tissue engineering offers, such as the use of biomaterials, the selection of subpopulations, and the addition in a temporal manner of growth factors.

It is important to emphasize that the understanding of the molecular mechanisms that governs the chondrocyte differentiation allows generating strategies that reflect a balance between the chondrocyte maturation and the containment in the development of fibrotic repair tissue.

Indeed, the therapeutic approaches for the repair and regeneration of joint tissue should consider as a goal mimic the osteochondral structure, which will result in an effective and safe clinical application.

Acknowledgements

The authors are supported by a grant from CONACyT; 233365-SALUD/2014.

Abbreviations

ACI	autologous chondrocyte implantation
AD-MSCs	adipose tissue-derived MSCs
ALP	alkaline phosphatase
ACT	autologous implantation of chondrocytes
BM-MSCs	bone marrow-derived stem cells
BMP	bone morphogenic protein
ECM	extracellular matrix
ERK1/2	extracellular signal-regulated kinase -1 and -2
FGF	fibroblast growth factor
FGFR	fibroblast growth factor receptor
GAGs	glycosaminoglycan
ICRS	International Cartilage Repair Society
IGF	insulin-like growth factor
LAP	latency-associated peptide
MMP13	matrix metalloproteinase 13
MSCs	mesenchymal stem cells
MAPK	mitogen-activated protein kinases
MEF2C	myocyte enhancer factor-2C
OA	osteoarthritis
PDGF	platelet-derived growth factor
PRP	platelet-rich plasma
PCL	polycaprolactone
PGA	polyglycolic acid
PLA	polylactic acid
PLCL	poly(L-lactide-co-caprolactone)
PLGA	poly(lactic-co-glycolic acid
PGs	proteoglycans
RUNX2	Runt-related transcription factor 2
SDSCS	synovium-derived MSCs
TI	tissue engineering
TGF-β	transforming growth factor-β
VEGF	vascular endothelial growth factor

Author details

Lizeth Fuentes-Mera[1*], Alberto Camacho[1,2], Nidia K. Moncada-Saucedo[1] and Víctor Peña-Martínez[3]

*Address all correspondence to: lizeth46@hotmail.com

1 Departamento de Bioquímica y Medicina Molecular, Universidad Autónoma de Nuevo León (UANL), Monterrey, México

2 Unidad de Neurociencias, Centro de Investigación y Desarrollo en Ciencias de la Salud, Universidad Autónoma de Nuevo León (UANL), Monterrey, México

3 Servicio de Ortopedia y Traumatología, Universidad Autónoma de Nuevo León (UANL), Monterrey, NL, México

References

[1] Longo UG, et al. Stem cells and gene therapy for cartilage repair. Stem Cells International. 2012;**2012**:1-9

[2] Chen CW, et al. Type I and II collagen regulation of chondrogenic differentiation by mesenchymal progenitor cells. Journal of Orthopaedic Research. 2005;**23**:446-453

[3] van der Kraan PM, et al, Interaction of chondrocytes, extracellular matrix and growth factors: Relevance for articular cartilage tissue engineering. Osteoarthritis and Cartilage. 2002;**10**:631-637

[4] Chen WH, et al. Tissue engineered cartilage using human articular chondrocytes immortalized by HPV-16 E6 and E7 genes. Journal of Biomedical Materials Research. Part A. 2006;**76**:512-520

[5] Verbruggen G, et al. Osteochondral repair in synovial joints. Current Opinion in Rheumatology. 2007;**19**:265-271

[6] Hunziker EB, et al. Articular cartilage repair: Basic science and clinical progress. A review of the current status and prospects. Osteoarthritis and Cartilage. 2002;**10**:432-463

[7] Kuo CK, et al. Cartilage tissue engineering: Its potential and uses. Current Opinion in Rheumatology. 2006;**18**:64-73

[8] Heng BC, et al. Directing stem cell differentiation into the chondrogenic lineage in vitro. Stem Cells. 2004;**22**:1152-1167

[9] Jadin KD, et al. Depth-varying density and organization of chondrocytes in immature and mature bovine articular cartilage assessed by 3D imaging and analysis. The Journal of Histochemistry and Cytochemistry. 2005;**53**:1109-1119

[10] Knight M, et al. Development of articular cartilage. Osteoarthritis and Cartilage. 2013;**21**:S2–S3

[11] Matsiko A, et al. Advanced strategies for articular cartilage defect repair. Materials (Basel). 2013;**6**:637-668

[12] Eyre D. Collagen of articular cartilage. Arthritis Research. 2002;**4**:30-35

[13] Sopena J, et al Estructura y función del cartílago articular. Portada: Armas Frente a la Patología Articular. 2000;**1**:24-26

[14] Chung C, et al. Engineering cartilage tissue. Advanced Drug Delivery Reviews. 2008; **60**:243-262

[15] Izadifar Z, et al. Strategic design and fabrication of engineered scaffolds for articular cartilage repair. Journal of Functional Biomaterials. 2012;**3**(4):799-838

[16] Minas T. Chapter 4: Patient evaluation, cartilage defect, and evidence putting it all together. Musculoskeletal Key

[17] Fox S, et al. The basic science of articular cartilage: Structure, composition, and function. Sports Health. 2009;**1**(6):461-468

[18] Roughley PJ. The structure and function of cartilage proteoglycans. European Cells & Materials. 2006;**12**:92-101

[19] Loeser RF. Integrins and chondrocyte-matrix interactions in articular cartilage. Matrix Biology. 2014;**39**:11-16

[20] Brodkin KR, et al. Chondrocyte phenotypes on different extracellular matrix monolayers. Biomaterials. 2004;**25**:5929-5938

[21] Dürr J, et al. Localization of beta 1-integrins in human cartilage and their role in chondrocyte adhesion to collagen and fibronectin. Experimental Cell Research. 1993;**207**: 235-244

[22] Enomoto M, et al. Beta 1 integrins mediate chondrocyte interaction with type I collagen, type II collagen, and fibronectin. Experimental Cell Research. 1993;**205**:276-285

[23] Bauer M, et al. Chondral lesions of the femoral condyles: A system of arthroscopic classification. Arthroscopy. 1988;**4**:97-102

[24] Peterson L, et al. Two- to 9-year outcome after autologous chondrocyte transplantation of the knee. Clinical Orthopaedics and Related Research. 2000;**374**:212-234

[25] Brittberg M, et al. Evaluation of cartilage injuries and repair. The Journal of Bone and Joint Surgery. American Volume. 2003;**85-A**(Suppl 2):58-69

[26] Mankin HJ. The response of articular cartilage to mechanical injury. The Journal of Bone and Joint Surgery. American Volume. 1982;**64**:460-466

[27] Lewis PB, et al. Basic science and treatment options for articular cartilage injuries. The Journal of Orthopaedic and Sports Physical Therapy. 2006;**36**(10):717-727

[28] Goldberg VM, et al. Biologic restoration of articular surfaces. Instructional Course Lectures. 1999;**48**:623-627

[29] Nehrer S, et al. Histologic analysis of tissue after failed cartilage repair procedures. Clinical Orthopaedics and Related Research. 1999;**365**:149-162

[30] Karnes J, et al. Cell therapy for the creation of cartilage and related clinical trials. In: Templeton NS, editor. Gene and Cell Therapy: Therapeutic Mechanisms and Strategies. 4th ed. Taylor & Francis/CRC Press; 2015. pp. 1123-1135

[31] Loeser RF, et al. Osteoarthritis: A disease of the joint as an organ. Arthritis and Rheumatism. 2012;**64**(6):1697-1707

[32] Burr DB, et al. Bone remodelling in osteoarthritis. Nature Reviews. Rheumatology. 2012;**8**(11):665-673

[33] Manferdini C, et al. Adipose-derived mesenchymal stem cells exert antiinflammatory effects on chondrocytes and synoviocytes from osteoarthritis patients through prostaglandin E2. Arthritis and Rheumatism. 2013;**65**(5):1271-1281

[34] Filardo G, et al. Mesenchymal stem cells for the treatment of cartilage lesions: From preclinical findings to clinical application in orthopaedics. Knee Surgery, Sports Traumatology, Arthroscopy. 2013;**21**(8):1717-1729

[35] Pérez-Silos V, et al. Mesenchymal stem cells subpopulations: Application for orthopedic regenerative medicine. Stem Cells International. 2016;**2016**:3187491

[36] Langer RS, et al. Tissue engineering: The challenges ahead. Scientific American. 1999;**280**: 86-89

[37] Spector M. Biomaterials-based tissue engineering and regenerative medicine solutions to musculoskeletal problems. Swiss Medical Weekly. 2008;**136**:293-301

[38] Marquass B, et al. A novel MSC-seeded triphasic construct for the repair of osteochondral defects. Journal of Orthopaedic Research. 2010;**12**:1586-1599

[39] Li JJ, et al. A biphasic scaffold based on silk and bioactive ceramic with stratified properties for osteochondral tissue regeneration. Journal of Materials Chemistry. B, Materials for Biology and Medicine. 2015;**3**(26):5361-5376

[40] Ho ST, et al. The evaluation of a biphasic osteochondral implant coupled with an electrospun membrane in a large animal model. Tissue Engineering. Part A. 2010;**16**(4):1123-1141

[41] Da H, et al. The impact of compact layer in biphasic scaffold on osteochondral tissue engineering. PLoS One. 2013;**8**(1):e54838

[42] Shen W, et al. Intra-articular injection of human meniscus stem/progenitor cells promotes meniscus regeneration and ameliorates osteoarthritis through stromal cell-derived factor-1/CXCR4-mediated homing. Stem Cells Translational Medicine. 2014;**3**(3):387-394

[43] Richardson JB, et al. Repair of human articular cartilage after implantation of autologous chondrocytes. Journal of Bone and Joint Surgery (British). 1999;**81**:1064-1068

[44] Saris DB, et al. Treatment of symptomatic cartilage defects of the knee: Characterized chondrocyte implantation results in better clinical outcome at 36 months in a randomized trial

compared to microfracture. The American Journal of Sports Medicine. 2009;**37**(Suppl. 1): 10S–19S

[45] Darling EM,.et al. Rapid phenotypic changes in passaged articular chondrocyte subpopulations. Journal of Orthopaedic Research. 2005;**23**:425-432

[46] Martinez I, et al. Redifferentiation of in vitro expanded adult articular chondrocytes by combining the hanging-drop cultivation method with hypoxic environment. Cell Transplantation. 2008;**17**:987-996

[47] Benya PD, et al. Dedifferentiated chondrocytes re-express the differentiated collagen phenotype when cultured in agarose gels. Cell. 1982;**30**(1):215-224

[48] Brittberg M, et al. Treatment of deep cartilage defects in the knee with autologous chondrocyte transplantation. The New England Journal of Medicine. 1994;**331**(14):889-895

[49] Somoza RA, et al. Chondrogenic differentiation of mesenchymal stem cells: Challenges and unfulfilled expectations. Tissue Engineering. Part B, Reviews. 2014;**20**:596-608

[50] Coleman RM, et al. Hydrogel effects on bone marrow stromal cell response to chondrogenic growth factors. Biomaterials. 2007;**28**:2077-2086

[51] Indrawattana N, et al. Growth factor combination for chondrogenic induction from human mesenchymal stem cell. Biochemical and Biophysical Research Communications. 2004;**320**:914-919

[52] Zhou GD, et al. In vivo chondrogenesis of BMSCs at non-chondrogenesis site by co-transplantation of BMSCs and chondrocytes with pluronic as biomaterial. Advanced Biomaterials VI. 2005;**288-289**:3-6

[53] Hennig T, et al. Reduced chondrogenic potential of adipose tissue derived stromal cells correlates with an altered TGFbeta receptor and BMP profile and is overcome by BMP-6. Journal of Cellular Physiology. 2007;**211**(3):682-691

[54] Zuk PA, et al. Human adipose tissue is a source of multipotent stem cells. Molecular Biology of the Cell. 2002;**13**:4279-4295

[55] Lai JH, et al. Interaction between osteoarthritic chondrocytes and adipose-derived stem cells is dependent on cell distribution in three-dimension and transforming growth factor-β3 induction. Tissue Engineering. Part A. 2015;**21**(5-6):992-1002

[56] Zhu S, et al. Programmed application of transforming growth factor β3 and Rac1 inhibitor NSC23766 committed hyaline cartilage differentiation of adipose-derived stem cells for osteochondral defect repair. Stem Cells Translational Medicine. 2014;**3**(10): 1242-1251

[57] Chiou M, et al. Mitogenic and chondrogenic effects of fibroblast growth factor-2 in adipose-derived mesenchymal cells. Biochemical and Biophysical Research Communications. 2006;**343**:644-652

[58] Estes BT, et al. Potent induction of chondrocytic differentiation of human adipose-derived adult stem cells by bone morphogenetic protein 6. Arthritis and Rheumatism. 2006;**54**:1222-1232

[59] Pei M, et al. Engineering of functional cartilage tissue using stem cells from synovial lining: A preliminary study. Clinical Orthopaedics and Related Research. 2008;**466**(8):1880-1889

[60] Jones BA, et al. Synovium-derived stem cells: A tissue-specific stem cell for cartilage engineering and regeneration. Tissue Engineering. Part B, Reviews. 2012;**18**(4):301-311

[61] Pei M, et al. Repair of full-thickness femoral condyle cartilage defects using allogeneic synovial cell-engineered tissue constructs. Osteoarthritis and Cartilage. 2009;**17**(6):714-722

[62] Sakaguchi Y, et al. Comparison of human stem cells derived from various mesenchymal tissues: Superiority of synovium as a cell source. Arthritis and Rheumatism. 2005;**52**(8):2521-2529

[63] Pei M, et al. Extracellular matrix deposited by synovium-derived stem cells delays replicative senescent chondrocyte dedifferentiation and enhances redifferentiation. Journal of Cellular Physiology. 2012;**227**:2163-2174

[64] Zhou C, et al. Gene expression profiles reveal that chondrogenic progenitor cells and synovial cells are closely related. Journal of Orthopaedic Research. 2014;**32**(8):981-988

[65] Arufe MC, et al. Chondrogenic potential of subpopulations of cells expressing mesenchymal stem cell markers derived from human synovial membranes. Journal of Cellular Biochemistry. 2010;**111**(4):834-845

[66] Arufe MC, et al. Differentiation of synovial CD-105+ human mesenchymal stem cells into chondrocyte-like cells through spheroid formation. Journal of Cellular Biochemistry. 2009;**108**(1):145-155

[67] Gullo F, et al. Prospective purification of a subpopulation of human synovial mesenchymal stem cells with enhanced chondro-osteogenic potency. Rheumatology. 2013;**52**(10): 1758-1768

[68] Burdan F, et al. Morphology and physiology of the epiphyseal growth plate. Folia Histochemica et Cytobiologica. 2009;**47**:5-16

[69] Chimal J, et al. Expression of N-cadherin, N-CAM, fibronectin and tenascin is stimulated by TGF-beta1, beta2, beta3 and beta5 during the formation of precartilage condensations. International Journal of Developmental Biology. 1999;**43**:59-67

[70] Goldring MB, et al. The control of chondrogenesis. Journal of Cellular Biochemistry. 2006;**97**:33-44

[71] Demoor M, et al. Cartilage tissue engineering: Molecular control of chondrocyte differentiation for proper cartilage matrix reconstruction. Biochimica et Biophysica Acta. 2014;**1840**:2414-2440

[72] Gaissmaier C, et al. Growth and differentiation factors for cartilage healing and repair. Injury. 2008;**39**:S88-96

[73] Chen Y, et al. Insulin-like growth factor-1 boosts the developing process of condylar hyperplasia by stimulating chondrocytes proliferation. Osteoarthritis and Cartilage. 2012;**20**:279-287

[74] Shanmugarajan T, et al. Growth factors and signaling pathways in the chondrogenic differentiation of mesenchymal stem cells. Tissue Engineering & Regenerative Medicine. 2011;**8**:292-299

[75] Fortier L, et al. The role of growth factors in cartilage repair. Clinical Orthopaedics and Related Research. 2011;**469**:2706-2715

[76] Vinatier C, et al. Cartilage engineering: A crucial combination of cells, biomaterials and biofactors. Trends in Biotechnology. 2009;**27**(5):307-314

[77] Haas AR, et al. Chondrogenic differentiation of murine C3H10T1/2 multipotential mesenchymal cells: II. Stimulation by bone morphogenetic protein-2 requires modulation of N-cadherin expression and function. Differentiation. 1999;**64**(2):77-89

[78] Fernández-Lloris R, et al. Induction of the Sry-related factor SOX6 contributes to bone morphogenetic protein-2-induced chondroblastic differentiation of C3H10T1/2 cells. Molecular Endocrinology. 2003;**17**(7):1332-1343

[79] Gründer T, et al. Bone morphogenetic protein (BMP)-2 enhances the expression of type II collagen and aggrecan in chondrocytes embedded in alginate beads. Osteoarthritis and Cartilage. 2004;**12**(7):559-567

[80] Shi S, et al. Growth factor transgenes interactively regulate articular chondrocytes. Journal of Cellular Biochemistry. 2013;**114**:908-919

[81] Aurich M, et al. Histological and cell biological characterization of dissected cartilage fragments in human osteochondritis dissecans of the femoral condyle. Archives of Orthopaedic and Trauma Surgery. 2006;**126**:606-614

[82] Zapata N, et al. Cultivo de tejido cartilaginoso articular: Acercamiento conceptual. Environmental Impact Assessment Review. 2007;**8**:117-129

[83] Akiyama H, et al. The transcription factor Sox9 has essential roles in successive steps of the chondrocyte differentiation pathway and is required for expression of Sox5 and Sox6. Genes & Development. 2002;**16**:2813-2828

[84] Glowacki J et al. Collagen scaffolds for tissue engineering. Biopolymers. 2008;**89**: 338-344

[85] Guilak F, et al. The pericellular matrix as a transducer of biomechanical and biochemical signals in articular cartilage. Ann N Y Acad Sci. 2006;**1068**:498-512

[86] Foss C, et al. Silk fibroin/hyaluronic acid 3D matrices for cartilage tissue engineering. Biomacromolecules. 2013;**14**(1):38-47

[87] Nagaya H, et al. Examination of synovial fluid and serum hyaluronidase activity as a joint marker in rheumatoid arthritis and osteoarthritis patients (by zymography). Annals of the Rheumatic Diseases. 1999;**58**(3):186-188

[88] Hori Y, et al. Modular injectable matrices based on alginate solution/microsphere mixtures that gel in situ and co-deliver immunomodulatory factors. Acta Biomaterialia. 2011;**5**:969-982

[89] Park H, et al. Facile control of RGD-alginate/hyaluronate hydrogel formation for cartilage regeneration. Carbohydrate Polymers. 2011;**86**:1107-1112

[90] Bhat A, et al. Alginate hydrogels containing cell-interactive beads for bone formation. FASEB Journal. 2013;**27**:4844-4852

[91] Stoop R. Smart biomaterials for tissue engineering of cartilage. Injury. 2008;**39**:S77-87

[92] Shen Y, et al. Biomaterial and mesenchymal stem cell for articular cartilage reconstruction. Current Stem Cell Research & Therapy. 2014;**9**(3):254-267

[93] Tavassoli A, et al. Bovine articular cartilage decellularized matrix as a scaffold for use in cartilage tissue engineering. Iranian journal of Veterinary Science and Technology. 2012;**4**:1-8

[94] Elder BD, et al. Extraction techniques for the decellularization of tissue engineered articular cartilage constructs. Biomaterials. 2009;**30**:3749-3756

[95] Diekman B et al. Chondrogenesis of adult stem cells from adipose tissue and bone marrow: Induction by growth factors and cartilage-derived matrix. Tissue Engineering. Part A. 2010;**16**:27-31

[96] Harvey A, et al. Proteomic analysis of the extracellular matrix produced by mesenchymal stromal cells: implications for cell therapy mechanism. PLoS One. 2013;**8**:11

[97] Caron MM, et al. Redifferentiation of dedifferentiated human articular chondrocytes: Comparison of 2D and 3D cultures. Osteoarthritis and Cartilage. 2012;**20**:1170-1178

[98] Oh SH, et al. Investigation of pore size effect on chondrogenic differentiation of adipose stem cells using a pore size gradient scaffold. Biomacromolecules. 2010;**127**:1948-1955

[99] Chen FH, et al. Technology insight: Adult stem cells in cartilage regeneration and tissue engineering. Nature Clinical Practice. Rheumatology. 2006;**2**:373-382

[100] O'Brien FJ, et al. The effect of pore size on cell adhesion in collagen-GAG scaffolds. Biomaterials. 2005;**26**:433-441

[101] Grigolo B, et al. Evidence for redifferentiation of human chondrocytes grown on a hyaluronan-based biomaterial (HYAff 11): Molecular, immunohistochemical and ultrastructural analysis. Biomaterials. 2002;**23**:1187-1195

[102] Hoch AI, et al. Concise review: Optimizing expansion of bone marrow mesenchymal stem/stromal cells for clinical applications. Stem Cells Translational Medicine. 2014;**3**:643-652

[103] Bush PG, et al. The osmotic sensitivity of rat growth plate chondrocytes in situ; clarifying the mechanisms of hypertrophy. Journal of Cellular Physiology. 2008;**214**(3):621-629

[104] Studer D, et al. Molecular and biophysical mechanisms regulating hypertrophic differentiation in chondrocytes and mesenchymal stem cells. European Cells & Materials. 2012;**24**:118-135

[105] Arnold MA, et al. MEF2C transcription factor controls chondrocyte hypertrophy and bone development. Developmental Cell. 2007;**12**(3):377-389

[106] Mueller MB, et al. Functional characterization of hypertrophy in chondrogenesis of human mesenchymal stem cells. Arthritis and Rheumatism. 2008;**58**(5):1377-1388

[107] Fosang AJ, et al. Degradation of cartilage aggrecan by collagenase-3 (MMP-13). FEBS Letters. 1996;**380**(1-2):17-20

[108] Shen G. The role of type X collagen in facilitating and regulating endochondral ossification of articular cartilage. Orthodontics & Craniofacial Research. 2005;**8**(1):11-17

[109] Van Donkelaar CC, et al. The PTHrP-Ihh feedback loop in the embryonic growth plate allows PTHrP to control hypertrophy and Ihh to regulate proliferation. Biomechanics and Modeling in Mechanobiology. 2007;**6**(1-2):55-62

[110] Tamamura Y, et al. Developmental regulation of Wnt/beta-catenin signals is required for growth plate assembly, cartilage integrity, and endochondral ossification. Journal of Biological Chemistry. 2005;**280**(19):19185-19195

[111] Woods A, et al. Rac1 signaling stimulates N-cadherin expression, mesenchymal condensation, and chondrogenesis. Journal of Biological Chemistry. 2007;**282**(32):23500-23508

[112] Zhu S, et al. Inhibition of Rac1 activity by controlled release of NSC23766 from chitosan microspheres effectively ameliorates osteoarthritis development in vivo. Annals of the Rheumatic Diseases. 2015;**74**(1):285-293

[113] Xiao J, et al. Regulation of chondrocyte proliferation through GIT1-Rac1-mediated ERK1/2 pathway by PDGF. Cell Biology International. 2014;**38**(6):695-701

[114] Long DL, et al. Rac1 is required for matrix metalloproteinase 13 production by chondrocytes in response to fibronectin fragments. Arthritis and Rheumatism. 2013;**65**(6):1561-1568

[115] Woods A, et al. Rac1 signaling regulates CTGF/CCN2 gene expression via TGFbeta/Smad signaling in chondrocytes. Osteoarthritis and Cartilage. 2013;**17**(3):406-413

[116] Hellingman CA, et al. Smad signaling determines chondrogenic differentiation of bone-marrow-derived mesenchymal stem cells: inhibition of Smad1/5/8P prevents terminal differentiation and calcification. Tissue Engineering. Part A. 2011;**17**:1157-1167

[117] Shintani N, et al. TGF-ss1 enhances the BMP-2-induced chondrogenesis of bovine synovial explants and arrests downstream differentiation at an early stage of hypertrophy. PLoS One. 2013;**8**:e53086

[118] Van Caam A, et al. The high affinity ALK1-ligand BMP9 induces a hypertrophy-like state in chondrocytes that is antagonized by TGFβ1. Osteoarthritis and Cartilage. 2015;**23**(6):985-995

[119] Prasadam I, et al. ERK-1/2 and p38 in the regulation of hypertrophic changes of normal articular cartilage chondrocytes induced by osteoarthritic subchondral osteoblasts. Arthritis and Rheumatism. 2010;**62**:1349-1360

[120] Pei M, et al. Repair of large animal partial-thickness cartilage defects through intraarticular injection of matrix-rejuvenated synovium-derived stem cells. Tissue Engineering. Part A. 2013;**19**(9-10):1144-1154

[121] Dexheimer V, et al. Differential expression of TGF-β superfamily members and role of Smad1/5/9-signalling in chondral versus endochondral chondrocyte differentiation. Scientific Reports. 2016;**6**:36655

[122] Correa D, et al. Sequential exposure to fibroblast growth factors (FGF) 2, 9 and 18 enhances hMSC chondrogenic differentiation. Osteoarthritis and Cartilage. 2015;**23**(3):443-453

[123] Lee JM, et al. PD98059-impregnated functional PLGA Scaffold for direct tissue engineering promotes chondrogenesis and prevents hypertrophy from mesenchymal stem cells. Tissue Engineering. Part A. 2014;**20**(5-6):982-991

[124] Oh CD, et al. Opposing role of mitogen-activated protein kinase subtypes, erk-1/2 and p38, in the regulation of chondrogenesis of mesenchymes. Journal of Biological Chemistry. 2000;**25**:5613

[125] Kim HJ, et al. The effects of ERK1/2 inhibitor on the chondrogenesis of bone marrow- and adipose tissue-derived multipotent mesenchymal stromal cells. Tissue Engineering. Part A. 2010;**16**:851

[126] Chen Z, et al. ERK1 and ERK2 regulate chondrocyte terminal differentiation during endochondral bone formation. Journal of Bone and Mineral Research. 2015;**30**(5):765-774

[127] Kim SH, et al. TGF-β3 encapsulated PLCL scaffold by a supercritical CO2-HFIP co-solvent system for cartilage tissue engineering. Journal of Controlled Release. 2015;**206**:101-107

[128] Gracitelli GC, et al. Surgical interventions (microfracture, drilling, mosaicplasty, and allograft transplantation) for treating isolated cartilage defects of the knee in adults. Cochrane Database of Systematic Reviews. 2016;**9**:CD010675

[129] Clar C, et al. Clinical and cost-effectiveness of autologous chondrocyte implantation for cartilage defects in knee joints: Systematic review and economic evaluation. Health Technology Assessment. 2005;**9**:iii–iv, ix–x, 1-82

[130] Randsborg P-H, et al. Focal cartilage defects in the knee—A randomized controlled trial comparing autologous chondrocyte implantation with arthroscopic debridement. BMC Musculoskeletal Disorders. 2016;**17**:117

[131] Mancò A, et al Microfracture versus microfracture and platelet-rich plasma: Arthroscopic treatment of knee chondral lesions. A two-year follow-up study. Joints. 2016;**4**:142-147

[132] Gormeli G, et al. Clinical effects of platelet-rich plasma and hyaluronic acid as an additional therapy for talar osteochondral lesions treated with microfracture surgery: A prospective randomized clinical trial. Foot & Ankle International. 2015;**36**:891-900

[133] Manunta AF, et al. The treatment of chondral lesions of the knee with the microfracture technique and platelet-rich plasma. Joints. 2013;**1**:167-170

[134] Koh Y-G, et al. Adipose-derived mesenchymal stem cells with microfracture versus microfracture alone: 2-year follow-up of a prospective randomized trial. Arthroscopy: The Journal of Arthroscopic & Related Surgery. 2016;**32**:97-109

[135] Pers Y-M, et al. Adipose mesenchymal stromal cell-based therapy for severe osteoarthritis of the knee: A phase I dose-escalation trial. Stem Cells Translational Medicine. 2016;**5**:847-856

[136] Jo CH, et al. Intra-articular injection of mesenchymal stem cells for the treatment of osteoarthritis of the knee: A proof-of-concept clinical trial. Stem Cells. 2014;**32**:1254-1266

[137] Vega A, et al. Treatment of knee osteoarthritis with allogeneic bone marrow mesenchymal stem cells: A randomized controlled trial. Transplantation. 2015;99:1681-1690

[138] Orozco L, et al. Treatment of knee osteoarthritis with autologous mesenchymal stem cells. Transplantation. 2013;**95**:1535-1541

[139] Saw K-Y, et al. Articular cartilage regeneration with autologous peripheral blood stem cells versus hyaluronic acid: A randomized controlled trial. Arthroscopy: The Journal of Arthroscopic & Related Surgery. 2013;**29**:684-694

[140] Lee MC, et al. A placebo-controlled randomised trial to assess the effect of TGF-ß1-expressing chondrocytes in patients with arthritis of the knee. The Bone & Joint Journal. 2015;**97**:924-932

[141] Cherian JJ, et al. Preliminary results of a phase II randomized study to determine the efficacy and safety of genetically engineered allogeneic human chondrocytes expressing TGF-β1 in patients with grade 3 chronic degenerative joint disease of the knee. Osteoarthritis and Cartilage. 2015;**12**:2109-2118

[142] Ha CW, et al. A multicenter, single-blind, Phase IIa clinical trial to evaluate the efficacy and safety of a cell-mediated gene therapy in degenerative knee arthritis patients. Human Gene Therapy. Clinical Development. 2015;**26**(2):125-130

[143] Zheng MH, et al. Matrix-induced autologous chondrocyte implantation (MACI): Biological and histological assessment. Tissue Engineering. 2007;**13**(4):737-746

[144] Kon E, et al. A novel nano-composite multi-layered biomaterial for treatment of osteochondral lesions: Technique note and an early stability pilot clinical trial. Injury. 2010;**41**:693-701

[145] Kon E, et al. Novel nano-composite multilayered biomaterial for osteochondral regeneration: A pilot clinical trial. The American Journal of Sports Medicine. 2011;**39**:1180-1190

[146] Panseri S, et al. Osteochondral tissue engineering approaches for articular cartilage and subchondral bone regeneration. Knee Surgery, Sports Traumatology, Arthroscopy. 2012;**20**:1182-1191

[147] Kon E, et al. Clinical results and MRI evolution of a nano-composite multilayered biomaterial for osteochondral regeneration at 5 years. The American Journal of Sports Medicine. 2014;**42**:158-165

[148] Kon E, et al. A one-step treatment for chondral and osteochondral knee defects: Clinical results of a biomimetic scaffold implantation at 2 years of follow-up. Journal of Materials Science: Materials in Medicine. 2014;**25**:2437-2444

[149] Getgood AMJ, et al. Evaluation of early-stage osteochondral defect repair using a biphasic scaffold based on a collagen–glycosaminoglycan biopolymer in a caprine model. The Knee. 2012;**19**:422-430

[150] Dell'Osso G, et al. The biphasic bioresorbable scaffold (Trufit®) in the osteochondral knee lesions: Long-term clinical and MRI assessment in 30 patients. Musculoskeletal Surgery. 2016;**100**:93-96

[151] Gelber PE, et al. Magnetic resonance evaluation of TruFit® plugs for the treatment of osteochondral lesions of the knee shows the poor characteristics of the repair tissue. The Knee. 2014;**21**:827-832

[152] Bekkers JEJ, et al. Articular cartilage evaluation after TruFit plug implantation analyzed by delayed gadolinium-enhanced MRI of cartilage (dGEMRIC). The American Journal of Sports Medicine. 2013;**41**:1290-1295

[153] De Girolamo L, et al. Regenerative approaches for the treatment of early OA. Knee Surgery, Sports Traumatology, Arthroscopy. 2016;**24**(6):1826-1835

[154] Filardo G, et al. Stem cells in articular cartilage regeneration. Journal of Orthopaedic Surgery and Research. 2016;**11**:42-56

[155] Caldwell KL, et al. Cell-based articular cartilage repair: The link between development and regeneration. Osteoarthritis and Cartilage. 2015;**23**(3):351-362

[156] Mollon B, et al. The clinical status of cartilage tissue regeneration in humans. Osteoarthritis and Cartilage. 2013;**21**:1824-1833

[157] Chiang H, et al. Clinical feasibility of a novel biphasic osteochondral composite for matrix-associated autologous chondrocyte implantation. Osteoarthritis and Cartilage. 2013;**21**:589-598

[158] Cui L, et al. Repair of articular cartilage defect in non-weight bearing areas using adipose derived stem cells loaded polyglycolic acid mesh. Biomaterials. 2009;**30**:2683-2693

[159] Cheng A, et al. Generating cartilage repair from pluripotent stem cells. Tissue Engineering. Part B, Reviews. 2013;**20**(4):257-266

[160] Diekman BO, et al. Cartilage tissue engineering using differentiated and purified induced pluripotent stem cells. Proceedings of the National Academy of Sciences of the United States of America. 2012;**109**:19172-19177

[161] Ko JY, et al. In vitro chondrogenesis and in vivo repair of osteochondral defect with human induced pluripotent stem cells. Biomaterials. 2014;**35**:3571-3581

[162] Mifune Y, et al. The effect of platelet-rich plasma on the regenerative therapy of muscle derived stem cells for articular cartilage repair. Osteoarthritis and Cartilage. 2013;**21**:175-185

[163] Levingstone TJ, et al. A biomimetic multi-layered collagen-based scaffold for osteochondral repair. Acta Biomaterialia. 2014;**10**:1996-2004

[164] Barr L, et al. The effect of recombinant human fibroblast growth factor-18 on articular cartilage following single impact load. Journal of Orthopaedic Research. 2014;**32**: 923-927

Chapter 9

Mesenchymal Stem Cells for Optimizing Bone Volume at the Dental Implant Recipient Site

Mustafa Ayna, Aydin Gülses, Jörg Wiltfang and Yahya Açil

Additional information is available at the end of the chapter

http://dx.doi.org/10.5772/intechopen.68514

Abstract

Inadequate bone volume at the implant recipient site presents a clinical challenge for many dental practitioners. To overcome these problems, several approaches have been developed and are currently used, including bone grafting strategies and distraction osteogenesis. Mesenchymal stem cells (MSCs) have gained their popularity within the last two decades, with regard to promising clinical results in improving the bone architecture at the implant recipient site. The aim of this chapter was to briefly outline the accessibility properties, differentiation capacities, isolation, and characterization of MSCs with regard to optimizing bone volume in dental implantology. Additionally, potential benefits and pitfalls are discussed in comparison with the conventional bone augmentation techniques.

Keywords: bone, dental, implantology, mesenchymal stem cells, platelet-rich plasma

1. Introduction

Dental implant therapies became an integral part of the daily dental practice. The success rate of implants is related to the correct position and angulation of implants in residual crest, so that height and thickness of bone augmentation can allow predictable results [1]. Therefore, the qualitative and quantitative characteristics of the surrounding tissues at the implant recipient site play a key role in the success of the procedure. Systemic diseases such as osteoporosis, changes in vitamin D metabolism, diabetes and adverse pregnancy outcomes, and local factors such as periodontitis, infections, pre-existing cysts or tumors, and traumatic extractions might result in the loss of both alveolar bone volume and quality and complicates the feasibility and long-term clinical outcomes of dental implant rehabilitation.

Various reconstructive surgical interventions could be necessary to regenerate bone defects prior to implant placement. In the literature, there are numerous clinical and experimental studies presenting techniques with different results that overcome the problems related to the insufficient bone volume at the edentulous alveolar ridge. Among these, the mostly performed surgical procedures to obtain bone augmentation are guided bone regeneration techniques via synthetic materials, xenografts or allografts, distraction osteogenesis of the alveolus, and the augmentation with autogenous bone blocks, which is thought to be the gold standard to obtain accurate bone volume and morphology with long-term predictable results. All techniques described above have their own advantages and pitfalls.

2. Conventional bone grafting strategies

2.1. Synthetic bone graft materials

A variety of artificial bone substitutes were used to reconstruct bone defects of the jaws. Synthetic bone grafts at most possess only osseoconductive characteristics and ideally should be biocompatible, show minimal fibrotic reaction, undergo remodeling, support new bone formation, and should have a similar strength or similar mechanical characteristics to that of the cortical/cancellous bone being replaced [2]. Availability of synthetic bone graft materials would eliminate the need for invasive graft-harvesting procedure and the dangers of pathogen transmission from immunogenic reaction to bank bone [2, 3]. In the maxillofacial reconstruction, the mostly used synthetic bone graft materials are:

- calcium phosphates,
- calcium phosphate cements,
- beta-tricalcium phosphates,
- synthetic hydroxyapatites,
- coralline hydroxyapatites, and
- bioactive glasses.

It is obvious that synthetic bone substitutes only have osteoconductive properties, and there is a need for improvement in their mechanical and degradation properties to ensure the replacement of the graft material with the living bone.

2.2. Allografts

The term allograft describes transplants between two subjects of the same species. Complications associated with the harvesting of autogenous bone have led to gain in their popularity as a treatment option in maxillofacial reconstruction. Allografts might offer the same characteristics as autograft; however, they do not present same osteogenic cells and therefore fulfill only the demand of osteoconductivity and serve mostly as a scaffold for new bone formation [2].

The advantages of allografts include availability and avoidance of morbidity associated with autogenous bone graft harvesting.

It is obvious that tissue safety is a major concern in transplantation. The major risk and disadvantage related to the use of allografts are the transmission of infectious agents from donor to recipient, which could result in microbial contamination from an infected donor, during collection of the tissue from donors or the environment and during processing of the tissues [4].

Viral transmission is a potential risk that is historically and serologically reported in association with allografts. Despite the exceedingly low risk, the transmission of human immunodeficiency virus (HIV-1) from seronegative cadaveric donors has reported in Refs. [4, 5]. During the history of allogenic tissue transfer, many sterilization techniques have been used to prevent infection through allografts which include gamma irradiation, ethylene oxide gas, thermal treatment with moist heat, beta-propiolactone, chemical processing, and antibiotic soaks [4]. Among these, gamma irradiation offers a clear advantage in terms of safety compared with other sterilization techniques.

2.3. Xenografts

Xenograft is a term used to describe a surgical graft of tissue from one species to an unlike species such as coral, bovine, and porcine and are used as calcified matrices generally. The processing of xenografts is reported to remove organic components such as cells and proteinaceous materials, leaving an inert absorbable bone scaffold, which assists in revascularization, osteoblast migration, and new bone formation [2, 6].

The use of xenografts has been demonstrated to be effective for increasing bone height and bone volume especially in sinus augmentation procedures (**Figure 1**). Xenogeneic bone is available in greater supply and larger sizes, and their physical properties are comparable to human cancellous bone [2, 6–8]. In the literature, it has been suggested that the resorption of xenografts and their replacement with new bone appears to be slow [9] and consideration must be given to the risk of cross contamination with bovine spongiform encephalopathy or porcine endogenous retroviruses [10].

2.4. Autografts

In the reconstruction of bony defects of the jaws, autogenous cancellous bone grafts are stated to be the most effective bone graft material considering their osteoinductive effects and predictable long-term results. Autogenous bone contains all of the elements necessary to promote vital bone formation, including mineral, collagen matrix, growth factors, and particularity vital cells [2]. Following the transplantation, few mature osteoblasts survive the procedure, but adequate numbers of precursor cells which have the osteogenic potential remain [2, 11]. Considering the bone volume needed, the donor sites for the reconstruction of the defects of the jaws are anterior or posterior iliac crest, mandibular ramus, mandibular symphysis, tibia, and parietal bone.

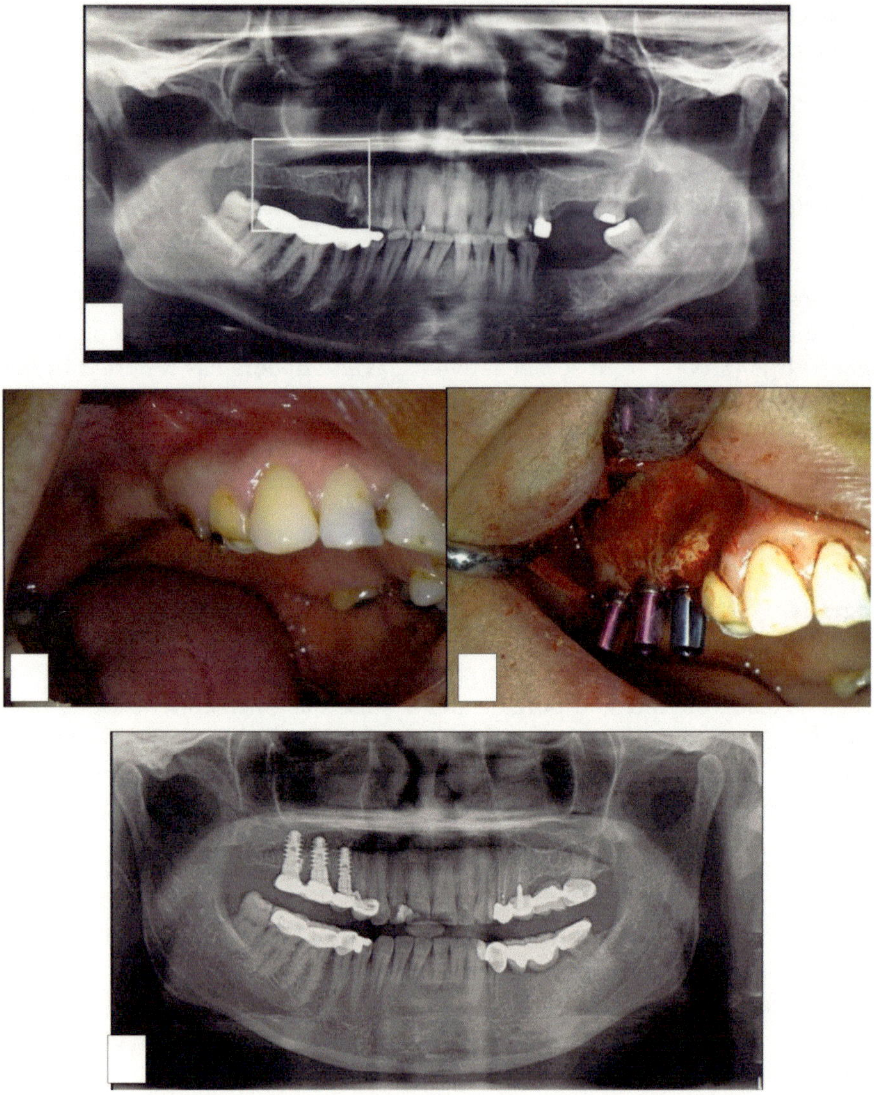

Figure 1. Augmentation of the atrophic posterior maxilla (a) The insufficent bone volume at the right posterior maxilla. (b) Intraoral clinical view (c) Sinus bone grafting with xenograft (Bioss®, Geistlich Germany). Implants were inserted simultaneously with sinus membrane elevation. (d) Panoramic radiograph after 1 year.

Limitations of the use of autogenous bone graft harvesting differ from the selected donor site, however, are mostly related to the so-called "donor site morbidity." The complications related to bone graft harvesting are [12]:

- increased operative time,
- limited availability and significant morbidity related to the intraoperative blood loss,
- wound complications,

- possible neurosensory alterations at the harvesting site, and
- chronic pain which is mainly attributed to the dissections performed during the graft harvesting.

An ultimate bone formation occurs only as a consequence of osteogenic/osteoinductive/osteoblastic cellular activity. In recent years, experimental studies followed by successfully clinical series have led to gaining the popularity of osteogenic precursors such as mesenchymal stem cells (MSCs) in bony reconstruction of the defects of the jaws and added a new dimension to the bone-gaining procedures in dental implantology.

The aim of this chapter was to describe the isolation methods regarding the MSCs used in oral implantology and briefly describe their clinical applications in peri-implantary surgical interventions.

3. Mesenchymal stem cells

Stem cells are unspecialized cells with the ability to proliferate and differentiate to multiple cell types when stimulated by both internal and external signals. They can be either embryonic stem cells, which are found in blastocysts or adult stem cells, which are called as pluripotent cells and can be found in bone marrow in the form of hematopoietic, endothelial, and MSCs.

The first successful isolation of bone marrow MSCs, then called colony-forming fibroblast-like cells, was described in 1968 by Friedenstein et al. [13] Today, MSCs are defined as nonhematopoietic progenitor cells that have the ability to differentiate into distinct mesodermal lineages (adipogenic, chondrogenic, osteogenic, or myogenic), which can produce bone, cartilage, fat, or fibrous connective tissue depending on their differentiation process [14].

Sources of MSCs in adult patients are [15, 16]:

- bone marrow [16] (**Figure 2**),
- peripheral blood [17],
- adipose tissue [18],
- muscle [19],
- periosteum [20],
- synovium [21], and
- teeth (perivascular niche of dental pulp and periodontal ligament) [22].

Since the first description of MSCs, various studies aimed to identify an ideal source for MSC harvesting. In 2006, Zhu et al. [23] have performed a study on the investigation of donor cell-related differences in tissue-engineered bone and examined bone marrow MSCs, alveolar bone cells, and periosteal cells for their in vivo potential to form bone.

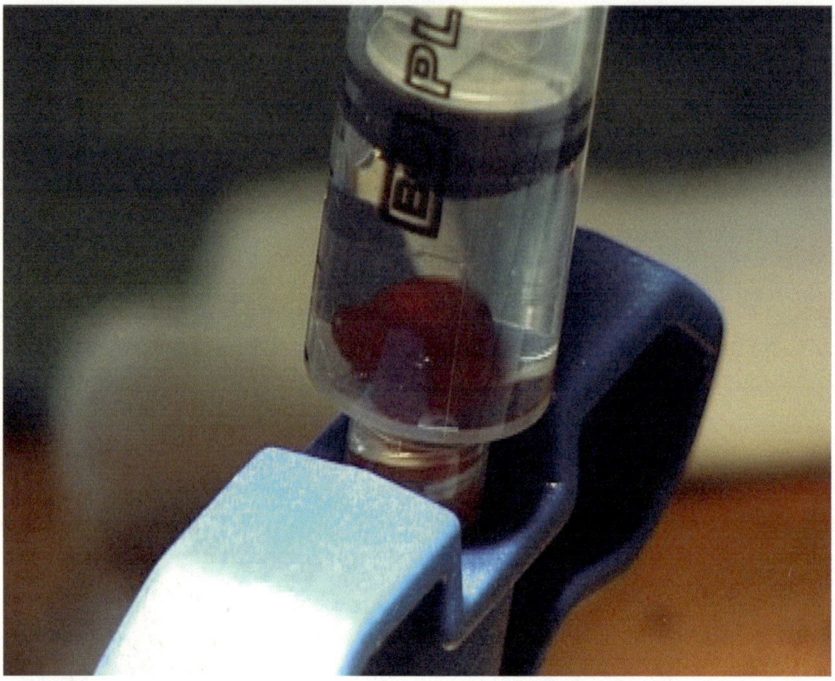

Figure 2. Bone marrow aspiration.

They have demonstrated that periosteal cells are the best choice for enhancing bone formation in tissue engineering of bone regeneration. In addition, recent studies showed a lower osteogenic differentiation potential of adipose tissue–derived stromal cells (ASCs) compared to bone marrow–derived mesenchymal stromal cells. According to Açil et al. [24], a careful reconsideration of the use of ASCs in bone tissue engineering application should be given.

3.1. Characterization of mesenchymal stem cells

Surface antigen expression, which allows for a rapid identification of a cell population, has been extensively used in experimental studies focusing on the identification of MSCs. For analysis of surface antigen expression, flow cytometry analysis and immunocytochemistry are efficient methods that reveal the marker profile of individual cells. In addition, fluorescence-activated cell sorting (FACS) is a valuable protocol for sorting isolation of MSCs. (**Figure 3**)

All techniques described above rely on both positive and negative selection by cell antigen surface markers, as well as physical properties of cells such as forward and side scatter characteristics [25].

According to the Mesenchymal and Tissue Stem Cell Committee of the International Society for Cellular Therapy [26], minimal criteria to define human MSCs are as follows:

- MSC must be plastic-adherent when maintained in standard culture conditions.

- MSC must express CD105, CD73, and CD90 and lack expression of CD45, CD34, CD14 or CD11b, CD79a, or CD19 and human leukocyte antigen-D-related (HLA-DR) surface molecules.
- MSC must differentiate to osteoblasts, adipocytes, and chondroblasts in vitro.

Since the first description of the above mentioned criteria by Mesenchymal and Tissue Stem Cell Committee of the International Society for Cellular Therapy in 2006, many studies have investigated the surface antigen expression of human MSCs in order to increase the confidence in their identification and verification. Lee et al. [27] have demonstrated that CD14, CD31, CD34, CD45, CD49d, CD49f, CD51, CD54, CD71, CD106, CD133, major histocompatibility complex (MHC II), cytokeratin, and desmin were absent from human MSCs, whereas CD13, CD29, CD44, CD59, CD73, CD90, CD105, CD166, MHC I,a-SMA, and vimentin were present on human MSCs. For human bone marrow stromal cells, common targets of negative antigene expression include CD2, CD3, CD11b/Integrin alpha M, CD14, CD15/Lewis X, CD16/$F_c$gamma RIII, CD19, CD38, CD56/NCAM-1, CD66b/CEACAM-8, CD123/IL-3 R alpha, and CD235a/Glycophorin.

For the positive selection of MSCs, CD271/NGF R, CD105/Endoglin, STRO-1, ganglioside GD2, and SUSD2 are relatively newly identified surface markers. In addition, STRO-1, CD271, CD200, ganglioside GD2, and frizzled-9 tissue non-specific alkaline phosphatase (TNAP) are suggested to be the latest markers used to verify MSC Identity [28, 29]. Identification of both positive and negative novel antigen surface markers would lead to modifications in the future.

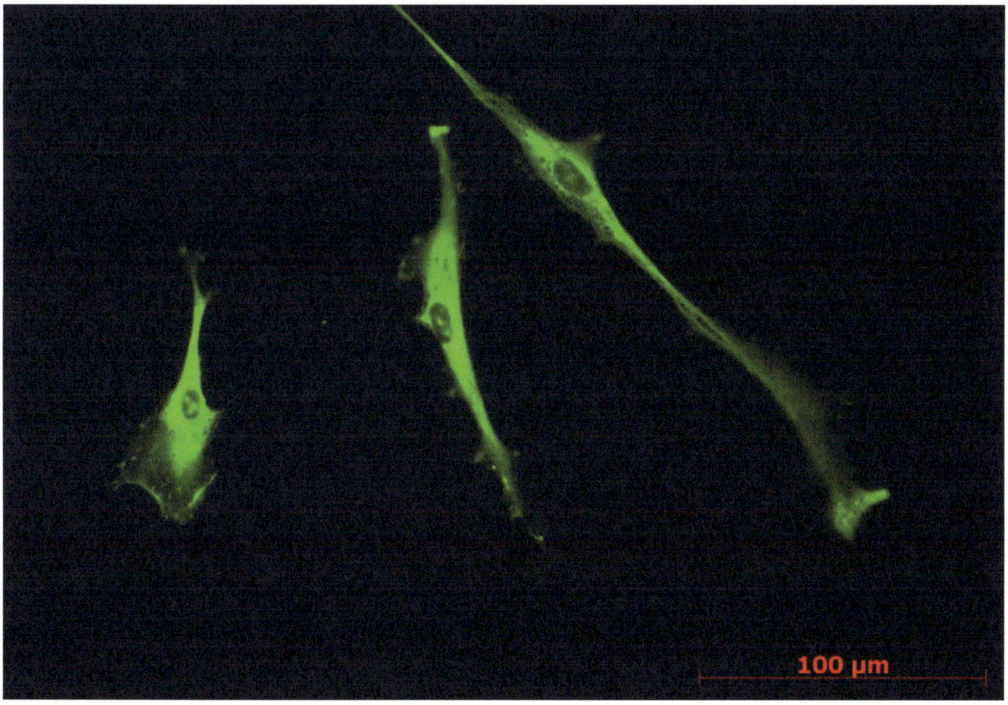

Figure 3. Morphology of MSCs obtained from bone marrow transfected with fluorescent protein. (Scale 100 µm).

4. Current concepts in mesenchymal stem cell harvest

4.1. Periodontal ligament

Shi et al. [14] have showed that periodontal ligament regeneration involves the recruitment of progenitor cells or stem cells, differentiating into either fibroblasts, cementoblasts, or osteoblasts, securing the teeth in the sockets between the cementum and adjacent alveolar bone. Seo et al. [30] have isolated stem cells from periodontal ligament for the first time and gave us new strategy to reconstruction of periodontium. According to Seo et al. [30], periodontal ligament stem cells (PDLSCs) share similar characteristics with other adult stem cells, including the ability to self-renew and multi-lineage differentiation potential. All these results suggested that PDLSCs might belong to a unique population of postnatal mesenchymal cells.

A literature survey could reveal that third molar teeth were mostly used for PDLSC isolation. Briefly [31], impacted third molars were surgically extracted, and periodontal dental ligament was gently scraped from the middle root surface. Coronal and apical portions of the ligament were not used in order to avoid contamination by gingival and pulpal cells. Periodontal dental ligament tissues were then minced then digested in a solution containing 3 mg/ml collagenase type I and 2.5 mg/ml dispase II for 1 h at 37°C. After digestion, tissue was seeded into culture flasks with alpha-modification of Eagle's Medium supplemented with 10% fetal bovine serum, 2-mM Glutamine, 100-U/mL penicillin and 100-μg/mL: streptomycin solution at 37°C in 5% CO_2 in a humidified atmosphere. After single cells were attached on the plastic bottom of the flask, non-adherent cells were removed by changing the medium [31].

Hakki et al. [32] have suggested that BMP-2, -6, and -7 are potent regulators of periodontal ligament stem cell gene expression and bio mineralization. BMPs with periodontal ligament stem cell isolated from periodontal ligament tissues provide a promising strategy for bone tissue engineering. According to a recent study performed by Açil et al. [51], BMP-7 triggers periodontal dental ligament cells to differentiate toward an osteoblast/cementoblast phenotype.

4.2. Adipose tissue

According to Açil et al. [24], ASCs could be easily isolated by using the modified technique that has been previously described by Zuk et al. [33]. Briefly description of the technique is; the adipose tissue, which could be obtained from liposuction procedures or from the subcutaneous tissue at the surgical access to the iliac crest during reconstructive maxillofacial surgical procedures.

Recent studies indicated a lower osteogenic differentiation potential of ASCs compared to bone marrow–derived mesenchymal stromal cells. As we have mentioned before, Açil et al. [24] have evaluated the effects of potent combinations of highly osteogenic BMPs in order to enhance the osteogenic differentiation potential of ASCs and indicated a restricted osteogenic differentiation potential of ASCs and suggest careful reconsideration of their use in bone tissue engineering applications.

4.3. Bone marrow

The superior iliac crest is usually preferred as a donor site due to its ease in access and trabecular structure. As described by Hernigou et al. [34] and later by Shapiro et al. [35], briefly, appropriate local anesthesia of the skin and subcutaneous soft tissues should be administered. Then, a 1-cm stab incision was performed over iliac crest. An 11-gauge, 11-cm Jamshidi needle was used to aspirate the bone marrow. Effort was taken to use a parallel approach, with the needle directed parallel to the iliac wing between the inner and outer tables, and the needle was subsequently withdrawn and repositioned [34, 35] (**Figure 4**). The marrow aspirates was then passed through a sterile filter into a separate compartment to remove particulate matter. The material was transferred for centrifugation resulting marrow cell concentration [34, 35].

Recent literatures have showed the potential benefits of using a cocktail of mononuclear cells without expanding them in vitro before reimplantation [36] (**Figure 5**). Therefore, there are also various systems developed for harvesting of MSCs from bone marrow. One of these is the bone marrow–derived MNCs isolation by synthetic poylsaccharid (FICOLL), technique, which is currently accepted as the gold standard [36, 37]. The FICOLL method might present a useful

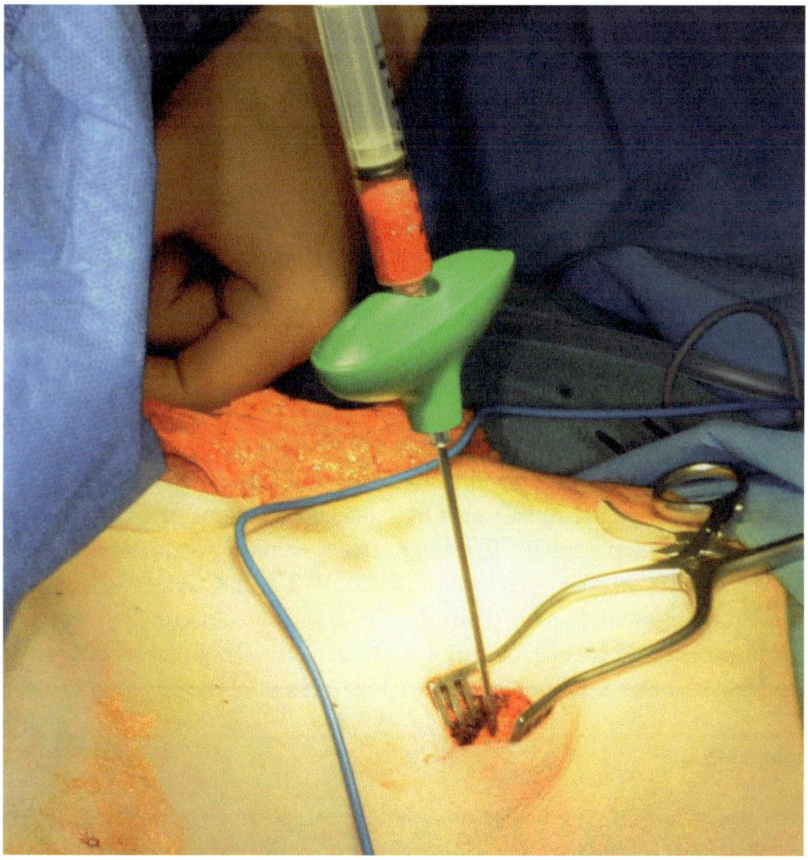

Figure 4. Bone marrow aspiration from the superior iliac crest.

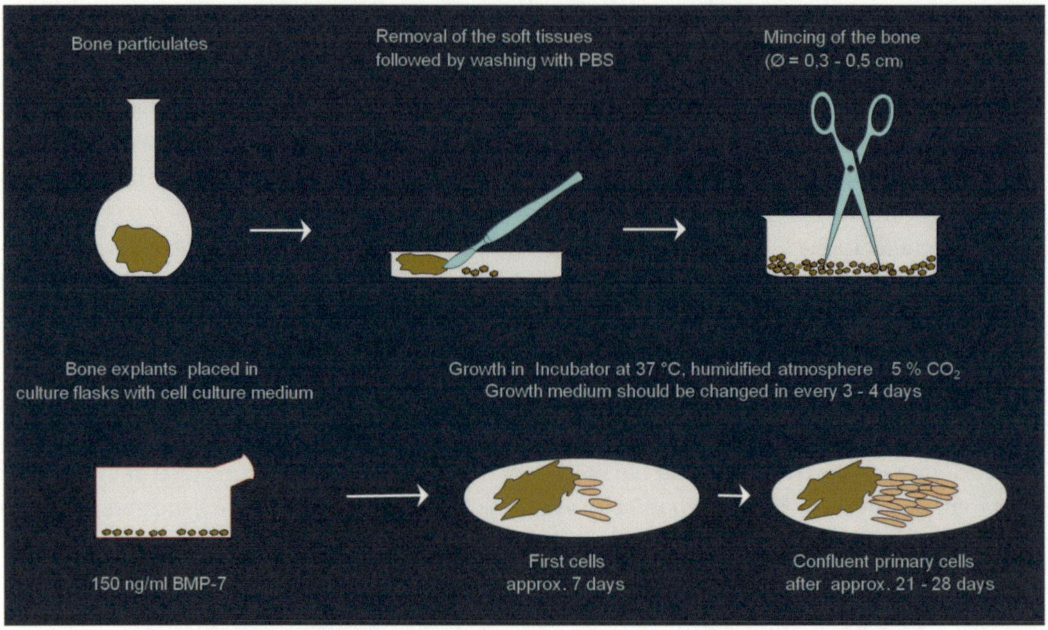

Figure 5. Diagrammatic illustration of the steps in osteoblast cell culture. Bone particles were obtained, the soft tissues were removed, and washed with PBS. The bone particles were minced and placed in culture flasks. After 3–4 weeks in incubation, cells have reached confluence.

technique for hospitals; however, the system is a time-consuming one, and a good manufacturing practice (GMP) laboratory is required. To ensure the clinical use in operating facilities without GMP possibilities, closed systems such as closed bone marrow aspirate concentrate (SmartPReP2 Bone Marrow Aspirate Concentrate System; BMAC; Harvest Technologies GmbH) system were developed [38]. Saurbier et al. [36] have compared new bone formation in maxillary sinus augmentation procedures using biomaterial associated with MSCs separated by FICOLL and BMAC and observed a higher proportion of hard tissue in the BMAC group.

Marx et al. [39] have compared the histologic parameters and outcomes of two types of grafts in large vertical maxillary defects: a composite graft of recombinant human bone morphogenetic protein-2/acellular collagen sponge (rhBMP-2/ACS), crushed cancellous freeze-dried allogeneic bone (CCFDAB), and platelet-rich plasma (PRP) and size-matched 100% autogenous grafts in 20 patients. According to their findings, the composite graft of rhBMP-2/ACS-CCFDAB-PRP regenerates bone in large vertical ridge augmentations as predictably as 100% autogenous graft with less morbidity, equal cost, and more viable new bone formation without residual nonviable bone particles but with more edema which might be attributed to the incisional release of the periosteum intraoperatively.

4.4. Peripheral blood

According to the material and methods of the experimental study performed by Sato et al. [40], peripheral blood could be obtained by jugular vein puncture, collected into syringes containing 0.5-ml sodium heparin and should be transported at 4°C to the laboratory within 3 h. To isolate peripheral blood-derived mononuclear cells, undiluted blood layered onto 12-ml Lympholyte in a 50-ml tube and centrifuged at 300 g for 40 min without braking [59]. The mononuclear cells were collected and washed twice with phosphate buffered saline (PBS) by centrifuging at 300 g for 5 min followed by an additional wash with PBS. After that, cells were re-suspended in culture medium which consists of Dulbecco's modified Eagle's medium with 5% separated autologous plasma, 10% fetal bovine serum and 10-μl/ml 100-units/ml Penicillin/Streptomycin solution. Subsequently, cells obtained from each 12 ml of blood were seeded onto a 100-mm^2 tissue culture dishes and incubated in a humidified atmosphere at 37°C with 5% CO_2. Nonadherent cells were removed by washing the mononuclear cells twice with PBS after 72 h of incubation. After 2 weeks, colonies of adherent fibroblast-like cells could be noticed. When the colony reached the approximate size of 5 cm^2, cells are detached and seeded in a new flask. The MSCs maintained in growth medium until ~70% confluence. The cells were then treated with 0.05% EDTA solution and could be cultured for subsequent passage in 100 mm^2 dishes at 7500 cells/cm^2 in base medium. This procedure was repeated as many times as possible.

Kassis et al. [41] evaluated the ability of fibrin microbeads (FMB) to separate human MSC from different sources other than bone marrow, with special emphasis on granulocyte colony-stimulating factor (G-CSF)–mobilized peripheral blood of healthy individuals. According to their material and methods, fibrin microbeads that bind matrix-dependent cells were produced from concentrated fibrinogen by a stirred heated oil emulsion technique and used to isolate MSC from the mononuclear fraction of mobilized peripheral blood of adult healthy human donors treated with G-CSF. Based on their results, FMB may have special advantage in isolating MSC from mobilized peripheral blood.

The isolation of MSCs from peripheral blood is a relatively new method with the main advantage of the ease in access, and further studies are needed to clarify the most appropriate technique. In addition, the introduction of platelet aggregates in oral and maxillofacial surgery has changed the approach toward extensive reconstruction of resorbed maxillae (**Figure 6**) and mandibles for implant reconstruction [42].

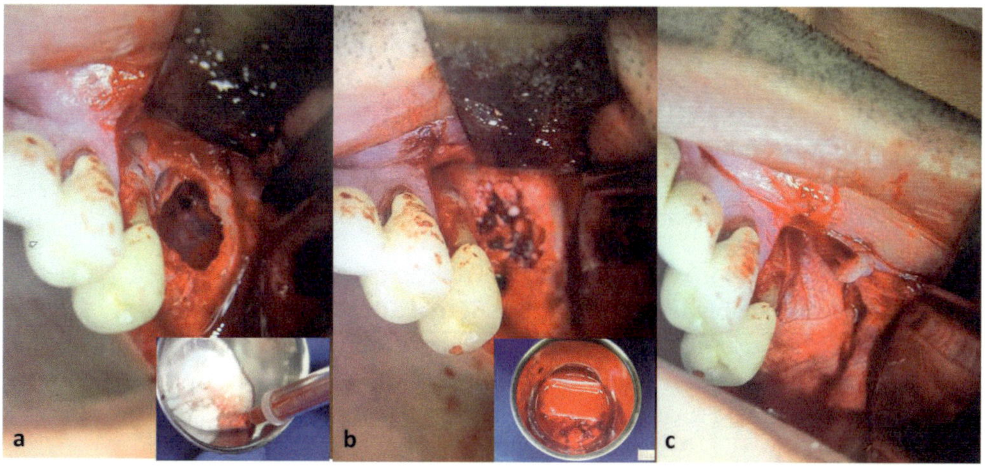

Figure 6. Second generation platelet aggregate (platelet rich fibrin) application in augmentation of the posterior maxilla. (a) Elevation of the membrane and preparation of PRF combined with Xenograft (Bioss®, Geistlich Germany) (b) Platelet rich fibrin (PRF) and Xenograft in situ. Preparation of the PRF membrane. (c) Placement of the membrane.

A possible role of platelet aggregates in local regulation of fracture healing and bone regeneration was attributed to the synergic effect of growth factors such as isomers of platelet-derived growth factor, transforming growth factor A1 and A2, insulin like growth factor > and A, and vascular endothelial growth factor. From this point of view, platelet aggregates could help in the differentiation and chemotactic and mitogenic stimulation of MSCs, which leads to an enhancement of bone repair and regeneration. Moreover, Marx [43] have recently confirmed that platelet rich plasma (PRF) contains an amount of 250×10^3 -400×10^3 per mL, which are positive for CD 44, CD 90, CD 105, and CD 34.

5. Mesenchymal stem cells in the reconstruction of the osseous defects of the jaws

Tissue-engineering therapy is a recent treatment modality in dental field to rehabilitate quantitative and qualitative properties of both soft and hard tissues with the use of cells with regenerative potential signaling molecules such as growth factors and a biocompatible matrix serving as a scaffold [44, 45].

During the past 2 decades, various experimental studies focusing on the osteogenic properties of MSCs have been performed. In 2001, Cooper et al. [46] have studied the relationship between bone sialoprotein (BSP) expression and osteocalcin expression with subsequent osteogenesis occurring in MSC-based implants and suggested that culture-expanded, cryopreserved human MSCs have osteogenic potential and demonstrated that implanted cell gene expression can reveal the early onset of bone formation.

In 2003, De Kok et al. [47] have evaluated MSC-based alveolar bone regeneration in a canine alveolar saddle defect model and observed that equivalent amounts of new bone were formed within the pores of the matrices loaded with autologous MSCs or MSCs from an unrelated donor, confirming the hypothesis that MSCs have the capacity to regenerate bone within craniofacial defects In addition, they have also stated that neither autologous nor allogeneic MSCs induced a systemic response by the host. Gutwald et al. [48] compared the osteogenic potential of mononuclear cells harvested from the iliac crest combined with bovine bone mineral (BBM) with that of autogenous cancellous bone alone and studied bilateral augmentations of the sinus floor in 6 adult sheep and reported that MSCs, in combination with BBM as the biomaterial, have the potential to form bone.

In the literature, there are also numerous studies focusing on the stimulating effects of various growth factors, most notably BMPs, on the osteogenic differentiation of MSCs [49, 50]. Açil et al. [51] have compared the most potent growth factors in regard to their osteoinductive potential and stated that the combined addition of BMP-2, BMP-6, and BMP-9 to the osteoinductive culture medium containing dexamethasone, β-glycerophosphate, and ascorbate-2-phosphate produces more potent osteoblast differentiation of human MSCs in vitro.

Following various experimental studies, the number of the clinical prospective studies has also increased steadily, and good cases of translational research from basic research to clinical application have arisen. In a groundbreaking study, Wiltfang et al. [52] have reconstructed a mandibular discontinuity defect after ablative surgery using the gastrocolic omentum as a bioreactor for heterotopic ossification via a titanium mesh cage filled with bone mineral blocks, infiltrated with 12 mg of recombinant human BMP2, and enriched with bone marrow aspirate. The scaffold was implanted into the gastrocolic omentum, and 3 months later, a free flap was harvested to reconstruct the mandibular defect. In vivo single-photon-emission computed tomography/computed tomography revealed bone remodeling and mineralization inside the mandibular transplant during prefabrication. They have reported that the quality of life of the patient significantly increased with acquisition of the ability to masticate and the improvement in pronunciation and aesthetics.

It is well known that MSCs can be directed to differentiate into an osteoblastic lineage in the presence of growth factors. Furthermore, platelet-rich plasma (PRP), which can be easily isolated from whole blood, was often used for bone regeneration, wound healing, and bone defect repair [53]. Marx [43] have stated that PRP contains an amount of 250×10^3 –400×10^3 per ml which are positive for CD 44, CD 90, CD 105, and CD 34.

Yamada et al. [54] investigated as basic research tissue-engineered bone regeneration using MSCs and PRP in a dog mandible model and confirmed the correlation between osseointegration in dental implants and the injectable bone. After that, same authors applied this injectable tissue-engineered bone to onlay plasty in the posterior maxilla or mandible in three human patient with simultaneous implant placement and reported stable and predictable results in terms of implant success [55]. In 2005, Ueda et al. [56] have used MSCs in a clinical study undertaken to evaluate the use of tissue-engineered bone, MSCs, platelet-rich plasma, and beta-tricalcium phosphate as grafting materials for maxillary sinus floor augmentation and proclaimed that tissue-engineered bone provided stable and predictable results in terms of implant success.

In order to increase the amount of available bone where dental implants must be placed, Filho Cerruti et al. [57] evaluated PRP and mononuclear cells (MNCs) from bone marrow aspirate and bone scaffold in 32 patients and have concluded that the process of healing observed in the patients was due to the presence of mesenchymal stem cell in MNC fraction in the bone grafts. Schmelzeisen et al. [58] reported a simplified method of using to regenerate hard tissue and suggested that bone marrow aspirate concentrate combined with a suitable biomaterial can form sufficient bone within 3 months for further implants to be inserted and at the same time minimize morbidity at the donor site. Similarly, Ricket et al. [59] have assessed whether differences occur in bone formation after maxillary sinus floor elevation surgery with bovine bone mineral mixed with autogenous bone or autogenous stem cells and stated that MSCs seeded on bovine bone mineral particles can induce the formation of a sufficient volume of new bone to enable the reliable placement of implants within a time frame comparable with that of applying either solely autogenous bone or a mixture of autogenous bone and bovine bone mineral particles.

Not only the defects at implant recipient sites, peri-implantar bone loss has also become a point of interest for some researchers, and efforts have been made over the last few decades to produce reliable and predictable methods to stimulate bone regeneration in bone defects resulting from peri-implant diseases [60]. Ribeiro et al. [61] have investigated the effect of bone marrow–derived cells associated with guided bone regeneration in the treatment of dehiscence bone defects around dental implants and suggested that bone marrow–derived cells provided promising results for peri-implantar bone regeneration, although the combined approach seems to be relevant, especially to bone formation out of the implant threads. Similarly, Kim et al. [62] evaluated the potential of periodontal ligament stem cells and bone marrow stem cells on alveolar bone regeneration in a canine peri-implant defect model and demonstrated the feasibility of using stem cell–mediated bone regeneration to treat peri-implant defects.

6. Conclusion

A growing number of studies indicate that stem applications are feasible protocols with clinically successful results in restoration of the bone architecture of the maxillofacial region. Composite grafts of MSCs, BMP, PRP, and bone graft combinations are able to achieve clinical results equivalent to autogenous grafts in large vertical ridge augmentations without donor bone harvesting.

Continued and extended experimental studies are needed to exactly determine the isolation, characterization, and differentiation properties of MSCs. In addition, development of chairside protocols would be beneficial in order to adapt MSC applications to the daily dental practice.

Abbreviations

ASC	Adipose tissue-derived stromal cells
BMAC	Bone marrow aspirate concentrate
BMP	Bone morphogenetic protein
CCFDAB	Crushed cancellous freeze-dried allogeneic bone
CD	Cluster of differentiation
CEACAM	Carcinoembryonic antigen-related cell adhesion molecule
FACS	Fluorescence-activated cell sorting
FMB	Fibrin microbeads
GMP	Good manufacturing practice
HIV	Human immunodeficiency virus
HLA-DR	Human leukocyte antigen-D-related
MHC	Major histocompatibility complex
MSC	Mesenchymal stem cell
NCAM	Neural cell adhesion molecule
PBS	Phosphate buffered solution

PDLSC	Periodontal ligament stem cell
PRF	Platelet rich plasma
PRP	Platelet-rich fibrin
rhBMP	Recombinant human bone morphogenetic protein
TNAP	Tissue non-specific Alkaline Phosphatase

Author details

Mustafa Ayna[1,3]*, Aydin Gülses[2], Jörg Wiltfang[3,4] and Yahya Açil[3,4]

*Address all correspondence to: praxis@dr-ayna.de

1 Danube University Krems, Krems, Austria

2 Ankara 2nd Division Public Hospital Association, Ankara, Turkey

3 Center for Oral Surgery and Implantology, Duisburg, Germany

4 Department of Oral and Maxillofacial Surgery, Christian Albrechts University, Kiel, Germany

References

[1] Baj A, Trapella G, Lauritano D, Candotto V, Mancini GE, et al. An overview on bone reconstruction of atrophic maxilla: Success parameters and critical issues. Journal of Biological Regulators and Homeostatic Agents. 2016;**30**:209-215. DOI: 0393-974x(2016)

[2] Moore WR, Graves SE, Bain GI. Synthetic bone graft substitutes. ANZ Journal of Surgery. 2001;**71**:354-361. DOI: 10.1046/j.1440-1622.2001.02128.x

[3] Desilets CP, Marden LJ, Patterson AL, Hollinger JO. Development of synthetic bone-repair materials for craniofacial reconstruction. The Journal of Craniofacial Surgery. 1990;**1**:150-153. DOI: 10.1097/00001665-199001030-00005

[4] Singh R, Singh D, Singh A. Radiation sterilization of tissue allografts: A review. World Journal of Radiology. 2016;**8**:355-369. DOI: 10.4329/wjr.v8.i4.355

[5] Simonds RJ, Holmberg SD, Hurwitz RL. Transmission of human immunodeficiency virus type 1 from a seronegative organ and tissue donor. New England Journal of Medicine. 1992;**326**:726-732. DOI: 10.1056/NEJM199203123261102

[6] Spector M. Anorganic bovine bone and ceramic analogs of bone mineral as implants to facilitate bone regeneration. Clinics in Plastic Surgery. 1994;**21**:437-444

[7] Hürzeler MB, Quinones CR, Kirsch A, Gloker C, Schubach P. Maxillary sinus augmentation using different grafting materials and dental implants in monkeys. Part I. Evaluation of anorganic bovine derived bone matrix. Clinical Oral Implants Research. 1997;**8**:476-486. DOI: 10.1034/j.1600-0501.1997.080606.x

[8] Yildirim M, Spiekermann H, Biesterfeld S, Edelhoff D. Maxillary sinus augmentation using xenogenic bone substitute material Bio-Oss in combination with venous blood. A histologic and histomorphometric study in humans. Clinical Oral Implants Research. 2000;**11**:217-229. DOI: 10.1034/j.1600-0501.2000.011003217.x

[9] Ayna M, Açil Y, Gulses A. Fate of a bovine-derived xenograft in maxillary sinus floor elevation after 14 years: Histologic and radiologic analysis. The International Journal of Periodontics & Restorative Dentistry. 2015;**35**:541-547. DOI: 10.11607/prd.2135

[10] Jacobsen G, Easter D. Allograft vs. Xenograft. Practical Considerations for Biologic Scaffolds. Available from: https://cme.ucsd.edu/biologicscaffolds/AllograftvsXenograftMonograph.pdf [Accessed:10-01-2017]

[11] Cypher TJ, Grossman JP. Biological principles of bone graft healing. Journal of Foot and Ankle Surgery. 1996;**35**:413-417. DOI: 10.1016/S1067-2516(96)80061-5

[12] Kurz LT, Garfin SR, Booth JR. Harvesting autogenous iliac bone grafts: A review of complications and techniques. Spine. 1989;**14**:1324-1331. DOI: 10.1097/00007632-198912000-00009

[13] Friedenstein AJ, Petrakova KV, Kurolesova AI, Frolova GP. Heterotopic of bone marrow. Analysis of precursor cells for osteogenic and hematopoietic tissues. Translantation. 1968;**6**:230-247

[14] Shi S, Bartold PM, Miura M, Seo BM, Robey PG, et al. The efficacy of mesenchymal stem cells to regenerate and repair dental structures. Orthodontics & Craniofacial Research. 2005;**8**:191-199. DOI: 10.1111/j.1601-6343.2005.00331.x

[15] Baena RR, Rizzo S, Graziano A, Lupi SM. Bone Regeneration in Implant Dentistry: Role of Mesenchymal Stem Cells, A Textbook of Advanced Oral and Maxillofacial Surgery Volume 3, Prof. Mohammad Hosein KM, editor, InTech; 2016. DOI: 10.5772/63346. Available from: http://www.intechopen.com/books/a-textbook-of-advanced-oral-and-maxillofacial-surgery-volume-3/bone-regene ration-in-implant-dentistry-role-of-mesenchymal-stem-cells

[16] Collins E, Gu F, Qi M, Molano I, Ruiz P, et al. Differential efficacy of human mesenchymal stem cells based on source of origin. Journal of Immunology. 2014;**193**:4381-4390. DOI: 10.4049/jimmunol.1401636

[17] Zuk PA, Zhu M, Ashjian P, De Ugarte DA, Huang JI, Mizuno H, et al. Human adipose tissue is a source of multipotent stem cells. Molecular Biology of the Cell. 2002;**13**:4279-4295. DOI: 10.1091/mbc.E02-02-0105

[18] Wang SJ, Jiang D, Zhang ZZ, Huang AB, Qi YS, et al. Chondrogenic potential of peripheral blood derived mesenchymal stem cells seeded on demineralized cancellous bone scaffolds.Scientific Reports. 2016;**6**:36400. DOI: 10.1038/srep36400

[19] Stanco D, Viganò M, Perucca Orfei C, Di Giancamillo A, Peretti GM, et al. Multidifferentiation potential of human mesenchymal stem cells from adipose tissue and hamstring tendons for musculoskeletal cell-based therapy. Regenerative Medicine. 2015;**10**:729-743. DOI: 10.2217/rme.14.92

[20] Ceccarelli G, Graziano A, Benedetti L, Imbriani M, Romano F, et al. Osteogenic potential of human oral-periosteal cells (PCs) isolated from different oral origin: An in vitro study. Journal of Cellular Physiology. 2016;**231**:607-612. DOI: 10.1002/jcp.25104

[21] De Bari C, Dell'Accio F, Tylzanowski P, Luyten FP. Multipotent mesenchymal stemcells from adult human synovial membrane. Arthritis and Rheumatism. 2001;**44**(8):1928-1942. DOI: 10.1002/1529-0131(200108)44:8<1928::AID-ART331>3.0.CO;2-P

[22] Ratajczak J, Bronckaers A, Dillen Y, Gervois P, Vangansewinkel T, et al. The neurovascular properties of dental stem cells and their importance in dental tissue engineering. Stem Cells International. 2016;**2016**:9762871. DOI: 10.1155/2016/9762871

[23] Zhu SJ, Choi BH, Huh JY, Jung JH, Kim BY, et al. A comparative qualitative histological analysis of tissue-engineered bone using bone marrow mesenchymal stem cells, alveolar bone cells, and periosteal cells. Oral Surgery, Oral Medicine, Oral Pathology, Oral Radiology, and Endodontics. 2006;**101**:164-169. DOI: 10.1016/j.tripleo.2005.04.006

[24] Açil Y, Ghoniem AA, Gülses A, Kisch T, Stang F, et al. Suppression of osteoblast-related genes during osteogenic differentiation of adipose tissue derived stromal cells. Journal of Cranio-Maxillo-Facial Surgery. 2017;**45**:33-38. DOI: 10.1016/j.jcms.2016.10.006

[25] Liu L, Cheung TH, Charville GW, Rando TA. Isolation of skeletal muscle stem cells by fluorescence-activated cell sorting. Nature Protocols. 2015;**10**:1612-1624. DOI: 10.1038/nprot.2015.110

[26] Dominici M, Le Blanc K, Mueller I, Slaper-Cortenbach I, Marini F, et al. Minimal criteria for defining multipotent mesenchymal stromal cells. The International Society for Cellular Therapy Position Statement. Cytotherapy. 2006;**8**:315-317. DOI: 10.1080/14653240600855905

[27] Lee T-C, Lee T-H, Huang Y-H, Chang N-K, Lin Y-J, et al. Comparison of surface markers between human and rabbit mesenchymal stem cells. PLoS One. 2014;**9**:e111390. DOI: 10.1371/journal.pone.0111390

[28] Markers & Methods to Verify Mesenchymal Stem Cell Identity, Potency, & Quality. Available from: https://www.rndsystems.com/resources/articles/markers-and-methods-verify-mesenchymal-stem-cell-identity-potency-and-quality [Accessed: 10-01-2017]

[29] Mödder UI, Roforth MM, Nicks KM, Peterson JM, McCready LK, et al. Characterization of mesenchymal progenitor cells isolated from human bone marrow by negative selection. Bone. 2012;**50**:804-810. DOI: 10.1016/j.bone.2011.12.014

[30] Seo BM, Miura M, Gronthos S, Bartold PM, Batouli S, et al. Investigation of multipotent postnatal stem cells from human periodontal ligament. Lancet. 2004;**364**:149-155. DOI: 10.1016/S0140-6736(04)16627-0

[31] Açil Y, Yang F, Gulses A, Ayna M, Wiltfang J, et al. Isolation, characterization and investigation of differentiation potential of human periodontal ligament cells and dental follicle progenitor cells and their response to BMP-7 in vitro. Odontology. 2016;**104**:123-135. DOI: 10.1007/s10266-015-0198-1

[32] Hakki SS, Bozkurt B, Hakki EE, Kayis SA, Turac G, et al. Bone morphogenetic protein-2, -6, and -7 differently regulate osteogenic differentiation of human periodontal ligament stem cells. Journal of Biomedical Materials Research Part B: Applied Biomaterials. 2014;**102**:119-130. DOI: 10.1002/jbm.b.32988

[33] Zuk PA, Zhu M, Mizuno H, Huang J, Futrell JW, et al. Multilineage cells from human adipose tissue: Implications for cell-based therapies. Tissue Engineering. 2001;**7**:211-228. DOI: 10.1089/107632701300062859

[34] Hernigou P, Poignard A, Beaujean F, Rouard H. Percutaneous autologous bone-marrow grafting for nonunions: Influence of the number and concentration of progenitor cells. The Journal of Bone and Joint Surgery. American Volume. 2005;**87**:1430-1437. DOI: 10.2106/JBJS.D.02215

[35] Shapiro SA, Kazmerchak SE, Heckman MG, Zubair AC, O'Connor MI. A prospective, single-blind, placebo-controlled trial of bone marrow aspirate concentrate for knee osteoarthritis. American Journal of Sports Medicine. 2017;**45**:82-90. DOI: 10.1177/0363546516662455

[36] Sauerbier S, Stricker A, Kuschnierz J, Bühler F, Oshima T, et al. In vivo comparison of hard tissue regeneration with human mesenchymal stem cells processed with either the FICOLL method or the BMAC method. Tissue Engineering. Part C, Methods. 2010;**16**:215-223. DOI: 10.1089/ten.TEC.2009.0269

[37] Assmus B, Honold J, Schächinger V, Britten MB, Fischer-Rasokat U, et al. Transcoronary transplantation of progenitor cells after myocardial infarction. New England Journal of Medicine. 2006;**355**:1222-1232. DOI: 10.1056/NEJMoa051779

[38] Wongchuensoontorn C, Liebehenschel N, Schwarz U, Schmelzeisen R, Gutwald R, et al. Application of a new chair-side method for the harvest of mesenchymal stem cells in a patient with nonunion of a fracture of the atrophic mandible–A case report. Journal of Cranio-Maxillo-Facial Surgery. 2009;**37**:155-161. DOI: 10.1016/j.jcms.2008.11.002

[39] Marx RE, Armentano L, Olavarria A, Samaniego J. rhBMP-2/ACS grafts versus autogenous cancellous marrow grafts in large vertical defects of the maxilla: An unsponsored randomized open-label clinical trial. The International Journal of Oral & Maxillofacial Implant. 2013;**28**:e243-e251. DOI: 10.11607/jomi.te04

[40] Sato K, Yamawaki-Ogata A, Kanemoto I, Usui A, Narita Y. Isolation and characterisation of peripheral blood-derived feline mesenchymal stem cells. Veterinary Journal. 2016;**216**:183-188. DOI: 10.1016/j.tvjl.2016.08.009

[41] Kassis I, Zangi L, Rivkin R, Levdansky L, Samuel S, et al. Isolation of mesenchymal stem cells from G-CSF-mobilized human peripheral blood using fibrin microbeads. Bone Marrow Transplantation. 2006;**37**:967-976. DOI: 10.1038/sj.bmt.1705358

[42] Gentile P, Bottini DJ, Spallone D, Curcio BC, Cervelli V. Application of platelet-rich plasma in maxillofacial surgery: Clinical evaluation. The Journal of Craniofacial Surgery. 2010;**21**:900-904. DOI: 10.1097/SCS.0b013e3181d878e9

[43] Marx R. Stem Cell Leased Regenerative Medicine. Available from: https://www.for.org/en/learn/videos/robert-marx-stem-cell-leased-regenerative-medicine [Accessed: 10-01-2017]

[44] Langer R, Vacanti JP, Vacanti CA, Atala A, Freed LE, et al. Tissue engineering: Biomedical applications. Tissue Engineering. 1995;**1**:151-161. DOI: 10.1089/ten.1995.1.151

[45] Misawa MYO, Huynh-Ba G, Villar GM, Villar CC. Efficacy of stem cells on the healing of peri-implant defects: Systematic review of preclinical studies. Clinical and Experimental Dental Research.. 2016;**2**:18-34. DOI: 10.1002/cre2.16

[46] Cooper LF, Harris CT, Bruder SP, Kowalski R, Kadiyala S. Incipient analysis of mesenchymal stem-cell-derived osteogenesis. Journal of Dental Research. 2001;**80**:314-320. DOI: 10.1177/00220345010800010401

[47] De Kok IJ, Peter SJ, Archambault M, van den Bos C, Kadiyala S, et al. Investigation of allogeneic mesenchymal stem cell-based alveolar bone formation: Preliminary findings. Clinical Oral Implants Research. 2003;**14**:481-489. DOI: 10.1034/j.1600-0501.2003.110770.x

[48] Gutwald R, Haberstroh J, Kuschnierz J, Kister C, Lysek DA, et al. Mesenchymal stem cells and inorganic bovine bone mineral in sinus augmentation: Comparison with augmentation by autologous bone in adult sheep. British Journal of Oral and Maxillofacial Surgery. 2010;**48**:285-290. DOI: 10.1016/j.bjoms.2009.06.226

[49] Khanna-Jain R, Agata H, Vuorinen A, Sándor GK, Suuronen R, et al. Addition of BMP-2 or BMP-6 to dexamethasone, ascorbic acid, and β-glycerophosphate may not enhance osteogenic differentiation of human periodontal ligament cells. Growth Factors. 2010;**28**:437-446. DOI: 10.3109/08977194.2010.495719

[50] Cruz AC, Silva ML, Caon T, Simões CM. Addition of bone morphogenetic protein type 2 to ascorbate and β-glycerophosphate supplementation did not enhance osteogenic differentiation of human adipose-derived stem cells. Journal of Applied Oral Science. 2012;**20**:628-635. DOI: 10.1590/S1678-77572012000600007

[51] Açil Y, Ghoniem AA, Wiltfang J, Gierloff M. Optimizing the osteogenic differentiation of human mesenchymal stromal cells by the synergistic action of growth factors. Journal of Cranio-Maxillo-Facial Surgery. 2014;**42**:2002-2009. DOI: 10.1016/j.jcms.2014.09.006

[52] Wiltfang J, Rohnen M, Egberts JH, Lützen U, Wieker H, et al. Man as a living bioreactor: Prefabrication of a custom vascularized bone graft in the gastrocolic omentum. Tissue Engineering. Part C, Methods. 2016;**22**:740-746. DOI: 10.1089/ten.TEC.2015.0501

[53] Fernandes G, Yang S. Application of platelet-rich plasma with stem cells in bone and periodontal tissue engineering. Bone Research. 2016;**13**:16036. DOI: 10.1038/boneres.2016.36

[54] Yamada Y, Ueda M, Naiki T, Takahashi M, Hata K, et al .Autogenous injectable bone for regeneration with mesenchymal stem cells and platelet-rich plasma: Tissue-engineered bone regeneration. Tissue Engineering. 2004;**10**:955-964. DOI: 10.1089/1076327041348284

[55] Yamada Y, Ueda M, Hibi H, Nagasaka T. Translational research for injectable tissue-engineered bone regeneration using mesenchymal stem cells and platelet-rich plasma: From basic research to clinical case study. Cell Transplantation. 2004;**13**:343-355. DOI: 10.3727/000000004783983909

[56] Ueda M, Yamada Y, Ozawa R, Okazaki Y. Clinical case reports of injectable tissue-engineered bone for alveolar augmentation with simultaneous implant placement. The International Journal of Periodontics & Restorative Dentistry. 2005;**25**:129-137. DOI: 10.1016/j.prosdent.2005.05.019

[57] Filho Cerruti H, Kerkis I, Kerkis A, Tatsui NH, da Costa Neves A, et al. Allogenous bone grafts improved by bone marrow stem cells and platelet growth factors: Clinical case reports. Artificial Organs. 2007;**31**:268-273. DOI: 10.1111/j.1525-1594.2007.00374.x

[58] Schmelzeisen R, Gutwald R, Oshima T, Nagursky H, Vogeler M, et al. Making bone II: Maxillary sinus augmentation with mononuclear cells—case report with a new clinical method. British Journal of Oral and Maxillofacial Surgery. 2011;**49**:480-482. DOI: 10.1016/j.bjoms.2010.06.020

[59] Rickert D, Sauerbier S, Nagursky H, Menne D, Vissink A, et al. Maxillary sinus floor elevation with bovine bone mineral combined with either autogenous bone or autogenous stem cells: A prospective randomized clinical trial. Clinical Oral Implants Research. 2011;**22**:251-258. DOI: 10.1111/j.1600-0501.2010.01981.x

[60] Larsson L, Decker AM, Nibali L, Pilipchuk SP, Berglundh T, et al. Regenerative medicine for periodontal and peri-implant diseases. Journal of Dental Research. 2016;**95**:255-266. DOI: 10.1177/0022034515618887

[61] Ribeiro FV, Suaid FF, Ruiz KG, Rodrigues TL, Carvalho MD, et al. Effect of autologous bone marrow-derived cells associated with guided bone regeneration or not in the treatment of peri-implant defects. International Journal of Oral and Maxillofacial Surgery. 2012;**41**:121-127. DOI: 10.1016/j.ijom.2011.06.025

[62] Kim SH, Kim KH, Seo BM, Koo KT, Kim TI, et al. Alveolar bone regeneration by transplantation of periodontal ligament stem cells and bone marrow stem cells in a canine peri-implant defect model: A pilot study. Journal of Periodontology. 2009;**80**:1815-1823. DOI: 10.1902/jop.2009.090249

Chapter 10

Stem Cell-Based Therapies for Osteoarthritis: From Pre-Clinical to Clinical Applications

Hechmi Toumi, Eric Lespessailles and Marija Mazor

Additional information is available at the end of the chapter

http://dx.doi.org/10.5772/intechopen.68176

Abstract

Although many surgical and pharmaceutical interventions are currently available for treating osteoarthritis (OA), restoration of normal cartilage function remains inefficient. In fact, because of the absence of vasculature within the articular cartilage (AC), the self-potential for regeneration is very poor. Recently, researchers and clinicians have been focusing on alternative methods for cartilage preservation and repair. It has been shown that AC contains a population of stem cells or progenitor cells, similar to those found in many other adult tissues that are thought to be involved in the maintenance of tissue homeostasis. In the present chapter, we review the current status of stem cells potential in the treatment of early OA and discuss the possible origin of these cells and the role they might have in cartilage repair. We also review the recent progress in the field of chondroprogenitors in cartilage.

Keywords: osteoarthritis, stem cells, chondrocytes, bone marrow, cartilage, progenitors

1. Introduction

Articular hyaline cartilage is a tissue whose mechanical properties allow joint movements with a low coefficient of friction and a high absorption of constraints. Degradation of hyaline cartilage, posttraumatic or degenerative, causes functional impairment of the joint, pain, and decreased quality of life. These conditions generally lead to the formation of the most common degenerative orthopedic disease such as osteoarthritis (OA). The OA involves gradual deterioration of cartilage and subchondral bone accompanied by chronic low-grade inflammation of the synovium. These pathological changes lead to destruction of the whole joint organ. Even it is agreed that OA affects entire joint articular cartilage, breakdown remains the principal characteristic of OA. Unfortunately, since cartilage is a neural tissue, the OA is generally diagnosed in more advanced stages when the majority of cartilage is already degraded. Thus, restoration of normal cartilage function in OA remains challenged despite many surgical and pharmaceutical interventions being currently available [1]. Several treatment options are available to support the knee articular cartilage injury. Painkillers and anti-inflammatory drugs are first prescribed in association with loss of weight or physiotherapy. When these options are not sufficient, intra-articular injections of corticosteroids, hyaluronic acid, or platelet-rich plasma (PRP) [2] represent non-surgical alternatives. Despite drugs used clinically to reduce pain and maintain joint movement, in many cases, surgical substitution with artificial implants is inevitable. A number of surgical treatment strategies are currently available for articular cartilage defect repair. The cartilage repair aims to restore the histological structure of the whole osteochondral structure so that it can restore the original mechanical and functional properties [3, 4]. Restorative procedures include abrasion chondroplasty, subchondral drilling, microfracture, and mosaicplasty arthroscopy. The procedure chosen will depend on the size of the lesion, its depth, the age of the patient, the nature of the symptoms, and the regulations in force in each country. Surgical possibilities routinely used to repair articular cartilage can be separated into three major groups; those conducting subchondral stimulation, reconstruction techniques which transplant mature cartilage, and finally cellular transplants which aim to create a favorable environment for cartilage healing [5]. Recently, both cartilage and bone marrow stromal cells (BMSCs), also known as bone marrow-derived "mesenchymal stem cells" and "mesenchymal stromal cells," with inherent chondrogenic differentiation potential appeared to present a potential for therapeutic use in cartilage regeneration. BMSCs are easy to isolate and expand in culture in an undifferentiated state for therapeutic use. Owing to their potential to modulate local microenvironment via anti-inflammatory and immunosuppressive functions, BMSCs have an additional advantage for allogeneic application.

2. Mesenchymal stem cells (MSC) in cartilage repair

2.1. Stem cells

Stem cells are the foundation cells for every organ, tissue, and cell in the body [6, 7]. They may be thought of as a blank microchip that can ultimately be programmed to perform any number of specialized tasks. This role is justified based on two key properties: (1) the ability to self-renew, dividing in a way to make copies of themselves and (2) the ability to differentiate, giving rise to the mature types of the cells that make up our organs and tissues [6, 7].

The stem cells can be generally divided into three groups: totipotent, pluripotent, and multipotent stem cells. Totipotent stem cells originate from the fertilized egg and give rise to the whole organism. These cells, through the process of proliferation and differentiation, become *pluripotent embryonic stem cells* that form three germ layers: ectoderm, mesoderm, and endoderm [8]. These three germ layers are the embryonic source of all cells of the body (adult organism consists of 200 different cells types). During embryonic development, stem cells become specialized, which makes them terminally differentiated with specific function and they are unable to be renewed [9, 10].

Yet, even in the specialized tissue, we can find a pool of cells referred to as "*adult*" or "*somatic*" *stem cells*, which replace injured and dead cells of certain tissue (blood, skin, liver, brain, etc.) [9, 10]. These cells are termed as multipotent as their potential is limited to produce some or all of the mature cell types within a particular tissue where they reside (tissue-specific stem cells) [9–11]. Yet, some of the adult stem cells are less differentiated than the others and can give rise to the several tissue types belonging to the same germ layer. These include hematopoietic stem cells as a source of both red and white blood cells and mesenchymal stem cells (MSC), which may be a potential source of the several mesodermal tissues [10–12].

Based on this, the focus of scientific research became the potential use of adult stem cells for tissue repair but also to generate new tissue under *in vitro* conditions for biological transplantation. The ability to obtain cells with proliferation and differentiation potential without sacrificing potential human life is a highly popular and hopeful tool for modern day researchers.

2.2. Phenotype and differentiation potential of MSC

The MSC cells are multipotent—self-renewing cells found in adult tissues, which can be *in vitro* differentiated and form adipocytes, fibroblast, osteocyte, and chondrocytes lineage [13, 14]. These cells had been primarily isolated in the early 1970s when Friedenstein et al. discovered that a specific number of fibroblastic cells isolated from bone marrow have the capacity to form colonies *in vitro* and under appropriate stimulating environmental conditions, small aggregates of bone, and cartilage [15, 16]. Over the years, it has become clear that MSC are not an exclusive feature of the bone marrow [17–19], but can also be isolated from other organs and tissues such as fat [20–22], skeletal muscles [23, 24], and synovium [25].

The isolation and characterization of MSC among the other cell types are based on their properties to adhere and grow on plastic, phenotype characteristics, and differentiation potential [26]. Over the last decades of research, significant effort has been made to establish phenotypic characterization of MSC. Despite all the effort, to date, there is no specific marker or combination of markers which will allow isolation of the homogeneous MSCs pool [27].

Nevertheless, it has been generally agreed that MSCs express specific surface antigens which involve: CD105 (endoglin—type I glycoprotein), CD73 (ecto-5′-nucleotidase), CD44 (HCAM—homing cell adhesion molecule), CD90 (cluster of differentiation 90 [Thy 1]), CD71 (cluster of differentiation 71) and Stro-1 as well as the adhesion molecules CD106 (vascular cell adhesion molecule [VCAM]-1), CD166 (activated leucocyte cell adhesion molecule [ALCAM]), intercellular adhesion molecule (ICAM)-1, neurogenic locus notch homolog protein 3 (NOTCH3), integrin alpha-11 (ITGA11), and CD29 [17, 26, 28–31]. However, they do not express the hematopoietic-specific markers CD79a, CD45, CD11, CD34, CD19, or CD14 and co-stimulatory molecules CD80, CD40, CD86, or the adhesion molecules CD31 (platelet/endothelial cell adhesion molecule [PECAM]-1), CD18 (leucocyte function-associated antigen-1 [LFA-1]), or CD56 (neuronal cell adhesion molecule-1) [26].

Hence, to confirm the presence of MSC and extract them among the other cell types, researchers use the different combinations of these markers.

Another way to identify supposed MSC population is by their differentiation capacity to bone, cartilage, and adipocyte tissue. Herein, MSC has to be cultured in the specific medium composed of the substituent known to stimulate and control these differentiations *in vivo*. These are mostly specific growth factors such as BMPs for osteocytes [32–34] and TGFs, BMPs, and FGFs for chondrocytes [35–38]. To optimize MSC differentiation, cells need to be put under the *in vivo*-like environment. Then MSC aimed to become osteocytes or chondrocytes will be cultured in 3D pellets [32–38] while differentiation to adipocytes will be performed in monolayer.

The fact that MSCs can be differentiated into several different cells types *in vitro* clearly makes MSC and MSC-like cells (progenitors) a promising cell source for tissue repair and regeneration. Moreover, MSCs are known to secrete a large number of growth factors (GFs), cytokines, and chemokines for mediating various functions including anti-inflammatory, anti-apoptotic, anti-fibrotic, angiogenic, mitogenic, and wound-healing through paracrine activity [27, 39, 40]. All these features are highly desired and support their candidature for therapeutic purpose.

2.3. MSC potential for cartilage repair

Current research into cartilage tissue engineering focuses on the use of adult MSCs as an alternative to autologous chondrocytes [41]. The advantage of MSC over chondrocytes is their ability to self-renew without loss of differentiation capacity. Likewise, MSCs may retain immunomodulatory activity in recipient tissue due to lack of human leucocyte antigen (HLA) class II expression [42, 43]. These properties make MSC promising for a diversity of clinical applications including *in vitro* development of the cartilage tissue and its transplantation into the joint defect.

To date, research has demonstrated that bone marrow, adipose, and synovial-derived MSCs are mostly relevant as MSC sources for cartilage repair [8].

2.3.1. Bone marrow-derived MSC in cartilage repair

2.3.1.1. In vitro studies

Since the Friedenstein study in the early 1970s to date, numerous reports confirmed the multipotency of MSC isolated from bone marrow (BMSC) [16, 44–48]. Although, they represent a minor fraction of the total nucleated cell population (1 MSC/5 $\times 10^3$ mononuclear cells), they could significantly increase their number through *in vitro* expansion [44, 49–51]. Sakaguchi et al. confirmed that BMSC potential to divide persists even after 10 *in vitro* passages [49]. This is a significant achievement as the high cells number is required to fill the cartilage defects. Note that, as opposed to chondrocytes, MSC retain chondrogenetic potential even after long monolayer expansion [46, 52]. When a sufficient cell number is reached, cells are placed in the differentiation-specific medium. The quality of BMSC-derived chondrocytes and the formed cartilage tissue is then estimated [46, 52].

The obtained tissue exhibited high positive staining for cartilage ECM components: glycosaminoglycans, collagen II, and lubricin [45–48]. Note that, however, positive staining was also obtained for the collagen X, which is well-known as a marker of hypertrophic chondrocytes and produces calcified cartilage [45].

In a comparative study of MSC isolated from versus tissues, BMSC showed greater chondrogenetic potential over the fetal lung MSC or placenta MSC [45, 46]. Nevertheless, BMSC-derived cartilage pellets exhibited significantly higher expression of collagen X than those derived from the two other sources [46]. Moreover, the capacity of BMSC to differentiate into chondrocytes was reduced by passaging of the cells [46]. This has been recently confirmed on the animal model [53]. The results showed that proliferative, differentiation, and metabolism profile of BMSC significantly decreases by age increase [53]. In the other comparative study from 2016, authors did not observe any preference in *in vitro* chondrogenesis among MSC derived from bone marrow, adipose tissue, and umbilical cord [54].

2.3.1.2. Pre-clinical studies in animal model

To investigate cartilage repair by MSC *in vivo*, most of these pre-clinical studies have been performed in rabbit models treated with MSCs combined with appropriated scaffold materials and environmental factors [55–57]. The histological outcomes confirmed formation of the hyaline cartilage-like tissue expressing collagen type II [55, 56, 58, 59] as well as collagen type I [55, 56, 58]. Note that, the latter is a marker of fibrocartilaginous tissue. However, compared to the traditional ACI, the MSC therapy of cartilage defect resulted in regenerated hyaline cartilage-like tissue and restored a smooth cartilage surface, while the chondrocyte-seeded constructs produced mostly fibrocartilage-like tissue with a discontinuous superficial cartilage contour [60].

This finding has been further tested in large animal models. The study on swine model confirmed the beneficial effect of MSC over the ACI [61, 62]. Moreover, ovine MSCs have been isolated from bone marrow, expanded, characterized, and injected with transforming growth factor (TGF) b3 in a fibrin clot [63]. Two months after implantation, histological analysis revealed chondrocyte-like cells surrounded by a hyaline-like cartilaginous matrix that was integrated to host cartilage [63, 64]. Similar findings had been observed in the *Cynomolgus macaque* OA-model. The 2 months postoperative evaluation confirmed regular surface integration with neighboring native cartilage, and reconstruction of trabecular subchondral bone in the BMSC filled defects [65].

Taken together, animal studies indicated that MSC may be a promising approach for cartilage repair. However, animal models could not completely mimic OA pathogenesis in humans. In human primary OA, disease generally develops as a result of disturbed cell homeostasis, which leads to misbalance in synthesis and degradation of cartilage and subchondral bone matrix. These pathological changes are widely spread in OA cartilage at advanced stages when OA is generally diagnosed. Unfortunately, at this stage of the disease, there is only a slight amount of normal cartilage left. In contrast, experimental OA induced by mechanical trauma represents cartilage lesion surrounded by healthy tissue. The implanted cartilage construct may interact differently with healthy tissue than with a damaged surrounding tissue. Thus, repair techniques performed on the OA experimental model may not be sufficient to predict outcomes of this technique in humans.

2.3.1.3. *Clinical studies*

The clinical reports of cartilage defects treated by bone marrow MSC did show promising results. The symptoms improvement was mostly expressed through the pain relief and progress in physical mobility [66, 67]. However, quality of regenerated tissue evaluated by MRI and histology vary with respect to the time elapsed since surgery [68–72].

Autologous BMSC embedded in a collagen gel were transplanted into articular cartilage defects and covered with autologous periosteum [68–71]. Six weeks follow-up revealed better arthroscopic and histological scores in the cell-transplanted compared to the cell-free control group [68]. The repaired defects were filled with hyaline-like cartilage tissue confirmed by positive Safranin O staining [71]. Moreover, pain and walking abilities have been improved significantly [69]. Nevertheless, 1-year follow-up analysis detected formation of fibrocartilaginous tissue instead of hyaline cartilage tissue in the repaired lesions [57, 70]. This has been further confirmed by a 5-year follow-up study, where in the first 6 months after surgery pain, walking, stairs climbing, patella crepitus, and flection contractures were all improved. However, after the 6 months, they started gradually to deteriorate [73].

In the comparative study of autologous BMSC and autologous chondrocyte implantation (ACI), it has been shown that older patients showed significantly lower improvement compared to the younger in the ACI group. Nevertheless, age did not make any difference for the patients treated by autologous BMSC [74]. This finding may indicate that cellular senescence downgraded chondrocytes molecular pathways that are involved in regulation of cell activity, which affected their ability to form functional cartilage tissue [75].

Yet, these results did not confirm significant improvement between ACI and MSC therapies [74, 76]. Moreover, the same as for ACI, being overweight and large lesion size are significant predictors of poor clinical and arthroscopic outcomes after MSC therapy [77, 78].

2.3.2. *Adipose tissue-derived MSC in cartilage repair*

2.3.2.1. *In vitro studies*

Even the BMSC were commonly investigated and used in treatment of cartilage defects, the harvesting of bone marrow is painful and followed by risk of wound infection. Moreover, the BMSC number in bone marrow is very low which requires extended *in vitro* expansion and may cause loss of cells regenerative potential [8]. Given that, the adipose tissue became a novel source of adult stem cells due to easier harvesting procedure from the wasted tissue after the liposuction treatment.

Moreover, the proportion of the AMSCs in adipose tissue is several times higher than of MSCs in bone marrow. Results have confirmed their potential for chondrogenesis, osteogenesis, adipogenesis, myogenesis, and some aspects of neurogenesis [79, 80].

Chondrogenesis of human AMSCs has shown significantly higher expression of chondrogenic markers after 1 week under appropriate conditions [81]. However, a significantly elevated expression of collagen type X was observed after 3 weeks of chondrogenic induction [41, 81]. The tendency of the AMSCs to differentiate in hypertrophic chondrocytes had been further confirmed by the other studies. These studies showed positive staining of the collagen I and X in newly formed tissue even after the stimulation with chondrogenic growth factors [82–84]. This indicates that the regulation of cellular activity by growth factors, scaffolds, and even gene therapy merits further investigation.

Compared to the BMSC, cartilage obtained from the adipose-derived MSC did not express significantly higher levels of hypertrophic markers: collagen X and MMP-13 [41]. The recent study from 2016 has emphasized that MSCs from bone marrow, adipose tissue, and umbilical cord share similar biological properties and that their chondrogenic potential does not vary [54].

Based on the *in vitro* studies, it is not clear if the AMSCs are the best choice for the cartilage repair. Even though their chondrogenic potential had been clearly justified, their predisposition to form hypertrophic and fibrous tissue should not be neglected.

2.3.2.2. Pre-clinical studies

In vitro studies on animal models demonstrated that adipocyte-derived MSCs were able to restore symptoms of OA-induced cartilage. The improvement had been observed macroscopically where cartilage lesion had been covered by repaired tissue and the surface was relatively smooth. The histological assessment revealed only a few fissures, few cracks, and an almost continuous superficial zone [85]. Another study showed that injected AMSC migrated to the synovial membrane and meniscus, however not in cartilage. Nevertheless, reduced OA progression had been observed [86]. The benefits obtained by AMSCs treatment could be due to a trophic mechanism of action by the release of growth factors and cytokines [86]. Taken together, these few pre-clinical studies are in favor of AMSCs-based cartilage repair.

2.3.3. Synovium-derived MSC in cartilage repair

2.3.3.1. In vitro studies

Another source of adult stem cells is synovium (synovium-derived stem cells (SDSC)). The comparative study of stem cells from five different sources (bone marrow, synovium, skeletal muscle, periosteum, and adipose tissue) confirmed that SDSC have proliferation and differentiation capacity similar to BMSC [49]. Moreover, the pellets derived from synovium were heavier than those from other tissues, because of their higher secretion of cartilage matrix [87–89]. This makes synovium-derived MSC potentially superior to bone marrow-derived MSC.

2.3.3.2. In vivo studies

The transplantation of the implant composed of MSC from different sources into the full-thickness cartilage defects of rabbits showed that synovium and bone marrow MSCs had greater *in vivo* chondrogenic potential than adipose and muscle MSCs [89]. Moreover, synovium MSCs had the advantage of the highest proliferation potential [90]. This study also noted that cartilage repair by synovium-derived MSC requires injection of a high number of these cells into the defect [90]. By contrast, another group reported that the aggregates with a high density of synovium-derived MSCs failed to regenerate cartilage due to cell death and nutrient deprivation into the core of the aggregates. Though, aggregates with relatively low-cartilage density successfully regenerated damaged tissue [91]. When compared to the healthy cartilage, tissue regenerated by constructs composed of the synovium-derived MSCs showed more fibrocartilage-like characteristics mostly in the superficial zone of the repair tissue [92].

This finding needs to be further confirmed by more *in vitro* and *in vivo* studies before introducing these cell types in clinical trials.

2.4. Regulation of the MSC chondrogenesis

It has been proposed that *in vitro* chondrogenic differentiation of MSCs mimics *in vivo* embryonic cartilage development. Hence, *in vitro* MSC expansion phase may correspond to the initial proliferation of mesenchymal cells before condensation. Switching over to the high-density MSC pellet cultures mimics the *in vivo* MSCs condensation steps and early stage chondrogenesis during embryonic development [93]. It has been shown that mechanical forces employed on the cell mass during chondrogenesis may promote the cells differentiation and secretion of the matrix-specific molecule. These biomechanical applications mimic the natural articular cartilage *in vivo* conditions [94, 95].

2.4.1. MSC isolation and in vitro culturing conditions

The MSC to be subjected to the cartilage formation first need to be isolated from their native tissue. To date, bone marrow, fat, and synovium tissue presents the most suitable sources of adult stem cells [8] with each tissue necessitating a specific isolation procedure [6]. BMSC are aspired by syringe from bone shafts, while ADMS are released and collected due to enzymatic digestion of the tissue [6]. Subsequently, these cells are *in vitro* expanded in order to obtain sufficient cell numbers for the following experimental procedures [6]. After the proliferation step, expanded cells need to be cultured under the 3D conditions in order to stimulate chondrogenesis. Thus, they are cultivated in micromass (pellets) or in scaffold materials, such as polymers, alginate beads, collagen sponges or hydrogels, and microspheres for 2–3 weeks in special chondrogenic medium enriched by growth factors [96]. Growth factors enhance expression of chondrocyte markers and support formation of cartilage tissue [35, 44, 97–99]. Moreover, hypoxic conditions seem to be the logical choice to stimulate chondrogenesis as it is present in *in vivo* articular tissue [100–104]. It has been shown that hypoxia induces expression of crucial genes for cartilage formation like SOX9, SOX6, and SOX5 as well as secretion of ECM molecules typical for hyaline cartilage [44, 100–104].

Reported *in vitro* conditions provide MSC differentiation to chondrocytes, nevertheless, do not stop chondrogenesis at the pre-hypertrophic stage, while cells undergo terminal differentiation to hypertrophic chondrocytes. These cells produce calcified instead of hyaline cartilage [105]. This remains crucial, a limitation in the formation of functional articular cartilage, as calcified cartilage has different biomechanical characteristics compared to hyaline cartilage [105, 106].

2.4.2. Role of growth factors in cartilage repair

Chondrogenic differentiation of MSCs is induced by various intrinsic and extrinsic factors [107]. Growth factors play the most important role in this process [107]. The importance of growth factors in the maintenance and production of cartilage *in vivo* had been explained previously. Hence, introduction of these factors in *in vitro* controlled chondrogenesis was the logical choice. Below are listed studies that clarified the importance of growth factors in treatment of cartilage defects with MSC. Keep in mind that TGF-β superfamily (TGF-β 1 & 2 and bone morphogenic proteins—BMPs), insulin-like growth factors (IGFs), and fibroblast growth factors (FGFs) are the major factors regulating chondrogenesis and synthesis of cartilage matrix.

Porcine MSCs encapsulated in agarose hydrogels after treatment with TGF-b3 increase the sulfated glycosaminoglycans in surrounding culture media, highlighting their role in cartilage ECM anabolism [35]. Moreover, the expression of BMP4 in transgenic MSC enhances their chondrogenesis in rat model through the positive regulation of main cartilage component, collagen type II [108]. Moreover, after 24 weeks, animals treated with BMP-4 showed significantly better cartilage repair than untreated animals [108]. Nevertheless, better results were obtained in chondrogenesis of MSC when TGF-b1, IGF-1, BMP-2, and BMP-7 were combined [36]. Also, intra-articular application of another growth factor, FGF-18-induced dose-dependent, increases the cartilage thickness of tibial plateau in rat OA model [37]. Similar effect to FGF-18 has FGF-2 which stimulates [38, 109] increase in glycosaminoglycan and collagen type II after its application on MSC culture in chondrogenic medium [38]. Overall, growth factors appear to be one of the main components in improving clinical cartilage regeneration, but they must be precisely combined and loaded on appropriate scaffold materials to simulate the conditions and three-dimensional (3D) structure most similar to the *in vivo* condition.

3. Chondroprogenitors in cartilage

3.1. Chondrogenesis

Chondrogenesis is a complex process that is initiated by MSC crowding and condensing on the bone-forming site, followed by maturation into terminally differentiated chondrocytes [110, 111]. This pathway is accompanied by stage-specific ECM production, synchronized by cellular interactions with the matrix, growth, and differentiation factors [110]. The latter initiate or suppress cellular signaling pathways and transcription of specific genes in a spatial-temporal manner [110, 111]. The anti-inflammatory and immunosuppressive properties of BMSCs suggest that these cells reduce inflammation in the joint. Moreover, BMSCs may initiate the repair process by differentiating into chondrocytes or by inducing proliferation and differentiation of the remaining healthy chondroprogenitor into mature chondrocytes or both. In addition, other features such as transcription factors, biological modulators, and extracellular matrix proteins expressed or produced by BMSCs may play an important role in enhancing cartilage formation.

Initially, MSCs express adhesion molecules including N-cadherin, N-CAM (Ncam1), tenascin-C (Tnc), and versican, which are involved in the compaction and condensation of MSCs regulated by different BMP factors [112]. Through progression of the condensation process, MSCs begin to express early cartilage markers collagen type II, aggrecan, and FGF receptor leading to chondrocytes progenitors stage of chondrogenesis [113]. Process of MSC condensation and chondrogenesis is triggered and positively regulated by major transcriptional factor, Sox 9. It is highly expressed in MSC before condensation and remains highly expressed in all stages of chondrogenesis through prechondrocytes to mature chondrocytes, while it is switched off when cells undergo hypertrophy [113, 114]. The formation of chondrocytes over osteocytes is regulated by combined action of Sox 9 and other transcriptional factors Pax/Nkx/Barx2, Sox 9 through inhibition of Runx2 (Cbfa1) as a domain transcriptional factor required for osteoblast differentiation [113, 115]. Moreover, Sox 9 positively regulates two other Sox family members Sox 5 and Sox 6, which play a significant role in activation of cartilage-specific genes: type II, IX, and XI collagen, aggrecan, and cartilage oligomeric matrix protein [114, 116, 117]. The role and spatio-temporal expression of Sox 5 and Sox 6 in chondrogenesis has been studied through single and double null mutations in mice model. Single gene deletion resulted in moderate skeletal abnormalities; however, double mutation induced animal death caused by systemic chondrodysplasia and skeletal deformity. These results indicate simultaneous action of these two transcription factors in formation of functional skeletal system. Nevertheless, in the double mutant low level of cartilage, specific extracellular matrix component was sustained by normal Sox 9 expression, but it was insufficient to support proper MSC differentiation and formation of cartilage [116]. This implies that synchronized action of Sox 5, 6, and 9 trios is required to maintain sufficient ECM component expression and normal matrix composition. Furthermore, these three genes promote the chondrogenesis by inhibition of hypertrophic and osteogenic differentiation [113]. Chondrocytes maturation to hypertrophic chondrocytes is repressed by Sox 9 modulation of the Wnt/beta-catenin signaling pathway with beta-catenin degradation or inhibition of beta-catenin transcriptional activity without affecting its stability [118]. In addition, Sox 5 and Sox 6 delay chondrocyte hypertrophy by down-regulating Ihh signaling, FGFR3, and Runx2 and up-regulating BMP6 [115].

Further maturation of chondrocytes is essential for the final remodeling of the cartilage into bone. Terminal chondrocytes differentiation into the hypertrophic chondrocytes is promoted by upregulation of Runx 2 and calcified cartilage markers collagen X and MMP13 [113, 117]. Later, hypertrophic and terminal chondrocytes express angiogenic factors, including VEGF, which provide the genesis for vascularization and formation of primary ossification centers within osteoblasts, osteocytes, and hematopoietic cells [119]. Equally, terminal chondrocytes undergo apoptosis by release of collagen types X and I and mineralization of the ECM [117]. Contrary to growth plate chondrogenesis, normal articular chondrocytes never undergo hypertrophic differentiation, except at the tidemark [113].

3.2. Chondroprogenitors potential in cartilage repair

Recent research reported the presence of MSC and their progenitors in cartilage itself [104]. These cells possess characteristics similar to stem cells isolated from other adult tissues involving proliferation and differentiation potential under appropriate *in vitro* conditions [120–123]. They were subjected to the process of isolation, expansion, and identification in order to confirm their stem cells phenotype previously established on MSC from other adult tissues [121–124]. To date, studies investigated the presence of these cells in normal and OA cartilage. Interestingly, several authors observed that OA cartilage contains higher number of mesenchymal progenitors compared to normal [122, 125–129].

Subpopulation of cells determined as cartilage progenitor cells (ACPCs) possess high-colony forming efficiency and express surface antigens specific to MSC (Notch 1, CD 105 & CD 166) [121–123]. Moreover, after the cultivation in specific chondrogenic medium, they showed capacity to differentiate into cartilage in 3D pellet cultures [130]. The expression of MSC markers and differentiation potential confirmed presence of multipotential mesenchymal progenitor cells in articular cartilage [122]. Comparative study of ACPCs and BMSCs revealed positive expression of adult stem cells markers (Notch 1, Stro 1, CD105, and CD 166) on both cell types. Nevertheless, chondrogenesis of BMSCs resulted in hypertrophic cartilage tissue confirmed by positive staining of collagen X, while this marker was not detected in tissue obtained from ACPCs [124]. Similar was reported by Alsalameh et al. where $CD105^+$ and $CD166^+$ cells showed no signs of hypertrophic chondrocytes and osteogenesis in chondrogenic micromass cultures after 3 weeks [128].

Likewise, cells positive for other markers that have been identified in MSC $CD9^+/CD90^+/CD166^+$ [131], $CD105^+/CD166^+$ [128], and Notch-1^+/Stro-1^+ [125] were capable of differentiating in chondrocytes and formed cartilage tissue *in vitro*. MCS differentiation into hypertrophic cartilage is the major limitation in hyaline functional cartilage production [105]. ACPCs may therefore be considered superior to MSCs from other tissues in cartilage repair [124, 125, 128, 129].

These results indicate the opportunity for using OA cartilage as a potential source of cells with cartilage-forming potential. Yet, further investigations are required to explore chondrogenesis regulation *in vitro*.

4. Conclusion

Based on self-repair and multilineage potentials, MCS provide hyaline cartilage regeneration opportunities. Studies on cartilage regeneration with adult mesenchymal stem cells have shown that BMSC are the most commonly used cell types to address cartilage regeneration. However, although short-term results appear satisfactory, hypertrophic chondrocyte and fibrocartilage formation emerge thereafter with hypertrophically differentiated MSC. Note that fibrocartilage provides a molecular pattern secreted by hypertrophic chondrocytes, leading to different biomechanical characteristics compared with hyaline cartilage.

Furthermore, harvesting bone marrow is a painful procedure with donor-site morbidity and risk of wound infection and sepsis. Hence, both AMSCs and synovium-derived stem cells have been considered as alternatives. However, results using these two cell lines have been similar to those obtained employing the bone marrow approach. In fact, although a high expression of chondrogenic markers was initially obtained, they appear to be expressed as collagen type X confirming the presence of hypertrophy.

Therefore, further investigations regarding the regulation of cellular activity by growth factors, scaffolds and even gene therapy remain viable options. Recently, one more potential source of MSC and progenitors for cartilage repair engineering from the cartilage itself has been tested. Cells isolated from the surface zone of articular cartilage have the capacity to differentiate into cartilage in 3D pellet culture. Moreover, no signs of hypertrophic chondrocytes and osteogenesis were observed. Thus, ACPCs could be considered more adequate than MSC in cartilage repair.

Abbreviations

OA	Osteoarthritis
AC	Articular cartilage
PRP	Platelet-rich plasma
ECM	Extra-cellular matrix
MSC	Mesenchymal stem cells
BMSCs	Bone marrow stromal cells
ACI	Autologous chondrocytes implantation
COMP	Cartilage oligometric matrix protein
TGF-β	Transforming growth factors-beta superfamily
IGFs	Insulin-like growth factors
FGFs	Fibroblast growth factors
BMPs	Bone morphogenetic proteins
ALK	Activin receptor like-kinase
IHH	Indian hedgehog protein
IRS	Insulin receptor-substrate family
FGF	Fibroblast growth factors
FGFR	Fibroblast growth factor receptor
CD105	Endoglin-type I glycoprotein
CD73	Ecto-5′-nucleotidase
CD90 (Thy)	Cluster of differentiation 90
CD106 (VCAM-1)	Vascular cell adhesion molecule-1
CD166 (ALCAM)	Activated leucocyte cell adhesion molecule
CD106 (ICAM-1)	Intercellular adhesion molecule-1
NOTCH	Neurogenic locus notch homolog protein
ITGA11	Integrin alpha-11

http://dx.doi.org/10.5772/intechopen.68176

CD31 (PECAM-1)	Platelet/endothelial cell adhesion molecule-1
CD18 (LFA-1)	Leucocyte function-associated antigen-1
CD56	Neuronal cell adhesion molecule-1
GFs	Growth factors
HLA	Human leucocyte antigen
BMSC	Bone marrow-derived mesenchymal stem cells
AMSC	Adipose-derived mesenchymal stem cells
ACPC	Articular cartilage progenitor cells

Author details

Hechmi Toumi*, Eric Lespessailles and Marija Mazor

*Address all correspondence to: hechmi.toumi@chro-orleans.fr

Service de Rhumatologie, Centre Hospitalier Régional d'Orléans, Collegium ST, University of Orleans, Orleans, France

References

[1] Lorenz H, Richter W. Osteoarthritis: Cellular and molecular changes in degenerating cartilage. Progress in Histochemistry and Cytochemistry. 2006;**40**(3):135-163

[2] Jayabalan P, Hagerty S, Cortazzo MH. The use of platelet-rich plasma for the treatment of osteoarthritis. The Physician and Sportsmedicine. 2014;**42**(3):53-62

[3] Eyre DR. Collagens and cartilage matrix homeostasis. Clinical Orthopaedics and Related Research. 2004;(427 Suppl): S118-S122

[4] Pearle AD, Warren RF, Rodeo SA. Basic science of articular cartilage and osteoarthritis. Clinics in Sports Medicine. 2005;**24**(1):1-12

[5] Versier G, Dubrana F, French Arthroscopy S. Treatment of knee cartilage defect in 2010. Orthopaedics and Traumatology Surgery and Research. 2011;**97**(8 Suppl):S140-S153

[6] Longo UG, et al. Stem cells and gene therapy for cartilage repair. Stem Cells International. 2012;**2012**:168385

[7] Alison MR, et al. An introduction to stem cells. The Journal of Pathology. 2002;**197**(4): 419-423

[8] Khan WS, Johnson DS, Hardingham TE. The potential of stem cells in the treatment of knee cartilage defects. Knee. 2010;**17**(6):369-374

[9] Wang L, et al. Progress in stem cells and regenerative medicine. Sheng Wu Gong Cheng Xue Bao. 2015;**31**(6):871-879

[10] Goodell MA, Rando TA. Stem cells and healthy aging. Science. 2015;**350**(6265):1199-1204

[11] Tuan RS, Boland G, Tuli R. Adult mesenchymal stem cells and cell-based tissue engineering. Arthritis Research & Therapy. 2003;**5**(1):32-45

[12] Raveh-Amit H, et al. Tissue resident stem cells: Till death do us part. Biogerontology. 2013;**14**(6):573-590

[13] Pittenger MF, et al. Multilineage potential of adult human mesenchymal stem cells. Science. 1999;**284**(5411):143-147

[14] Liu ZJ, Zhuge Y, Velazquez OC. Trafficking and differentiation of mesenchymal stem cells. Journal of Cellular Biochemistry. 2009;**106**(6):984-991

[15] Friedenstein AJ. Stromal mechanisms of bone marrow: Cloning in vitro and retransplantation in vivo. Haematology and Blood Transfusion. 1980;**25**:19-29

[16] Friedenstein AJ, Chailakhjan RK, Lalykina KS. The development of fibroblast colonies in monolayer cultures of guinea-pig bone marrow and spleen cells. Cell and Tissue Kinetics. 1970;**3**(4):393-403

[17] Delorme B, et al. Specific plasma membrane protein phenotype of culture-amplified and native human bone marrow mesenchymal stem cells. Blood. 2008;**111**(5):2631-2635

[18] Yang M, Zhang H, Gangolli R. Advances of mesenchymal stem cells derived from bone marrow and dental tissue in craniofacial tissue engineering. Current Stem Cell Research & Therapy. 2014;**9**(3):150-161

[19] Gottipamula S, et al. Isolation, expansion and characterization of bone marrow-derived mesenchymal stromal cells in serum-free conditions. Cell and Tissue Research. 2014;**356**(1):123-135

[20] Laschke MW, et al. In vitro osteogenic differentiation of adipose-derived mesenchymal stem cell spheroids impairs their in vivo vascularization capacity inside implanted porous polyurethane scaffolds. Acta Biomaterialia. 2014;**10**(10):4226-4235

[21] Wu AY. Autologous fat transfer with in-situ mediation (AIM): A novel and compliant method of adult mesenchymal stem cell therapy. Journal of Translational Medicine. 2013;**11**:136

[22] Ozpur MA, et al. Generation of skin tissue using adipose tissue-derived stem cells. Plastic and Reconstructive Surgery. 2016;**137**(1):134-143

[23] Gao X, et al. A comparison of bone regeneration with human mesenchymal stem cells and muscle-derived stem cells and the critical role of BMP. Biomaterials. 2014;**35**(25):6859-6870

[24] Radtke CL, et al. Characterization and osteogenic potential of equine muscle tissue- and periosteal tissue-derived mesenchymal stem cells in comparison with bone marrow- and adipose tissue-derived mesenchymal stem cells. American Journal of Veterinary Research. 2013;**74**(5):790-800

[25] Liu H, et al. Comparison of cellular responses of mesenchymal stem cells derived from bone marrow and synovium on combined silk scaffolds. Journal of Biomedical Materials Research A. 2015;**103**(1):115-125

[26] Dominici M, et al. Minimal criteria for defining multipotent mesenchymal stromal cells. The International Society for Cellular Therapy position statement. Cytotherapy. 2006;**8**(4):315-317

[27] Szala S, Wisniewska E, Czapla J. Mesenchymal stromal cells. Postepy Higieny I Medycyny Doswiadczalnej (Online). 2014;**68**:1287-1298

[28] Simmons PJ, Torok-Storb B. Identification of stromal cell precursors in human bone marrow by a novel monoclonal antibody. *STRO-1*. Blood. 1991;**78**(1):55-62

[29] Halfon S, et al. Markers distinguishing mesenchymal stem cells from fibroblasts are downregulated with passaging. Stem Cells and Development. 2011;**20**(1):53-66

[30] Rozemuller H, et al. Prospective isolation of mesenchymal stem cells from multiple mammalian species using cross-reacting anti-human monoclonal antibodies. Stem Cells and Development. 2010;**19**(12):1911-1921

[31] De Schauwer C, et al. Markers of stemness in equine mesenchymal stem cells: A plea for uniformity. Theriogenology. 2011;**75**(8):1431-1443

[32] Kim TH, et al. In vitro and in vivo evaluation of bone formation using solid freeform fabrication-based bone morphogenic protein-2 releasing PCL/PLGA scaffolds. Biomedical Materials. 2014;**9**(2):025008

[33] Moeinzadeh S, Jabbari E. Morphogenic peptides in regeneration of load bearing tissues. Advances in Experimental Medicine and Biology. 2015;**881**:95-110

[34] McMahon MS. Bone morphogenic protein 3 signaling in the regulation of osteogenesis. Orthopedics. 2012;**35**(11):920

[35] Thorpe SD, et al. The response of bone marrow-derived mesenchymal stem cells to dynamic compression following TGF-beta3 induced chondrogenic differentiation. Annals of Biomedical Engineering. 2010;**38**(9):2896-2909

[36] An C, et al. IGF-1 and BMP-2 induces differentiation of adipose-derived mesenchymal stem cells into chondrocytes-like cells. Annals of Biomedical Engineering. 2010;**38**(4):1647-1654

[37] Davidson D, et al. Fibroblast growth factor (FGF) 18 signals through FGF receptor 3 to promote chondrogenesis. The Journal of Biological Chemistry. 2005;**280**(21):20509-20515

[38] Moore EE, et al. Fibroblast growth factor-18 stimulates chondrogenesis and cartilage repair in a rat model of injury-induced osteoarthritis. Osteoarthritis & Cartilage. 2005;**13**(7):623-631

[39] Hocking AM, Gibran NS. Mesenchymal stem cells: Paracrine signaling and differentiation during cutaneous wound repair. Experimental Cell Research. 2010;**316**(14):2213-2219

[40] Doorn J, et al. Therapeutic applications of mesenchymal stromal cells: Paracrine effects and potential improvements. Tissue Engineering Part B Reviews. 2012;**18**(2):101-115

[41] Ronziere MC, et al. Chondrogenic potential of bone marrow—and adipose tissue-derived adult human mesenchymal stem cells. BioMedical Materials and Engineering. 2010;**20**(3):145-158

[42] De Miguel MP, et al. Immunosuppressive properties of mesenchymal stem cells: Advances and applications. Current Molecular Medicine. 2012;**12**(5):574-591

[43] Castro-Manrreza ME, Montesinos JJ. Immunoregulation by mesenchymal stem cells: Biological aspects and clinical applications. Journal of Immunology Research. 2015;**2015**:394917

[44] Freyria AM, Mallein-Gerin F. Chondrocytes or adult stem cells for cartilage repair: the indisputable role of growth factors. Injury. 2012;**43**(3):259-265

[45] Boyette LB, et al. Human bone marrow-derived mesenchymal stem cells display enhanced clonogenicity but impaired differentiation with hypoxic preconditioning. Stem Cells Translational Medicine. 2014;**3**(2):241-254

[46] Bernardo ME, et al. Human mesenchymal stem cells derived from bone marrow display a better chondrogenic differentiation compared with other sources. Connective Tissue Research. 2007;**48**(3):132-140

[47] Nakagawa Y, et al. Cartilage derived from bone marrow mesenchymal stem Cells expresses lubricin in vitro and in vivo. PLoS One. 2016;**11**(2):e0148777

[48] Murdoch AD, et al. The development of a mature collagen network in cartilage from human bone marrow stem cells in Transwell culture. Matrix Biology. 2016;**50**:16-26

[49] Sakaguchi Y, et al. Comparison of human stem cells derived from various mesenchymal tissues: Superiority of synovium as a cell source. Arthritis and Rheumatism. 2005;**52**(8):2521-2529

[50] Norambuena GA, Khoury M, Jorgensen C. Mesenchymal stem cells in osteoarticular pediatric diseases: An update. Pediatric Research. 2012;**71**(4 Pt 2):452-458

[51] Lubis AM, Lubis VK. Adult bone marrow stem cells in cartilage therapy. Acta Medica Indonesiana. 2012;**44**(1):62-68

[52] Heng BC, Cao T, Lee EH. Directing stem cell differentiation into the chondrogenic lineage in vitro. Stem Cells. 2004;**22**(7):1152-1167

[53] Fafian-Labora J, et al. Influence of age on rat bone-marrow mesenchymal stem cells potential. Scientific Reports. 2015;**5**:16765

[54] Danisovic L, et al. Comparative analysis of mesenchymal stromal cells from different tissue sources in respect to articular cartilage tissue engineering. General Physiology and Biophysics. 2016;**35**(2):207-14.

[55] Chang F, et al. Repair of large full-thickness articular cartilage defects by transplantation of autologous uncultured bone-marrow-derived mononuclear cells. Journal of Orthopaedic Research. 2008;**26**(1):18-26

[56] Yan H, Yu C. Repair of full-thickness cartilage defects with cells of different origin in a rabbit model. Arthroscopy. 2007;**23**(2):178-187

[57] Bai T, et al. Experimental research on repair of rabbit articular cartilage deffects with composite of autologous cell-carriers. Zhongguo Xiu Fu Chong Jian Wai Ke Za Zhi. 2008;**22**(4):487-491

[58] Wakitani S, et al. Mesenchymal cell-based repair of large, full-thickness defects of articular cartilage. The Journal of Bone and Joint Surgery American. 1994;**76**(4):579-592

[59] Ding Z, Huang H. Mesenchymal stem cells in rabbit meniscus and bone marrow exhibit a similar feature but a heterogeneous multi-differentiation potential: Superiority of meniscus as a cell source for meniscus repair. BMC Musculoskeletal Disorders. 2015;**16**:65

[60] Tay LX, et al. Treatment outcomes of alginate-embedded allogenic mesenchymal stem cells versus autologous chondrocytes for the repair of focal articular cartilage defects in a rabbit model. The American Journal of Sports Medicine. 2012;**40**(1):83-90

[61] Li WJ, et al. Evaluation of articular cartilage repair using biodegradable nanofibrous scaffolds in a swine model: A pilot study. Journal of Tissue Engineering and Regenerative Medicine. 2009;**3**(1):1-10

[62] Dutton AQ, et al. Enhancement of meniscal repair in the avascular zone using mesenchymal stem cells in a porcine model. The Journal of Bone and Joint Surgery British. 2010;**92**(1):169-175

[63] Mrugala D, et al. Phenotypic and functional characterisation of ovine mesenchymal stem cells: Application to a cartilage defect model. Annals of the Rheumatic Diseases. 2008;**67**(3):288-295

[64] Dorotka R, et al. Repair of articular cartilage defects treated by microfracture and a three-dimensional collagen matrix. Biomaterials. 2005;**26**(17):3617-3629

[65] Araki S, et al. Improved quality of cartilage repair by bone marrow mesenchymal stem cells for treatment of an osteochondral defect in a cynomolgus macaque model. Acta Orthopaedica. 2015;**86**(1):119-126

[66] Vega A, et al. Treatment of knee osteoarthritis with allogeneic bone marrow mesenchymal stem cells: A randomized controlled trial. Transplantation. 2015;**99**(8):1681-1690

[67] Orozco L, et al. Treatment of knee osteoarthritis with autologous mesenchymal stem cells: A pilot study. Transplantation. 2013;**95**(12):1535-1541

[68] Wakitani S, et al. Human autologous culture expanded bone marrow mesenchymal cell transplantation for repair of cartilage defects in osteoarthritic knees. Osteoarthritis & Cartilage. 2002;**10**(3):199-206

[69] Wakitani S, et al. Autologous bone marrow stromal cell transplantation for repair of full-thickness articular cartilage defects in human patellae: Two case reports. Cell Transplantation. 2004;**13**(5):595-600

[70] Wakitani S, et al. Repair of articular cartilage defects in the patello-femoral joint with autologous bone marrow mesenchymal cell transplantation: Three case reports involving nine defects in five knees. Journal of Tissue Engineering and Regenerative Medicine. 2007;**1**(1):74-79

[71] Kuroda R, et al. Treatment of a full-thickness articular cartilage defect in the femoral condyle of an athlete with autologous bone-marrow stromal cells. Osteoarthritis & Cartilage. 2007;**15**(2):226-231

[72] Kim YS, et al. Assessment of clinical and MRI outcomes after mesenchymal stem cell implantation in patients with knee osteoarthritis: A prospective study. Osteoarthritis & Cartilage. 2016;**24**(2):237-245

[73] Davatchi F, et al. Mesenchymal stem cell therapy for knee osteoarthritis: 5 years follow-up of three patients. International Journal of Rheumatic Diseases. 2016;**19**(3):219-225

[74] Nejadnik H, et al. Autologous bone marrow-derived mesenchymal stem cells versus autologous chondrocyte implantation: An observational cohort study. The American Journal of Sports Medicine. 2010;**38**(6):1110-1116

[75] Burrage PS, Mix KS, Brinckerhoff CE. Matrix metalloproteinases: Role in arthritis. Frontiers in Bioscience. 2006;**11**:529-543

[76] Giannini S, et al. Cartilage repair evolution in post-traumatic osteochondral lesions of the talus: from open field autologous chondrocyte to bone-marrow-derived cells transplantation. Injury. 2010;**41**(11):1196-1203

[77] Kim YS, et al. Mesenchymal stem cell implantation in osteoarthritic knees: Is fibrin glue effective as a scaffold? The American Journal of Sports Medicine. 2015;**43**(1):176-185

[78] Kim YS, Choi YJ, Koh YG. Mesenchymal stem cell implantation in knee osteoarthritis: an assessment of the factors influencing clinical outcomes. The American Journal of Sports Medicine. 2015;**43**(9):2293-2301

[79] Estes BT, et al. Isolation of adipose-derived stem cells and their induction to a chondrogenic phenotype. Nature Protocols. 2010;**5**(7):1294-1311

[80] Guilak F, et al. Nicolas Andry Award: Multipotent adult stem cells from adipose tissue for musculoskeletal tissue engineering. Clinical Orthopaedics and Related Research. 2010;**468**(9):2530-2540

[81] Hamid AA, et al. Characterization of human adipose-derived stem cells and expression of chondrogenic genes during induction of cartilage differentiation. Clinics (Sao Paulo). 2012;**67**(2):99-106

[82] Diekman BO, Estes BT, Guilak F. The effects of BMP6 overexpression on adipose stem cell chondrogenesis: Interactions with dexamethasone and exogenous growth factors. Journal of Biomedical Materials Research A. 2010;**93**(3):994-1003

[83] Diekman BO, et al. Chondrogenesis of adult stem cells from adipose tissue and bone marrow: Induction by growth factors and cartilage-derived matrix. Tissue Engineering Part A. 2010;**16**(2):523-533

[84] Mehlhorn AT, et al. Differential effects of BMP-2 and TGF-beta1 on chondrogenic differentiation of adipose derived stem cells. Cell Proliferation. 2007;**40**(6):809-823

[85] Wang W, et al. Human adipose-derived mesenchymal progenitor cells engraft into rabbit articular cartilage. International Journal of Molecular Sciences. 2015;**16**(6):12076-12091

[86] Desando G, et al. Intra-articular delivery of adipose derived stromal cells attenuates osteoarthritis progression in an experimental rabbit model. Arthritis Research & Therapy. 2013;**15**(1):R22

[87] Yoshimura H, et al. Comparison of rat mesenchymal stem cells derived from bone marrow, synovium, periosteum, adipose tissue, and muscle. Cell and Tissue Research. 2007;**327**(3):449-462

[88] Shirasawa S, et al. In vitro chondrogenesis of human synovium-derived mesenchymal stem cells: Optimal condition and comparison with bone marrow-derived cells. Journal of Cellular Biochemistry. 2006;**97**(1):84-97

[89] Isobe Y, et al. Comparison of human mesenchymal stem cells derived from bone marrow, synovial fluid, adult dental pulp, and exfoliated deciduous tooth pulp. International Journal of Oral and Maxillofacial Surgery. 2016;**45**(1):124-131

[90] Koga H, et al. Comparison of mesenchymal tissues-derived stem cells for in vivo chondrogenesis: Suitable conditions for cell therapy of cartilage defects in rabbit. Cell and Tissue Research. 2008;**333**(2):207-215

[91] Suzuki S, et al. Properties and usefulness of aggregates of synovial mesenchymal stem cells as a source for cartilage regeneration. Arthritis Research & Therapy. 2012;**14**(3):R136

[92] Ando W, et al. Detection of abnormalities in the superficial zone of cartilage repaired using a tissue engineered construct derived from synovial stem cells. European Cells & Materials. 2012;**24**:292-307

[93] Mamidi MK, et al. Mesenchymal stromal cells for cartilage repair in osteoarthritis. Osteoarthritis & Cartilage. 2016;**24**(8):1307-16

[94] Li D, et al. Role of mechanical factors in fate decisions of stem cells. Regenerative Medicine. 2011;**6**(2):229-240

[95] Wang YK, Chen CS. Cell adhesion and mechanical stimulation in the regulation of mesenchymal stem cell differentiation. Journal of Cellular and Molecular Medicine. 2013;**17**(7):823-832

[96] Ishii I, et al. Healing of full-thickness defects of the articular cartilage in rabbits using fibroblast growth factor-2 and a fibrin sealant. The Journal of Bone and Joint Surgery British. 2007;**89**(5):693-700

[97] Chubinskaya S, et al. Effects induced by BMPS in cultures of human articular chondrocytes: Comparative studies. Growth Factors. 2008;**26**(5):275-283

[98] Danisovic L, Varga I, Polak S. Growth factors and chondrogenic differentiation of mesenchymal stem cells. Tissue and Cell. 2012;**44**(2):69-73

[99] Sekiya I, et al. Comparison of effect of BMP-2, -4, and -6 on in vitro cartilage formation of human adult stem cells from bone marrow stroma. Cell and Tissue Research. 2005;**320**(2):269-276

[100] Markway BD, et al. Enhanced chondrogenic differentiation of human bone marrow-derived mesenchymal stem cells in low oxygen environment micropellet cultures. Cell Transplantation. 2010;**19**(1):29-42

[101] Khan WS, Adesida AB, Hardingham TE. Hypoxic conditions increase hypoxia-inducible transcription factor 2alpha and enhance chondrogenesis in stem cells from the infrapatellar fat pad of osteoarthritis patients. Arthritis Research & Therapy. 2007;**9**(3):R55

[102] Khan WS, et al. Bone marrow-derived mesenchymal stem cells express the pericyte marker 3G5 in culture and show enhanced chondrogenesis in hypoxic conditions. Journal of Orthopaedic Research. 2010;**28**(6):834-840

[103] Merceron C, et al. Differential effects of hypoxia on osteochondrogenic potential of human adipose-derived stem cells. American Journal of Physiology Cell Physiology. 2010;**298**(2):C355-364

[104] Murphy CL, et al. Hypoxia. HIF-mediated articular chondrocyte function: Prospects for cartilage repair. Arthritis Research & Therapy. 2009;**11**(1):213

[105] van Osch GJ, et al. Cartilage repair: Past and future—lessons for regenerative medicine. Journal of Cellular and Molecular Medicine. 2009;**13**(5):792-810

[106] Pacifici M, et al. Hypertrophic chondrocytes. The terminal stage of differentiation in the chondrogenic cell lineage? Annals of the New York Academy of Sciences. 1990;**599**:45-57

[107] Kolf CM, Cho E, Tuan RS. Mesenchymal stromal cells. Biology of adult mesenchymal stem cells: Regulation of niche, self-renewal and differentiation. Arthritis Research & Therapy. 2007;**9**(1):204

[108] Kuroda R, et al. Cartilage repair using bone morphogenetic protein 4 and muscle-derived stem cells. Arthritis and Rheumatism. 2006;**54**(2):433-442

[109] Stewart AA, et al. Effect of fibroblast growth factor-2 on equine mesenchymal stem cell monolayer expansion and chondrogenesis. American Journal of Veterinary Research. 2007;**68**(9):941-945

[110] DeLise AM, Fischer L, Tuan RS. Cellular interactions and signaling in cartilage development. Osteoarthritis & Cartilage. 2000;**8**(5):309-334

[111] Shimizu H, Yokoyama S, Asahara H. Growth and differentiation of the developing limb bud from the perspective of chondrogenesis. Development Growth & Differentiation. 2007;**49**(6):449-454

[112] Zuscik MJ, et al. Regulation of chondrogenesis and chondrocyte differentiation by stress. The Journal of Clinical Investigation. 2008;**118**(2):429-438

[113] Lefebvre V, Smits P. Transcriptional control of chondrocyte fate and differentiation. Birth Defects Research Part C Embryo Today. 2005;**75**(3):200-212

[114] Akiyama H, et al. The transcription factor Sox9 has essential roles in successive steps of the chondrocyte differentiation pathway and is required for expression of Sox5 and Sox6. Genes & Development. 2002;**16**(21):2813-2828

[115] Zhou G, et al. Dominance of SOX9 function over RUNX2 during skeletogenesis. Proceedings of the National Academy of Sciences of the United States of America. 2006;**103**(50):19004-19009

[116] Smits P, et al. The transcription factors L-Sox5 and Sox6 are essential for cartilage formation. Developmental Cell. 2001;**1**(2):277-290

[117] Vinatier C, et al. Cartilage engineering: A crucial combination of cells, biomaterials and biofactors. Trends in Biotechnology. 2009;**27**(5):307-314

[118] Topol L, et al. Sox9 inhibits Wnt signaling by promoting beta-catenin phosphorylation in the nucleus. The Journal of Biological Chemistry. 2009;**284**(5):3323-3333

[119] Carlevaro MF, et al. Vascular endothelial growth factor (VEGF) in cartilage neovascularization and chondrocyte differentiation: Auto-paracrine role during endochondral bone formation. Journal of Cell Science. 2000;**113**(Pt 1):59-69

[120] O'Sullivan J, et al. Mesenchymal chondroprogenitor cell origin and therapeutic potential. Stem Cell Research & Therapy. 2011;**2**(1):8

[121] Dowthwaite GP, et al. The surface of articular cartilage contains a progenitor cell population. Journal of Cell Science. 2004;**117**(Pt 6):889-897

[122] Hiraoka K, et al. Mesenchymal progenitor cells in adult human articular cartilage. Biorheology. 2006;**43**(3-4):447-454

[123] Chang HX, et al. Age-related biological characterization of mesenchymal progenitor cells in human articular cartilage. Orthopedics. 2011;**34**(8):e382-388

[124] McCarthy HE, et al. The comparison of equine articular cartilage progenitor cells and bone marrow-derived stromal cells as potential cell sources for cartilage repair in the horse. Veterinary Journal. 2012;**192**(3):345-351

[125] Grogan SP, et al. Mesenchymal progenitor cell markers in human articular cartilage: Normal distribution and changes in osteoarthritis. Arthritis Research & Therapy. 2009;**11**(3):R85

[126] Pretzel D, et al. Relative percentage and zonal distribution of mesenchymal progenitor cells in human osteoarthritic and normal cartilage. Arthritis Research & Therapy. 2011;**13**(2):R64

[127] Nelson L, et al. Evidence of a viable pool of stem cells within human osteoarthritic cartilage. Cartilage. 2014;**5**(4):203-214

[128] Alsalameh S, et al. Identification of mesenchymal progenitor cells in normal and osteoarthritic human articular cartilage. Arthritis and Rheumatism. 2004;**50**(5):1522-1532

[129] Fickert S, Fiedler J, Brenner RE. Identification of subpopulations with characteristics of mesenchymal progenitor cells from human osteoarthritic cartilage using triple staining for cell surface markers. Arthritis Research & Therapy. 2004;**6**(5):R422-R432

[130] Khan IM, et al. Clonal chondroprogenitors maintain telomerase activity and Sox9 expression during extended monolayer culture and retain chondrogenic potential. Osteoarthritis & Cartilage. 2009;**17**(4):518-528

[131] Ozbey O, et al. Characterization f colony-forming cells in adult human articular cartilage. Acta Histochemica. 2014;**116**(5):763-770

Chapter 11

Prerequisites for Mesenchymal Stem Cell Transplantation in Spinal Cord Injury

Sherif M. Amr

Additional information is available at the end of the chapter

http://dx.doi.org/10.5772/intechopen.69554

Abstract

We have aimed at distinguishing obligatory prerequisites for mesenchymal stem cell transplantation in spinal cord injury from those prerequisites which are unnecessary or are prerequisites that have to be further investigated. Obligatory prerequisites include the following. First, the site of injury is extensively gliotic, constituting an unsuitable medium for stem cell transplantation. It has to be dissolved by neurolyzing agents, chondroitinase ABC as an example. Second, stem cells need a suitable biomaterial scaffold for their proper integration. Third, the biomaterial scaffold necessitates a tissue filler harboring stem cells, other cells and neurotrophic factors in a combinatorial approach. Fourth, the efficiency of mesenchymal stem cells themselves has to be increased (by reducing oxidative stress-induced apoptosis, by hypoxic preconditioning, by modulating the extracellular matrix and by other measures). Prerequisites that have to be further investigated include the ideal source, mode, quantity, time point and number of injections of mesenchymal stem cells; which growth factors and cells to be used in the combinatorial approach; transforming mesenchymal stem cells into motor neuron-like cells or Schwann cells; increasing the homing effect of stem cells and how to establish a continuous drug and cell delivery system.

Keywords: spinal cord injury, mesenchymal stem cells, scaffolds, nerve grafting, neurotrophic factors, chondroitinase ABC, continuous drug delivery systems

1. Introduction

Traumatic spinal cord injury results usually from cervical and lumbar fractures; it may be associated with complete paraplegia. Regeneration after such an injury is fairly limited mainly due to the inhibitory milieu (the gliosis) within the spinal cord. Cellular therapeutic strategies may overcome this milieu by neuroprotection, immunomodulation, axon regeneration, neuronal relay formation and myelin regeneration [1]. Clinically, in a meta-analysis on cellular therapy in traumatic spinal cord injury in humans published in 2012 [2], the authors reviewed eight bone marrow mesenchymal and hematopoietic stem cell studies, two olfactory ensheathing cell studies, one Schwann cell study and one fetal neurogenic tissue study. Three of these were Grade III and nine Grade IV level of evidence. It was concluded that improved preclinical studies and prospective, controlled clinical trials were needed. Nevertheless, ever since, the number of clinical trials have been increased. Mesenchymal stem cells, in particular, are easy to isolate, can be rapidly expanded in culture and can be cryopreserved without loss of potency [3, 4]. Clinical reports on their use have varied, starting from documenting their safety [5, 6] up to limited clinical efficacy [7], even partial or complete efficacy [8–11].

The aim of this review is to distinguish necessary prerequisites for effective mesenchymal stem cell transplantation in spinal cord injuries from those prerequisites which are unnecessary or are prerequisites that have to be further investigated.

2. Establishing a suitable niche

2.1. Dissolving the gliosis

Axonal regeneration following spinal cord injury is limited not only because central nervous system neurons have a poor intrinsic capacity for growth but also because injured axons encounter a series of inhibitory factors that are non-permissive for growth. These include myelin inhibitors [Nogo-A, MAG108 (myelin-associated glycoprotein) and OMgp109 (oligodendrocyte myelin glycoprotein)]; chondroitin sulfate proteoglycans (neurocan, versican, aggrecan, brevican, phosphacan and NG2); semaphorins and ephrins. In the central nervous system, laminin is replaced by netrins [12–15].

2.1.1. Chondroitinase ABC

Chondroitinase ABC [16–18] has improved recovery of function in synergy with mesenchymal stromal cells without [19] or with the addition of an acellular nerve allograft [20] or in synergy with brain-derived neurotrophic factor (BDNF) secreting mesenchymal stem cells [21]. Chondroitinase ABC should be thermostabilized with the sugar trehalose to reduce its temperature-dependent loss of activity [22]; it should be injected in high doses (50 or 100 IUs) [23–25], at multiple times [26–29] and be combined with cell transplantation and growth factor infusion [30, 31].

2.1.2. Other measures to overcome the gliosis

In a rat model of spinal cord contusion injury [32], infused *sialidase* has acted robustly throughout the spinal cord gray and white matter, whereas chondroitinase ABC activity has been more intense superficially, thus raising the possible consideration that it might be superior to chondroitinase ABC. Blocking myelin-associated inhibitors with *Nogo-A monoclonal antibodies or with Nogoreceptor competitive agonist peptide (NEP1-40)* has been shown to increase axonal regeneration [33]. Bone marrow mesenchymal stem cells with Nogo-66 receptor gene silencing have been used for repair of spinal cord injury [34]. *Blocking Rho-A with Rho inhibitor 'cethrin'* might overcome its effect; a synthetic membrane-permeable peptide mimetic of the protein tyrosine phosphatase σ, wedge domain can bind to tyrosine phosphatase σ and relieve chondroitin sulfate proteoglycan-mediated inhibition [35]. Chondroitin sulfate proteoglycans inhibition of phosphoinositide 3-kinase (PI3K) signaling is reversed by *cell permeable phosphopeptide (PI3Kpep)* [36]; *rolipram,* a phosphodiesterase4 inhibitor, can increase intracellular cAMP levels [33]; *taxol,* a microtubule-stabilizing agent, increases neurite outgrowth [37, 38].

2.1.3. Emerging role of heparin in lysing the gliosis

There is an emerging role of heparin in lysing of the gliosis, as reviewed elsewhere [39]. Both unfractionated and low molecular weight heparins have a fibrolytic (gliolytic) effect, can modulate astrocyte function and are used as lumen fillers. Astrocytes release a variety of trophic factors. These trophic factors include nerve growth factor, basic fibroblast growth factor, transforming growth factor-β, platelet-derived growth factor, brain-derived neurotrophic factor, ciliary neurotrophic factor and others. Astrocyte stress response and trophic effects are mediated by the fibroblastic growth factor family member, on which heparin exerts a profound influence [40–42].

2.2. Providing a suitable scaffold, both to bridge the gap and to harbor the cells

2.2.1. Biomaterial scaffolds in spinal cord injury

Biomaterial scaffolds in spinal cord injury have been reviewed elsewhere [43, 44]. Mesenchymal stromal cells have been grown onto *fibrin* scaffolds [45, 46]. The survival and neural differentiation of human bone marrow stromal cells have been tested on fibrin versus fibrin platelet-rich plasma scaffolds. The results have shown a clear superiority of platelet-rich plasma scaffolds, mainly after BDNF administration [47]. Mesenchymal stem cells have also been grown onto *collagen* scaffolds [48]. Rat adipose-derived stem cells have differentiated into olfactory ensheathing cell-like cells on collagen scaffolds by co-culturing with olfactory ensheathing cells [49]. *Acellular* spinal cord scaffolds [50, 51] and acellular muscle bioscaffolds [52] seeded with bone marrow stromal cells have promoted functional recovery in spinal cord-injured rats. Electro-acupuncture has been found to promote the survival and differentiation of transplanted bone marrow mesenchymal stem cells pre-induced with neurotrophin-3 and retinoic acid in *gelatin* sponge scaffold after rat spinal cord transaction [53]. Human bone marrow mesenchymal stem cells and endometrial stem cells have been found to differentiate better into motor neurons on electrospun *poly(ε-caprolactone)* scaffolds [54]. Nogo-66 receptor gene-silenced cells have been transplanted in a poly(D,L-lactic-co-glycolic acid) scaffold for the treatment of spinal cord injury [55]. Bone marrow mesenchymal stem cells seeded in *chitosan*-alginate scaffolds [56] and biodegradable chitin conduit tubulation combined with bone marrow mesenchymal stem cell transplantation have reduced glial scar and cavity formation in spinal cord injury [57]. In a comparative study investigating the efficacy of allogeneic mesenchymal stem cell transplantation via simple intralesional injection versus the use of a poly (lactic-co-glycolic acid) scaffold or a chitosan scaffold, higher mesenchymal stem cell engraftment rates have been reported in the scaffold groups, particularly, in the chitosan scaffold group [58].

Injectable extracellular matrix *hydrogels* have been used as scaffolds for spinal cord injury repair [59]. Matrix metalloproteinase-sensitive, hyaluronic acid-based biomimetic hydrogel scaffolds containing brain-derived neurotrophic factor have been implanted [60]. Cell-seeded alginate hydrogel scaffolds have promoted directed linear axonal regeneration in the injured rat spinal cord [61]. Multichannel polymer scaffolds fabricated from positively charged oligo[poly(ethylene glycol)fumarate] hydrogel and loaded with either syngeneic Schwann cells or mesenchymal stem cells derived from enhanced green fluorescent protein transgenic rats have been successfully implanted into rat spinal cords following T9 complete transection [62]. Highly superporous poly(2-hydroxyethyl methacrylate) scaffolds with oriented pores [63] and highly superporous cholesterol-modified poly(2-hydroxyethyl methacrylate) scaffolds have been developed for spinal cord injury repair [64].

Three-dimensional culture can mimic the stem cell niche compared to conventional two-dimensional culture. Bone marrow-derived mesenchymal stem cells cultured in three-dimensional collagen scaffold have exhibited distinctive features including significantly enhancing neurotrophic factor secretions and reducing macrophage activations challenged by lipopolysaccharide [65]. A polyhydroxybutaryl-hydroxyvinyl-based three-dimensional scaffold for a tissue engineering and cell-therapy combinatorial approach for spinal cord injury regeneration has been developed [66]. A three-dimensional biomimetic hydrogel has been implemented to deliver factors secreted by human mesenchymal stem cells in spinal cord injury [67]. Bone marrow mesenchymal stem cells in a three-dimensional gelatin sponge scaffold have attenuated inflammation, have promoted angiogenesis and have reduced cavity formation in experimental spinal cord injury [68].

2.2.2. Prerequisites for the use of biomaterial scaffolds in spinal cord injury

Biomaterial scaffolds should be biocompatible, non-toxic, chemically stable, of known absorption and degradation kinetics matching the degree of in vivo cell/tissue growth and should have adequate surface for cell access, proliferation and cell differentiation [69, 70]. They *should meet macroengineering requirements* being of proper form [71, 72], design (shape) [73] and size (diameter) [74]. They should be supplied with macrogrooves [43, 75, 76] and have a wall thickness of 0.6 mm, a porosity of 80% and a pore size range of 10–40 μm [77–79]. They *should meet microengineering requirements,* microgrooves directing axonal growth [80–87]. Prestretch-induced surface anisotropy has been beneficial in enhancing axon alignment, growth and myelination [88]. Also, filament inclusion has been more effective for bridging long nerve defect gaps [43, 89, 90]; Schwann cell migration over gaps exceeding 18 mm is superior in the presence of filaments. Yoshii et al. [91, 92] have tested collagen microfilaments with diameters of 20 μm to repair long gaps (20 or 30 mm) in the rat sciatic nerve. Increasing fiber number (4000 versus 2000 filaments) has enhanced nerve regeneration. Thus, increasing the whole filament surface area by increasing their number and reducing their diameter (increased surface area-to-volume ratio) is also critical [89, 93, 94].

Scaffolds should fulfill nearly the same mechanical conditions of the recipient spinal cord, exerting incremental tensile forces on intact cord segments to promote axonal regeneration while unloading gliotic segments to reduce gliosis and harbor cellular transplants (**Figure 1a and b**). A scaffold should

http://dx.doi.org/10.5772/intechopen.69554

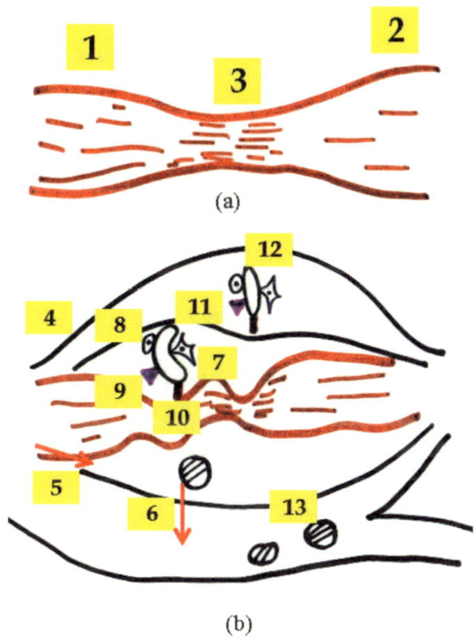

Figure 1. (a) How a spinal cord lesion looks like; (1) cranial spinal cord; (2) rostral spinal cord and (3) the gliotic segment. (b) A biomaterial scaffold (4) should fulfill nearly the same mechanical conditions of the recipient spinal cord, exerting incremental tensile forces (5—arrows) on intact cord segments to promote axonal regeneration while unloading gliotic segments (6—arrows) to reduce gliosis and harbor cellular transplants. In addition, it should meet macro- and microengineering requirements; it should provide adequate space for the interplay and manipulation of the different molecular pathways for axonal regeneration through lumen filling technology and it should meet requirements based on spatial distribution of neurotrophic factor gradients. Lumen filling technology allows for the incorporation and gradual local release of stem cells (7), accessory cells (8), molecular growth factors (e.g. BDNF, neurotrophin-3, etc.) (9) and neurolyzing agents (e.g. chondroitinase ABC) (10), either by combining them with a growth-supporting matrix in the lumen (11), by crosslinking (12) them to nerve conduit walls or by using microspheres (13) to deliver them. Growth-supporting matrices (11) in the lumen include hydrogel-forming collagen, fibrin, laminin, alginate, heparin and heparin sulfate. A natural and low-toxicity crosslinking agent (12), genipin, is commonly used.

possess sufficient toughness to resist compression or collapse, yet still be flexible and suturable [95]. A brittle scaffold that sustains little or no plastic deformation before fracture might break hampering axonal progression.

A scaffold should have an elastic modulus comparable with that of the recipient spinal cord. To approach appropriate mechanical properties, one strategy has been to form polymer composites with biopolymers such as chitosan [96], a polymer which has been established as being "softer" and biocompatible. The role of mechanical compliance in directing cell fate and function is a critical issue in material design [97–99]. A low elasticity and hierarchically aligned fibrillar fibrin hydrogel fabricated through electrospinning and concurrent molecular self-assembly process has been tested. Matrix stiffness and aligned topography have instructed stem cell neurogenic differentiation and rapid neurite outgrowth [100].

Scaffolds should provide adequate space for the interplay and manipulation of the different molecular pathways for axonal regeneration [80, 81, 101–103].

To provide adequate space and adherence for cells and molecules, biomaterial polymer nerve scaffolds should be porous [43]. Currently, ideal scaffolding should have 80–90% porosity with a pore size of 50–250 μm. Its pores should be interconnected so as to provide physical support to cells and guide their proliferation and differentiation, also facilitating neovascularization [69, 104]. The porous structure can be stabilized by adding glutaraldehyde, polyethylene glycol, heparin or collagen, allowing the structure to become more resistant and to maintain elasticity. A natural and low-toxicity cross-linking agent, genipin, has been used to immobilize nerve growth factor, a neurotrophic factor, onto chitosan-based neural scaffolds to generate a novel nerve graft, which has been beneficial for peripheral nerve repair [105]. A novel method has been introduced for standardized microcomputed tomography-guided evaluation of scaffold properties in bone and tissue research [106].

Scaffolds should provide adequate space for lumen fillers Methods of lumen filling allow for incorporation of cells and molecular factors either by combining them with a *growth-supporting matrix* in the lumen, by crosslinking them to nerve conduit walls or by using microspheres to deliver them [107]. *Growth-supporting matrices in the lumen* include hydrogel-forming collagen, fibrin, laminin, alginate, heparin, and heparin sulfate.

Scaffolds should meet requirements based on spatial distribution of neurotrophic factor gradients.

Spatial molecular concentration gradients of nerve growth factor [108] and laminin [43, 109, 110] promote axonal sprouting. Thus, axonal growth can be hypothetically made to bridge the whole length of the neural gap by seeding the scaffolds with multiple nerve growth factor/laminin spatial concentration gradients [111].

3. Optimizing the therapeutic effect of mesenchymal stem cell transplantation

3.1. The ideal source for mesenchymal stem cells

Mesenchymal stem cells reside not only in various tissues of mesenchymal origin (e.g. bone marrow, adipose tissue, skin and peripheral blood) but also in perinatal sources (e.g. umbilical cord blood, umbilical cord matrix or Wharton's jelly, amniotic fluid and placenta) [112].

In a comparative study using mesenchymal stem cells extracted from both bone marrow and adipose tissue for spinal cord injury, animals receiving adipose tissue cells have presented higher levels of tissue brain-derived neurotrophic factor, increased angiogenesis, higher number of preserved axons and a decrease in the number of macrophages, suggesting the superiority of mesenchymal stem cells extracted from adipose tissue [113]. In another study, however, no difference has been found between animals receiving mesenchymal stem cells derived from bone marrow or adipose tissue, whether in terms of axonal regeneration, neuroprotection or functional recovery [114].

Mesenchymal stem cells obtained from perinatal sources can proliferate more rapidly and extensively than adult mesenchymal stem cells and are easily obtained after normal and cesarean births, with low risk of viral contamination. They may be used for allogenic transplantation because they act by suppressing immune response and are, therefore, considered non-immunogenic cells [112].

In a study comparing mesenchymal stem cells derived from fat, bone marrow, Wharton's jelly and umbilical cord blood for treating spinal cord injuries, dogs have been treated with only matrigel or matrigel mixed with each type of mesenchymal stem cells. Although there have been no significant differences in functional recovery among the mesenchymal stem cell groups, application of umbilical cord stem cells has led to more nerve regeneration, neuroprotection and less inflammation compared to other mesenchymal stem cells [115].

Central nervous system pericytes (perivascular stromal cells) have recently gained significant attention. These cells not only display a mesenchymal stem cell phenotype in vitro but also have similar in vivo immunomodulatory effects after spinal cord injury that are more potent than those of non-central nervous system tissue-derived cells [116].

3.2. Increasing the efficiency of mesenchymal stem cells and their influence on spinal cord regeneration

3.2.1. *Influence of mesenchymal stem cells on spinal cord regeneration in general*

Present around blood vessels, mesenchymal stem cells respond more readily to tissue damage [3]. The transdifferentiation capacity of mesenchymal stem cells into neuronal and glial lineages has been debated; transplanted mesenchymal stem cells do not differentiate into a neuronal fate, even if they display weak expression of NeuN (a neuronal marker) [3]. Mesenchymal stem cell-based cell therapy, even when applied during the chronic phase of spinal cord injury, leads to changes in a number of structural and functional parameters, all of which indicate improved recovery [117]. Mesenchymal stem cells promote repair in the injured cord by secreting growth factors that overcome the inhibitory environment of the lesion. These cells have anti-inflammatory, immunomodulatory, vascular promoting oxidative stress reducing and neuroprotective effects. They can secrete trophic factors thus exerting a paracrine effect that can stimulate axon regeneration contributing to functional recovery enhancement [112, 118]. Human mesenchymal stem/stromal cells suppress spinal inflammation in mice with contribution of pituitary adenylate cyclase-activating polypeptide [119]. Intrathecal transplantation of mesenchymal stem cells activates extracellular adjusting protein kinase1 and 2 in the spinal cord following ischemia reperfusion injury, partially improving spinal cord function and inhibiting apoptosis in rats [120].

Measures to increase the *efficiency of mesenchymal stem cells* include the following. *Replacing fetal bovine serum* has been proposed as a gold standard for human cell propagation [121]. *Mechanical fibrinogen-depletion* has been found to support heparin-free mesenchymal stem cell propagation in human platelet lysate [122]. A combination of *electroacupuncture* and grafted mesenchymal stem cells overexpressing tyrosine kinase C has been found to improve remyelination and function in demyelinated spinal cord of rats [123]. Arginine decarboxylase is a rate-limiting enzyme of agmatine synthesis and is known to exist in the central nervous system of mammals. Arginine decarboxylase-secreting human mesenchymal stem cells have been found to be more suitable candidates than human mesenchymal stem cell for stem cell therapy after spinal cord injury [124]. Heme oxygenase-1 is a stress-responsive enzyme that modulates immune response and oxidative stress associated with spinal cord injury. Functional recovery after spinal cord injury has been promoted by transplantation of mesenchymal stem cells overexpressing heme oxygenase-1 [125]. *Hypothermia* is known to improve the microenvironment of the injured spinal cord in a number of ways. Neural cell transplantation has promoted the recovery of hind limb function in rats, and a combination treatment with hypothermia has produced synergistic effects [126]. *Extracorporeal shock wave* can introduce alteration of microenvironment in cell therapy for chronic spinal cord injury [127].

3.2.2. Peculiarities of bone marrow stromal cells in spinal cord regeneration

Bone marrow stromal cell transplantation has been shown to overcome the gliosis [3]. They have been reported to enhance neuronal protection and cellular preservation *via* reduction in injury-induced sensitivity to mechanical trauma. They can attenuate astrocyte reactivity and chronic microglia/macrophage activation. They have been found to infiltrate primarily into the ventrolateral white matter tracts, spreading to adjacent segments rostrocaudal to the injury epicenter. However, bone marrow stromal cell transplantation present certain issues. Migration beyond the injection site after intraspinal delivery is limited and inter-donor variability in efficacy and immunomodulatory potency might affect clinical outcome [4].

Measures to increase the *efficiency of bone marrow mesenchymal stem cells* include mainly measures to reduce oxidative stress-induced apoptosis, hypoxic preconditioning, measures to modulate the extracellular matrix and other measures.

Studies have demonstrated that the inhibition of the Notch1 pathway in bone marrow mesenchymal stem cells contributes to the differentiation of these cells. Research findings that certain antioxidants induce bone marrow mesenchymal stem cells to differentiate into neuronal cells suggest that bone marrow mesenchymal stem cell differentiation is related to the level of reactive oxygen species in cells. After bone marrow mesenchymal stem cell induction with the antioxidant β-mercaptoethanol, Western blotting and immunofluorescence have revealed gradual increases in the expression of Nestin (a neural stem cell-specific protein) and neuron-specific enolase but decreases in Notch1 expression. The decreased expression levels of Notch1 have correlated positively with changes in reactive oxygen species [128]. The effects of a calpain inhibitor (MDL28170) on increasing survival of bone marrow mesenchymal stem cells transplanted into the injured rat spinal cord have been investigated. The protective effects of MDL28170 on survival of bone marrow mesenchymal stem cells have inhibited the activation of calpain and stress-induced apoptosis [129]. Treatment with bone marrow mesenchymal stem cells combined with plumbagin may alleviate spinal cord injury by affecting oxidative stress, inflammation, apoptosis and the activation of the Nrf2 pathway [130]. Polydatin, a glucoside of resveratrol, has been reported to possess potent antioxidative effects and can used in combination with bone marrow mesenchymal stem cell for the treatment of spinal cord injury. Polydatin significantly protects bone marrow mesenchymal stem cell against apoptosis due to its antioxidative effects and the regulation of Nrf 2/ARE pathway [131]. Carvedilol, a nonselective β-adrenergic receptor blocker, has been reported to exert potent anti-oxidative activities. It has been shown that carvedilol protects cell death of H_2O_2-induced bone marrow mesenchymal stem cells partly through PI3K-Akt pathway, suggesting its use in combination with bone marrow mesenchymal stem cells to improve cell survival in oxidative stress microenvironments [132].

Hypoxic preconditioning effectively increases the survival rate of bone marrow mesenchymal stem cells following transplantation and increases their protective effect on injured tissues. Hypoxic preconditioning has upregulated the expression of hypoxia-inducible factor 1α in spinal cord tissues [133].

Cytokines and extracellular matrix can trigger various types of neural differentiation. To highlight the current understanding of their effects on neural differentiation of human bone marrow-derived multipotent progenitor cells, extracellular matrix proteins, tenascin-cytotactin, tenascin-restrictin and chondroitin sulfate, with the cytokines, nerve growth factor/brain-derived neurotrophic factor/retinoic acid, have been incorporated to induce transdifferentiation of human bone marrow-derived multipotent progenitor cells. Greater amounts of neuronal morphology have appeared in cultures incorporated with tenascin-cytotactin and tenascin-restrictin than those with chondroitin sulfate. It has been suggested that the combined use of tenascin-cytotactin, nerve growth factor /brain-derived neurotrophic factor/retinoic acid and human bone marrow-derived multipotent progenitor cells offers a new feasible method for nerve repair [134]. Fibronectin secreted by mesenchymal stem cells in the early stage has been found to accumulate on gelatin sponge scaffolds and promote neurite elongation of neuronal differentiating mesenchymal stem cells as well as nerve fiber regeneration after spinal cord injury [135].

Transplanted bone mesenchymal stem cells can be mobilized by erythropoietin toward lesion sites following spinal cord injury [136]. Propofol injection combined with bone marrow mesenchymal stem cell transplantation has improved electrophysiological function in the hindlimb of rats with spinal cord injury than monotherapy [137]. Combining bone marrow stromal cells with green tea polyphenols has attenuated the blood-spinal cord barrier permeability in rats with compression spinal cord injury [138]. Bone marrow stromal cells transplantation combined with ultrashortwave therapy has promoted functional recovery in spinal cord injury in rats [139].

Microtubule-associated protein 1B plays an important role in axon guidance and neuronal migration. Phosphatidylinositol 3-kinase and extracellular signal-regulated kinase 1/2 in bone marrow mesenchymal stem cells have been found to modulate the phosphorylation of microtubule-associated protein 1B via a cross-signaling network and have affected the migratory efficiency of bone marrow mesenchymal stem cells towards injured spinal cord [140]. Administration of valproic acid potentiates the therapeutic effect of mesenchymal stem cell therapy [141]. Interleukin-8 enhances the angiogenic potential of human bone marrow mesenchymal stem cells by increasing vascular endothelial growth factor production [142].

3.2.3. *Peculiarities of adipose-derived stem cells in spinal cord regeneration*

Human mesenchymal cells from adipose tissue have deposited laminin and have promoted regeneration of injured spinal cord in rats [143–146]. Transplanted during the acute and subacute phases after spinal cord injury, they have enabled the remodulation and regeneration of the lesion site, decreasing the importance of transplantation time in the treatment of spinal cord injury [145]. Chondroitinase ABC-adipose-derived stem cells constructed using lentiviral vector transfection have stably expressed chondroitinase ABC, and chondroitinase ABC expression has significantly enhanced their migratory capacity [146]. Cytoplasmic extracts prepared from adipose tissue stromal cells have *inhibited H_2O_2-mediated apoptosis* of cultured spinal cord-derived neural progenitor cells and have improved cell survival. *Predifferentiation* of adipose tissue-derived stromal cells has promoted the protection of denuded axons and cellular repair. Such predifferentiated cells and hematopoietic stem cells have been successfully infused intrathecally [143]. Nevertheless, no evidence points to the superiority of neural differentiated adipose tissue-derived stromal over undifferentiated ones. *Allogenic* adipose-derived stem cells have improved neurological function in a canine model. All of the former evidence, however, is contradicted by a study in a rat C3–C4 hemisection in which adipose tissue-derived stromal cell transplantation has significantly reduced sprouting of the descending serotonergic fibers at the injured site [147].

Hypoxic preconditioning of adipose tissue-derived mesenchymal stem cells has increased their survival. Cotransplantation of such cells with engineered neural stem cells has improved both cell survival and gene expression of the engineered neural stem cells [4].

3.2.4. Peculiarities of human umbilical cord blood-derived mesenchymal stem cells in spinal cord regeneration

Human umbilical cord blood-derived mesenchymal stem cells (whether Wharton's jelly mesenchymal stem cells or human umbilical cord perivascular cells) may reverse spinal cord injury pathophysiology by *downregulating apoptotic genes and secreting neurotrophic factors* in few days; they may *transdifferentiate* toward neuronal and oligodendroglial phenotypes [3]. Intrathecal transplantation of human amniotic mesenchymal stem cells has promoted functional recovery in a rat model of traumatic spinal cord injury [148] and in a chronic constrictive nerve injury model [149]. Placental mesenchymal stromal cells have rescued ambulation in ovine myelomeningocele [150]. Umbilical cord-derived mesenchymal stem cell therapy for neurological disorders may act via inhibition of mitogen-activated protein kinase pathway-mediated apoptosis [115]. Through the effect on glial cells(suppression of activated astrocytes and microglia), proinflammatory (Interleukin-1β and Interleukin-17A) and anti-inflammatory cytokines (anti-inflammatory cytokine Interleukin-10), intrathecal injection of human umbilical cord-derived mesenchymal stem cells has ameliorated neuropathic pain in rats [151]. Also, neurotrophic factors have been expressed in the injured spinal cord after transplantation of human-umbilical cord blood stem cells in rats [152].

Preconditioning of umbilical cord mesenchymal stem cells in physioxic environment can enhance the regenerative properties of these cells in the treatment of rat spinal cord injury. In a study on umbilical cord, mesenchymal stem cells pretreated with either atmospheric normoxia (21% O_2) or physioxia (5% O_2) have grown faster, whereas physioxia has upregulated the expression of trophic and growth factors, including hepatocyte growth factor, brain-derived neurotrophic factor and vascular endothelial growth factor. This has been associated with a significant increase in axonal preservation and a decrease in the number of caspase-3+ cells and ED-1+ macrophages [153].

Calcitonin gene-related peptide, a neural peptide synthesized in spinal cord, contributes to homing of human umbilical cord mesenchymal stem cells. The PI3K/Akt and p38MAPK signaling pathways have played a critical role in the calcitonin gene-related peptide-induced chemotactic migration of human umbilical mesenchymal stem cells [154].

Lavandula angustifolia has neuroprotective effects; it has potentiated the functional and cellular recovery with human umbilical mesenchymal stem cell treatment in rats after spinal cord injury [155]. The combined treatment with methylprednisolone and amniotic membrane mesenchymal stem cells after spinal cord injury in rats has potentiated the anti-inflammatory and anti-apoptotic effect of mesenchymal stem cell transplantation [156]. The neuroprotective effects of conditioned medium from cultured human CD34(+) cells have been similar to those of human CD34(+) cells and the conditioned medium has been found to enhance the neuroprotective effects of 17β-estradiol in rat spinal cord injury [157].

3.3. Inducing the transformation of mesenchymal stem cells into motor neuron-like cells or Schwann cells

A third method for optimizing the therapeutic effect of mesenchymal stem cell transplantation is inducing their transformation into motor neuron-like cells or Schwann cells [158–169]. Their differentiation into *motor neuron-like cells* has been induced through a pre-induction step using β-mercaptoethanol followed by 4 days of induction with retinoic acid and sonic hedgehog [158]. Motor neuron axonal sprouting has been induced by adding different concentrations of a nerve growth factor to the differentiation media. In another study [159], such cells have been tested for 2′,3′-cyclic-nucleotide-3′-phosphodiesterase and microtubule-associated protein 2, as well as to glial fibrillary acidic protein and beta III tubulin. Cells have been injected percutaneously into the spinal cord of paraplegic dogs for two times separated by a 21-day interval. *Optimal culture conditions* have been investigated as to the production of neural cells and neural stem cells [160]. β-Mercaptoethanol has been used as the main inducer of the neurogenesis pathway. Three types of neural markers have been used: nestin as the immaturation stage marker, neurofilament light chain as the early neural marker, and microtubule-associated protein 2 as the maturation marker. Results have shown that the best exposure time for the production of neural stem cells is 6 hours. It has also been demonstrated that LY294002, a small molecule inhibitor of phosphatidylinositol 3-kinase (PI3K)/Akt signal pathway, can promote neuronal differentiation of mesenchymal stem cells cultured on polycaprolactone/collagen scaffolds [161]. Similarly, microRNA-124 has promoted bone marrow mesenchymal stem cell differentiation into neurogenic cells for accelerating recovery in the spinal cord injury [166, 169]. *Such induced motor neuron-like cells* have promoted axonal regeneration into the injured spinal cord, whether derived from bone marrow [162, 163, 168], human chorion [164] and placenta [167]. Their in vivo tracking by magnetic resonance has been possible in rabbit models of spinal cord injury [169].

3.4. Mode, quantity and number of injections; time point for injection age and donor variation; allo- and xenotransplantation

The mode, quantity and number of injections may influence the therapeutic effect of mesenchymal stem cell transplantation

3.4.1. *Mode of injection*

All methods for stem cell transplantation (intravenous, intrathecal, intramedullary, intranasal or skeletal muscle injection) are based on the homing effect, the ability of implanted stem cells to move to the injured area [170–180]. Mesenchymal progenitor cells have been injected *intravenously* in two models of cervical spinal cord injury, unilateral C5 contusion and complete unilateral C5 hemisection. Cells have been isolated from green fluorescence protein-luciferase transgenic mice and have been injected via the tail vein at D1, D3, D7, D10, or D14. Transplanted cells have been tracked via postmortem bioluminescence imaging. Cells have been found to accumulate in the lungs, irrespective of the time of injection or injury model. It has been proposed that they modulate the immune system via the lungs through secreted immune mediators [173]. The antioxidant and anti-inflammatory effects of intravenously injected adipose-derived mesenchymal stem cells have been proven in dogs with acute spinal cord injury [174]. Diffuse and persistent blood-spinal cord barrier disruption after contusive spinal cord injury has recovered following intravenous infusion of bone marrow mesenchymal stem cells [177]. Intravenous mesenchymal stem cell therapy has been effective after recurrent laryngeal nerve injury [179]. In a meta-analysis, the efficacy of intravenous bone marrow mesenchymal stem cell transplantation in spinal cord injury has been investigated. It has been concluded that the therapeutic window of intravenous bone marrow mesenchymal stem cell transplantation is wide [180]. The feasibility and safety of *intrathecal* transplantation of autologous bone marrow mesenchymal stem cells have been investigated in horses [175]. The *intranasal* delivery of bone marrow stromal cells to spinal cord lesions has been successfully tried out [176]. Stem cell injection in the *hindlimb* skeletal muscle has enhanced neurorepair in mice with spinal cord injury [178].

Although intrathecal is more effective than intravenous injection, it needs large stem cell numbers. Subarachnoid adhesions may prevent the cells from reaching the target site. The homing effect is absent in the chronic stage of spinal cord injury. Therefore, direct intramedullary injection into the injured site is the most effective method for delivering stem cells. Intramedullary injection proximal to the injured area is ideal for stem cell survival, but is hampered by volume effects caused by high tissue pressure and subsequent normal spinal cord damage. On the contrary, large volumes can be injected into the cavity area at the injured level. Injecting into the contused cavity may lead to resolution of the glial scar and may bridge for axonal regeneration. Therefore, Park et al. [171, 172] have injected into both the normal proximal spinal cord and the injured area. In addition, subdural stem cells have been applied in the hope the homing effect has been reinduced because of intramedullary injection.

3.4.2. *Quantity, number and time point for mesenchymal stem cell transplantation*

3.4.2.1. *Quantity and number*

Diversity of lesion models, animal types and route of cell administration influence the quantity of mesenchymal stem cells administered. Cell survival and enhancement in locomotor performance have been observed both after intravenous injection of one million cells in a volume of 0.5 mL of DMEM in a model of balloon compressive injury in rats and after transplantation of 600,000 cells in a volume of 6 μL directly into the injury site after contusive injury in rats [112]. Other studies have advocated intrathecal administration from 100×10^6 up to 230×10^6 cells followed by an additional 30×10^6 cell administration at 3 months [5], or the administration of two or three intrathecal injections with a median of 1.2×10^6 mesenchymal stem cells/kg body weight [6]. In a phase III clinical trial, limited efficacy has been proven after injecting 1.6 × 10 autologous mesenchymal stem cells into the intramedullary area at the injured level and 3.2 × 10 autologous mesenchymal stem cells into the subdural space. Single mesenchymal stem cell application to intramedullary and intradural space has had a very weak therapeutic effect compared to multiple injections [7]; partial efficacy has been demonstrated in other trials [8–11]. Continuous improvement after multiple mesenchymal stem cell transplantations has been observed in a patient with complete spinal cord injury [181]. Multiple injections of human umbilical cord-derived mesenchymal stromal cells through the tail vein have improved microcirculation and the microenvironment in a rat model of radiation myelopathy [182].

3.4.2.2. Time point

Acute phase is defined as the first three days after spinal cord injury and chronic phase is defined as more than 12 months after spinal cord injury. Subacute phase is defined as the period between acute and chronic phase. In the acute phase, reactive oxygen-free radicals, excitatory transmitters, inflammatory molecules and hypoxia caused by hypoperfusion are cytotoxic to implanted stem cells. In the chronic phase, glial scar tissue acts as a physical barrier to axonal regrowth. Thus, it is difficult for implanted stem cells to survive in chronic spinal cord injury. In contrast, in the subacute phase, the inflammatory response is reduced and the glial scar formation has not formed. Therefore, the subacute phase seems to be an optimal phase in the respect of timing of stem cell application [170]. Experimentally, bone marrow-derived stem cells have been infused intravenously 10 weeks after spinal cord injury [183].

3.4.3. Age and donor variation, allo- and xenotransplantation

3.4.3.1. Age and donor variation

The potency of mesenchymal stem cells exhibits significant age and donor variation [3, 184–186]. A robust potency assay has been established based on pooling responder leukocytes to minimize individual immune response variability. It has highlighted significant donor variation of human mesenchymal stem/progenitor cell immune modulatory capacity and extended radioresistance [184, 185].

3.4.3.2. Allo- and xenotransplantation

The neuroprotective and immunomodulatory effects of *xenotransplantation* of adipose tissue mesenchymal stem cells in Lewis rats after lumbar ventral root avulsion have been proven [187]. The therapeutic effects of autologous and *allogenic* bone marrow-derived mesenchymal stem cell transplantation have been established in canine spinal cord injury [188]. Immunosuppression of allogenic mesenchymal stem cells transplantation after spinal cord injury may improve graft survival [189].

3.4.4. Evaluating the therapeutic effect of mesenchymal stem cell transplantation

Although neurological evaluation of the spinal cord injured patient is usually conducted according to the International Standards for Neurological Classification of Spinal Cord Injury recommended by the American Spinal Cord Injury Association, it should be confirmed by electrophysiological studies (somatosensory evoked potentials and motor evoked potentials) and magnetic resonance imaging studies. Magnetic resonance imaging findings after stem cell therapy include widening of cord diameter, blurring of intramedullary cavity margin and appearance of fiber-like streak pattern in the injured spinal cord. Diffusion tensor imaging can perform accurate visualization and assessment of white matter tracts and is useful for the prediction of neurological recovery in spinal cord injury patients. Fiber continuity on diffusion tensor imaging not seen before stem cell therapy may be an indicator of axonal regeneration in stem cell therapy. Cell labeling techniques for *in vivo* visualization using biological indicators or contrast agents have helped monitoring the status of the transplanted stem cells in the body (survival, migration and exact location of implanted stem cells). Typical examples are supermagnetic iron oxide particle monitoring using magnetic resonance imaging and radionuclide monitoring using positron emission tomography or single-photon emission computed tomography [170, 190, 191].

4. Supplying neurotrophic factors and accessory cells

A combinatorial approach has been agreed upon for effective treatment of spinal cord injury [192–208].

The combination of *neurotrophic factors* such as BDNF and neurotrophin-3 has enhanced axonal regeneration and myelination [193]. Cyclic adenosine monophosphate (a neuronal stimulator) and neurotrophin-3 (neurotrophic factor) have been injected 5 days prior to a C4 transection at L4 to precondition the dorsal root ganglion soma. Bone marrow mesenchymal stem cells have been transplanted 7 days post injury. The effect of bone marrow mesenchymal stem cells on spinal cord regeneration has been augmented by modifying them to either express human brain-derived neurotrophic factor (BDNF) in an acute injury or neurotrophin-3 in a chronic injury model, by prestimulating them to secrete neurotrophic factors, e.g. by pretreating them with Schwann cell differentiating factors [3]. In an attempt to generate mesenchymal-derived differentiated neural cells expressing nerve growth factor or neurotrophin-3, mesenchymal stem cells have been infected with recombinant lentiviruses that express nerve growth factor both to induce their neural lineage genes and as a combinatorial approach [194]. Magnetic targeting of neurotrophin-3 gene-transfected bone marrow mesenchymal stem cells via lumbar puncture has enhanced their delivery to the site of injury and has significantly improved functional recovery and nerve regeneration compared to transplanting neurotrophin-3 gene-transfected bone marrow mesenchymal stem cells without magnetic targeting system [195, 196]. Pulsed electromagnetic field exposure near the injured site and for 8 hours per day over 4 weeks has been suggested as a suitable protocol for directing the cells to the site of injury [197]. Electro-acupuncture has promoted the survival and differentiation of transplanted bone marrow mesenchymal stem cells pre-induced with neurotrophin-3 and retinoic acid in gelatin sponge scaffold after rat spinal cord transection [53, 198].

A combination of other trophic factors, including epidermal growth factor, fibroblast growth factor type 2 and platelet-derived growth factor have enhanced the survival of implanted cells. Likewise has been the addition of granulocyte macrophage-colony stimulating factor [4, 170]. Co-transplantation of bone marrow-derived mesenchymal stem cells and nanospheres containing FGF-2 has improved cell survival and neurological function in the injured rat spinal cord [199]. Human ciliary neurotrophic factor overexpressing stable bone marrow stromal cells have proved effective in a rat model of traumatic spinal cord injury [200]. Bone marrow mesenchymal stem cells combined with minocycline have improved spinal cord injury in a rat model [201]. Propofol has enhanced the therapeutic effect of bone marrow mesenchymal stem cell transplantation on spinal cord injury in rats [202].

The addition of accessory cells includes combining mesenchymal stem cells with neural progenitor cells [3], neural crest stem cells [203], olfactory ensheathing cells [204, 205] or Schwann cells [207, 208]. The effects of mesenchymal stem cell and olfactory ensheathing cell transplantation at early or delayed time after a spinal cord contusion injury in the rat have been compared. Mesenchymal stem cell grafting seems a better option than olfactory ensheathing cell grafting [206].

5. Establishing a continuous drug and cell delivery system

In spinal cord injury, the gap is usually extensive and associated with excessive scarring. The axonal growth cone would thus take years to reach the distal spinal cord. Consequently, the factors mentioned before have to be replenished continually.

This can take place through an intrathecal (possibly extradural) continuous cell and drug delivery system (catheter) [39, 209]. Catheter-related complications include tension headache, meningitis, fibrous track formation, catheter slippage, difficult catheter insertion and catheter blockage. Microsphere, nanosphere and nanoshell technology may help keep the catheter patent, dissolve fibrosis and replenish molecules and cells [43, 210–215]. Co-transplantation of bone marrow-derived mesenchymal stem cells and nanospheres containing FGF-2 has improved cell survival and neurological function in the injured rat spinal cord [199]. Controlling surface tension as well as hydrophobic and hydrophilic properties of the conduit lumen and the microspheres may help us fulfill the three aims described previously. One method to achieve the latter aim is using magnetic nanoparticle-incorporated human bone marrow-derived mesenchymal stem cells exposed to pulsed electromagnetic fields [190, 191, 197] (**Figure 2**).

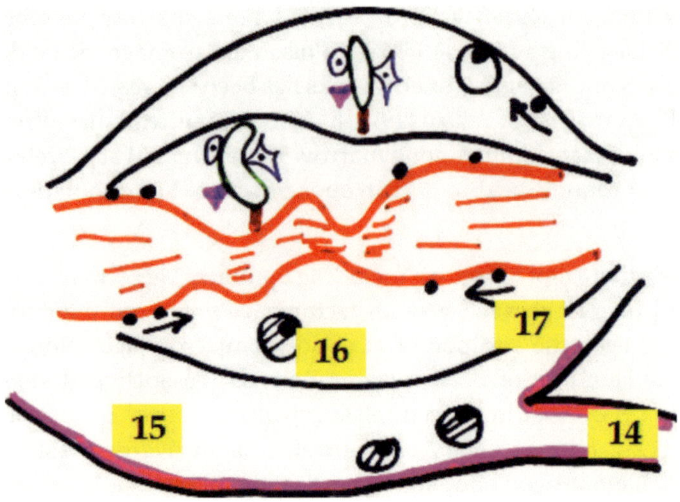

Figure 2. An intrathecal continuous cell and drug delivery system (catheter) (14) allows for the replenishment of stem cells, accessory cells, molecular growth factors and neurolyzing agents. To avoid catheter-related complications, it had better be lined with a biomaterial used for vascular grafts (15). Hydrophobic microsphere, nanosphere and nanoshell technology may also help keep the catheter patent, dissolve fibrosis and replenish molecules and cells. Magnetic nanoparticles (16) incorporated into microspheres may help guide the latter to the gliotic segment. After their release from microspheres, magnetic nanoparticles may be made to attach to the scaffold and to the intact cord segments and to apply tension on them (17—arrows), thus promoting axonal regeneration and enhancing engraftment and differentiation of transplanted cells.

6. Conclusion

We have attempted to identify the prerequisites for effective mesenchymal stem cell transplantation in spinal cord injuries. These fall into three categories (**Table 1**). The first category comprises those prerequisites, on which the literature is united. Research workers are thus obliged to follow them or provide a reasonable explanation for having not followed them.

The literature is unanimous on the following: (1) the gliosis has to be dissolved prior to mesenchymal stem cell transplantation (e.g. through chondroitinase ABC in high doses (50 or 100 IUs) and at multiple times); (2) a suitable scaffold has to be used; this scaffold should meet both macro- and microengineering requirements and should provide adequate space for lumen fillers; (3) the efficiency of mesenchymal stem cells themselves has to be increased (by reducing oxidative stress-induced apoptosis, by hypoxic preconditioning, by modulating the extracellular matrix and by other measures); (4) a combinatorial approach including growth factors, cellular transplants and neurolyzing agents has to be followed.

There are many issues, however, on which the literature is still not united. These fall into the second category. Among others, they include (1) the ideal source for mesenchymal stem cells, mode, quantity, time point and number of injections; (2) which growth factors and cells to be used in the combinatorial approach; (3) optimizing the therapeutic effect of mesenchymal stem cell transplantation by inducing their transformation into motor neuron-like cells or Schwann cells; (4) increasing the homing effect of stem cells (by calcitonin gene-related peptide). In the third category, more research has to be stimulated, e.g. as to how to establish a continuous drug and cell delivery system.

1. Establishing a suitable niche

1.1. Dissolving the gliosis

Category I (prerequisites, on which the literature is united)

Chondroitinase ABC in high doses (50 or 100 IUs) and at multiple times (at 0, 1, 2 and 4 weeks)

Category II (prerequisites, on which the literature is still not united)

- Heparins, sialidase
- Blocking myelin-associated inhibitors with Nogo-A monoclonal antibodies or with Nogoreceptor competitive agonist peptide (NEP1-40)
- Blocking Rho-A with Rho inhibitor 'cethrin'
- A synthetic membrane-permeable peptide mimetic of the protein tyrosine phosphatase σ can bind to protein tyrosine phosphatase σ and relieve proteoglycan-mediated inhibition
- Cell permeable phosphopeptide (PI3Kpep) reverses proteoglycans inhibition of phosphoinositide 3-kinase signaling in axons.
- Rolipram, a phosphodiesterase4 inhibitor, can increase intracellular cAMP levels
- Improving blood vessel formation might reduce cell death and promote angiogenesis within the injury zone
- Taxol, a microtubule-stabilizing agent, increases neurite outgrowth

1.2. Providing a suitable scaffold, both to bridge the gap and to harbor the cells

Category I (prerequisites, on which the literature is united)

- Scaffolds should meet macro- and microengineering requirements
- Scaffolds should fulfill the same mechanical conditions of the recipient spinal cord
- Scaffolds should provide adequate space for the different molecular pathways for axonal regeneration; they should be of ideal porosity
- Scaffolds should provide adequate space for lumen fillers
- Scaffolds should meet requirements based on spatial distribution of neurotrophic factor gradients

2. Optimizing the therapeutic effect of mesenchymal stem cell transplantation

2.1. The ideal source for mesenchymal stem cells

Category II (prerequisites, on which the literature is still not united)

Compared to stem cells of other mesenchymal origin (e.g. bone marrow, adipose tissue, skin), umbilical cord stem cells are superior

2.2. Increasing the efficiency of mesenchymal stem cells

Category I (prerequisites, on which the literature is united)

- Reducing oxidative stress-induced apoptosis
- Hypoxic preconditioning
- Modulating the extracellular matrix

Category II (prerequisites, on which the literature is still not united)

- Measures to reduce oxidative stress-induced apoptosis (arginine decarboxylase expressing cells; heme oxygenase-1 expressing cells; calpain inhibitor MDL28170; plumbagin; polydatin, a glucoside of resveratrol; carvedilol, a nonselective β-adrenergic receptor blocker)
- Measures during stem cell culture (replacing fetal bovine serum, mechanical fibrinogen-depletion)
- Measures during grafting (electroacupuncture, hypothermia, extracorporeal shock wave, propofol, green tea polyphenols, ultrashortwave therapy, valproic acid, IL-8)
- Measures increasing the homing effect and mobilization of stem cells (calcitonin gene-related peptide, erythropoietin)

2.3. Inducing the transformation of mesenchymal stem cells into motor neuron-like cells or Schwann cells

Category II (prerequisites, on which the literature is still not united)

2.4. Mode, quantity and number of injections; time point for injection; age and donor variation; allo- and xenotransplantation

Category I (prerequisites, on which the literature is united): intramedullary injection; injection during the subacute phase

Category II (prerequisites, on which the literature is still not united): all other issues

3. Supplying neurotrophic factors and accessory cells

Category I (prerequisites, on which the literature is united)

A combinatorial approach, including growth factors, cellular transplants and neurolyzing agents, has to be followed

Category II (prerequisites, on which the literature is still not united)

Which growth factors (epidermal growth factor, fibroblast growth factor type 2, platelet-derived growth factor, riluzole, minocycline, granulocyte-colony stimulating factor, BDNF, neurotrophin-3) and cells (embryonic stem cells, neural stem cells, induced pluripotent stem cells, neural crest stem cells, mesenchymal stromal cells, Schwann cells, olfactory ensheathing cells or macrophages) to be used in combination

4. Establishing a continuous drug and cell delivery system

Category III (prerequisites defective in the literature)

Table 1. Prerequisites for effective mesenchymal stem cell transplantation in spinal cord injuries.

List of abbreviations

Akt	Protein kinase B (PKB), a serine/threonine-specific protein kinase
BDNF	Brain-derived neurotrophic factor
cAMP	Cyclic adenosine monophosphate
DMEM	Dulbecco's Modified Eagle Medium ED-1+ macrophages: antibody against cellular marker CD68 macrophages
FGF-2	Fibroblast growth factor type 2
LY294002	Morpholine-containing chemical compound that is a potent inhibitor of numerous proteins, and a strong inhibitor of phosphoinositide 3-kinases
MAG108	Myelin-associated glycoprotein
MDL28170	Calpain inhibitor III
NEP1-40	Nogoreceptor competitive agonist peptide
NeuN	Feminizing locus on X-3, Fox-3, Rbfox3, or hexaribonucleotide binding protein-3
NG2	Neural/glial antigen 2
Nogo-A	Reticulon-4, neurite outgrowth inhibitor
Nrf 2	Nuclear factor (erythroid-derived 2)-like 2, also known as NFE2L2
Nrf 2/ARE pathway	The transcription factor Nrf2 (NF-E2-related factor 2) binds to the ARE, a cis-acting element called the antioxidant responsive element
OMgp109	Oligodendrocyte myelin glycoprotein
PI3K	Phosphatidylinositol 3-kinase
PI3Kpep	Cell permeable phosphopeptide: p38MAPK P38 mitogen-activated protein kinases
Rho-A	ras homolog gene family, member A

Author details

Sherif M. Amr

Address all correspondence to: sherifamrh@hotmail.com

The Department of Orthopaedics and Traumatology, Cairo University, Egypt

References

[1] Assinck P, Duncan GJ, Hilton BJ, Plemel JR, Tetzlaff W. Cell transplantation therapy for spinal cord injury. Nature Neuroscience. 2017;**20**(5):637-647

[2] Harrop JS, Hashimoto R, Norvell D, Raich A, Aarabi B, Grossman RG, Guest JD, Tator CH, Chapman J, Fehlings MG. Evaluation of clinical experience using cell-based therapies in patients with spinal cord injury: A systematic review. Journal of Neurosurgery: Spine. 2012 Sep;**17**(1 Suppl):230-246. DOI: 10.3171/2012.5.AOSPINE12115.Review. PubMed PMID: 22985383

[3] Doulames VM, Marquardt LM, Jayaram B, Plant CD and Plant GW. Stem Cell Therapies for Cervical Spinal Cord Injury, Recovery of Motor Function Following Spinal Cord Injury. Heidi F, editor, InTech; 2016. DOI: 10.5772/63580. Available from: http://www.intechopen.com/books/recovery-of-motor-function-following-spinal-cord-injury/stem-cell-therapies-for-cervical-spinal-cord-injury

[4] Dasari VR, Veeravalli KK, Dinh DH. Mesenchymal stem cells in the treatment of spinal cord injuries: A review. World Journal of Stem Cells. 2014 Apr 26;**6**(2):120-133. DOI: 10.4252/wjsc.v6.i2.120. Review. PubMed PMID: 24772239; PubMed Central PMCID: PMC3999770

[5] Vaquero J, Zurita M, Rico MA, Bonilla C, Aguayo C, Montilla J, Bustamante S, Carballido J, Marin E, Martinez F, Parajon A, Fernandez C, Reina LD, Neurological Cell Therapy Group. An approach to personalized cell therapy in chronic complete paraplegia: The Puerta de Hierro phase I/II clinical trial. Cytotherapy. 2016 Aug;**18**(8):1025-1036. DOI: 10.1016/j.jcyt.2016.05.003. PubMed PMID: 27311799

[6] Satti HS, Waheed A, Ahmed P, Ahmed K, Akram Z, Aziz T, Satti TM, Shahbaz N, Khan MA, Malik SA. Autologous mesenchymal stromal cell transplantation for spinal cord injury: A phase I pilot study. Cytotherapy. 2016 Apr;**18**(4):518-522. DOI: 10.1016/j.jcyt.2016.01.004. PubMed PMID: 26971680

[7] Oh SK, Choi KH, Yoo JY, Kim DY, Kim SJ, Jeon SR. A phase III clinical trial showing limited efficacy of autologous mesenchymal stem cell therapy for spinal cord injury. Neurosurgery. 2016 Mar;**78**(3):436-447; discussion 447. DOI: 10.1227/NEU.0000000000001056. PubMed PMID: 26891377

[8] Hur JW, Cho TH, Park DH, Lee JB, Park JY, Chung YG. Intrathecal transplantation of autologous adipose-derived mesenchymal stem cells for treating spinal cord injury: A

human trial. The Journal of Spinal Cord Medicine. 2015 Jul 24. [Epub ahead of print] PubMed PMID: 26208177

[9] Miao X, Wu X, Shi W. Umbilical cord mesenchymal stem cells in neurological disorders: A clinical study. Indian Journal of Biochemistry & Biophysics. 2015 Apr;**52**(2):140-146. PubMed PMID: 26118125

[10] Mendonça MV, Larocca TF, de Freitas Souza BS, Villarreal CF, Silva LF, Matos AC, Novaes MA, Bahia CM, de Oliveira Melo Martinez AC, Kaneto CM, Furtado SB, Sampaio GP, Soares MB, dos Santos RR. Safety and neurological assessments after autologous transplantation of bone marrow mesenchymal stem cells in subjects with chronic spinal cord injury. Stem Cell Research & Therapy. 2014 Nov 17;**5**(6):126. DOI: 10.1186/scrt516. PubMed PMID: 25406723; PubMed Central PMCID: PMC4445989

[11] Cheng H, Liu X, Hua R, Dai G, Wang X, Gao J, An Y. Clinical observation of umbilical cord mesenchymal stem cell transplantation in treatment for sequelae of thoracolumbar spinal cord injury. Journal of Translational Medicine. 2014 Sep 12;**12**:253. DOI: 10.1186/s12967-014-0253-7. PubMed PMID: 25209445; PubMed Central PMCID:PMC4172930

[12] Bradbury EJ, McMahon SB. Spinal cord repair strategies: Why do they work? Nature Reviews Neuroscience. 2006;**7**:644-653

[13] Thuret S, Moon LDF, Gage FH. Therapeutic interventions after spinal cord injury. Nature Reviews Neuroscience. 2006;**7**:628-643

[14] Harel NY, Strittmatter SM. Can regenerating axons recapitulate developmental guidance during recovery from spinal cord injury? Nature Reviews Neuroscience. 2006;**7**:603-616

[15] Yiu G, He Z. Glial inhibition of CNS axon regeneration. Nature Reviews Neuroscience. 2006;**7**:617-627

[16] Yamagata T, Saito H, Habuchi O, Suzuki S. Purification and properties of bacterial chondroitinases and chondrosulfatases. Journal of Biological Chemistry. 1968;**243**:15231535. [PubMed: 5647268]

[17] Bradbury EJ, Carter LM. Manipulating the glial scar: Chondroitinase ABC as a therapy for spinal cord injury. Brain Research Bulletin. 2011;**84**:306316. [PubMed: 20620201]

[18] Tom VJ, Sandrow-Feinberg HR, Miller K, Santi L, Connors T, Lemay MA, Houle JD. Combining peripheral nerve grafts and chondroitinase promotes functional axonal regeneration in the chronically injured spinal cord. The Journal of Neuroscience. 2009;**29**:14881-14890. [PMC free article] [PubMed]

[19] Lee SH, Kim Y, Rhew D, Kuk M, Kim M, Kim WH, Kweon OK. Effect of the combination of mesenchymal stromal cells and chondroitinase ABC on chronic spinal cord injury. Cytotherapy. 2015 Oct;**17**(10):1374-1383. DOI: 10.1016/j.jcyt.2015.05.012. PubMed PMID: 26188966

[20] Wang Y, Jia H, Li WY, Guan LX, Deng L, Liu YC, Liu GB. Molecular examination of bone marrow stromal cells and chondroitinase ABC-assisted acellular nerve allograft

for peripheral nerve regeneration. Experimental and Therapeutic Medicine. 2016 Oct;**12**(4):1980-1992. PubMed PMID: 27698684; PubMed Central PMCID: PMC5038205

[21] Xiong LL, Li Y, Shang FF, Chen SW, Chen H, Ju SM, Zou Y, Tian HL, Wang TH, Luo CZ, Wang XY. Chondroitinase administration and pcDNA3.1-BDNF-BMSC transplantation promote motor functional recovery associated with NGF expression in spinal cord-transected rat. Spinal Cord. 2016 Jun 28. DOI: 10.1038/sc.2016.55. [Epub ahead of print] PubMed PMID: 27349609

[22] Lee H, McKeon RJ, Bellamkonda RV. Regenerative medicine special feature: Sustained delivery of thermostabilized chABC enhances axonal sprouting and functional recovery after spinal cord injury. Proceedings of the National Academy of Sciences USA. 2009 Nov 2

[23] Hunanyan AS, García-Alías G, Alessi V, Levine JM, Fawcett JW, Mendell LM, Arvanian VL. Role of chondroitin sulfate proteoglycans in axonal conduction in Mammalian spinal cord. The Journal of Neuroscience. 2010 Jun 9;**30**(23):7761-7769. DOI: 10.1523/JNEUROSCI.4659-09.2010

[24] Cheng CH, Lin CT, Lee MJ, Tsai MJ, Huang WH, Huang MC, Lin YL, Chen CJ, Huang WC, Cheng H. Local delivery of high-dose chondroitinase ABC in the sub-acute stage promotes axonal outgrowth and functional recovery after complete spinal cord transection. PLoS One. 2015 Sep 22;**10**(9):e0138705. DOI: 10.1371/journal.pone.0138705. eCollection 2015

[25] Shields LB, Zhang YP, Burke DA, Gray R, Shields CB. Benefit of chondroitinase ABC on sensory axon regeneration in a laceration model of spinal cord injury in the rat. Surgical Neurology. 2008 Jun;**69**(6):568-577; discussion 577. DOI: 10.1016/j.surneu.2008.02.009

[26] Iseda T, Okuda T, Kane-Goldsmith N, Mathew M, Ahmed S, Chang YW, Young W, Grumet M. Single, high-dose intraspinal injection of chondroitinase reduces glycosaminoglycans in injured spinal cord and promotes corticospinal axonal regrowth after hemisection but not contusion. Journal of Neurotrauma. 2008 Apr;**25**(4):334-349. DOI: 10.1089/neu.2007.0289

[27] Caggiano AO, Zimber MP, Ganguly A, Blight AR, Gruskin EA. Chondroitinase ABCI improves locomotion and bladder function following contusion injury of the rat spinal cord. Journal of Neurotrauma. 2005 Feb;**22**(2):226-239

[28] Mondello SE, Jefferson SC, Tester NJ, Howland DR. Impact of treatment duration and lesion size on effectiveness of chondroitinase treatment post-SCI. Experimental Neurology. 2015 May;**267**:64-77. DOI: 10.1016/j.expneurol.2015.02.028

[29] Ma Y, Liu M, Li Y. Secretion of bacterial chondroitinase ABC from bone marrow stromal cells by glycosylation site mutation: A promising approach for axon regeneration. Medical Hypotheses. 2011 Nov;**77**(5):914-916. DOI: 10.1016/j.mehy.2011.08.010. Epub 2011 Aug 31

[30] Wilems TS, Pardieck J, Iyer N, Sakiyama-Elbert SE. Combination therapy of stem cell derived neural progenitors and drug delivery of anti-inhibitory molecules for spinal

cord injury. Acta Biomaterialia. 2015 Sep 15. DOI: 10.1016/j.actbio.2015.09.018. PII: S1742-7061(15)30112-4. [Epub ahead of print]

[31] Kumar P, Choonara YE, Modi G, Naidoo D, Pillay V. Multifunctional therapeutic delivery strategies for effective neuro-regeneration following traumatic spinal cord injury. Current Pharmaceutical Design. 2015;**21**(12):1517-1528

[32] Mountney A, Zahner MR, Sturgill ER, Riley CJ, Aston JW, Oudega M, Schramm LP, Hurtado A, Schnaar RL. Sialidase, chondroitinase ABC, and combination therapy after spinal cord contusion injury. Journal of Neurotrauma. 2013 Feb 1;**30**(3):181-190. DOI: 10.1089/neu.2012.2353

[33] McCreedy DA, Sakiyama-Elbert SE. Combination therapies in the CNS: Engineering the environment. Neuroscience Letters. 2012 Jun 25;**519**(2):115-121. DOI: 10.1016/j.neulet.2012.02.025. Epub 2012 Feb 17. Review. PMID: 22343313

[34] Li Z, Zhang Z, Zhao L, Li H, Wang S, Shen Y. Bone marrow mesenchymal stem cells with Nogo-66 receptor gene silencing for repair of spinal cord injury. Neural Regeneration Research. 2014 Apr 15;**9**(8):806-814. DOI: 10.4103/1673-5374.131595. PubMed PMID: 25206893; PubMed Central PMCID: PMC4146260

[35] Lang BT, Cregg JM, DePaul MA, Tran AP, Xu K, Dyck SM, Madalena KM, Brown BP, Weng YL, Li S, Karimi-Abdolrezaee S, Busch SA, Shen Y, Silver J. Modulation of the proteoglycan receptor PTPσ promotes recovery after spinal cord injury. Nature. 2015 Feb 19;**518**(7539):404-408. DOI: 10.1038/nature13974. Epub 2014 Dec 3. PubMed PMID: 25470046; PubMed Central PMCID: PMC4336236

[36] Silver L, Michael JV, Goldfinger LE, Gallo G. Activation of PI3K and R-Ras signaling promotes the extension of sensory axons on inhibitory chondroitin sulfate proteoglycans. Developmental Neurobiology. 2014 Sep;**74**(9):918-933. DOI: 10.1002/dneu.22174. Epub 2014 Mar 27

[37] Tetzlaff W, Alexander SW, Miller FD, Bisby MA. Response of facial and rubrospinal neurons to axotomy: Changes in mRNA expression for cytoskeletal proteins and GAP-43. The Journal of Neuroscience. 1991;**11**:2528-2544; [PubMed]

[38] Sengottuvel V, Leibinger M, Pfreimer M, Andreadaki A, Fischer D. Taxol facilitates axon regeneration in the mature CNS. The Journal of Neuroscience. 2011;**31**:2688-2699 [PubMed]

[39] Amr SM. Bridging Defects in Chronic Spinal Cord Injury Using Peripheral Nerve Grafts: From Basic Science to Clinical Experience, Recovery of Motor Function Following Spinal Cord Injury. Heidi F, editor. InTech; 2016. DOI: 10.5772/64211. Available from: http://www.intechopen.com/books/recovery-of-motor-function-following-spinal-cord-injury/bridging-defects-in-chronic-spinal-cord-injury-using-peripheral-nerve-grafts-from-basic-science-to-clinical-experience

[40] Billings PC, Pacifici M. Interactions of signaling proteins, growth factors and other proteins with heparan sulfate: Mechanisms and mysteries. Connective Tissue Research. 2015;**56**(4):272-280. DOI: 10.3109/03008207.2015.1045066

[41] Olczyk P, Mencner Ł, Komosinska-Vassev K. Diverse roles of heparan sulfate and heparin in wound repair. BioMed Research International. 2015;**2015**:549417. DOI: 10.1155/2015/549417. Epub 2015 Jul 7. Review. PubMed PMID: 26236728; PubMed Central PMCID: PMC4508384

[42] Santos-Silva A, Fairless R, Frame MC, Montague P, Smith GM, Toft A, Riddell JS, Barnett SC. FGF/heparin differentially regulates Schwann cell and olfactory ensheathing cell interactions with astrocytes: A role in astrocytosis. The Journal of Neuroscience. 2007 Jul 4;**27**(27):7154-7167

[43] Madigan NN, McMahon S, O'Brien T, Yaszemski MJ, Windebank AJ. Current tissue engineering and novel therapeutic approaches to axonal regeneration following spinal cord injury using polymer scaffolds. Respiratory Physiology & Neurobiology. 2009 Nov 30;**169**(2):183-199. DOI: 10.1016/j.resp.2009.08.015. Epub 2009 Sep 6. Review. PubMed PMID: 19737633; PubMed Central PMCID: PMC2981799

[44] Straley KS, Foo CW, Heilshorn SC. Biomaterial design strategies for the treatment of spinal cord injuries. Journal of Neurotrauma. 2010 Jan;**27**(1):1-19. DOI: 10.1089/neu.2009.0948. Review. PubMed PMID: 19698073; PubMed Central PMCID: PMC2924783

[45] Hyatt AJ, Wang D, van Oterendorp C, Fawcett JW, Martin KR. Mesenchymal stromal cells integrate and form longitudinally-aligned layers when delivered to injured spinal cord via a novel fibrin scaffold. Neuroscience Letters. 2014 May 21;**569**:12-17. DOI: 10.1016/j.neulet.2014.03.023. PubMed PMID: 24680849; PubMed Central PMCID: PMC4015360

[46] Liu J, Chen Q, Zhang Z, Zheng Y, Sun X, Cao X, Gong A, Cui Y, He Q, Jiang P. Fibrin scaffolds containing ectomesenchymal stem cells enhance behavioral and histological improvement in a rat model of spinal cord injury. Cells, Tissues, Organs. 2013;**198**(1):35-46. DOI: 10.1159/000351665. PubMed PMID: 23774080

[47] Zurita M, Otero L, Aguayo C, Bonilla C, Ferreira E, Parajón A, Vaquero J. Cell therapy for spinal cord repair: Optimization of biologic scaffolds for survival and neural differentiation of human bone marrow stromal cells. Cytotherapy. 2010 Jul;**12**(4):522-537. DOI: 10.3109/14653241003615164. PubMed PMID: 20465485

[48] Cholas R, Hsu HP, Spector M. Collagen scaffolds incorporating select therapeutic agents to facilitate a reparative response in a standardized hemiresection defect in the rat spinal cord. Tissue Engineering Part A. 2012 Oct;**18**(19-20):2158-2172. DOI: 10.1089/ten.TEA.2011.0577. PubMed PMID: 22827732

[49] Wang B, Han J, Gao Y, Xiao Z, Chen B, Wang X, Zhao W, Dai J. The differentiation of rat adipose-derived stem cells into OEC-like cells on collagen scaffolds by co-culturing with OECs. Neuroscience Letters. 2007 Jun 29;**421**(3):191-196. PubMed PMID: 17574753

[50] Chen J, Zhang Z, Liu J, Zhou R, Zheng X, Chen T, Wang L, Huang M, Yang C, Li Z, Yang C, Bai X, Jin D. Acellular spinal cord scaffold seeded with bone marrow stromal cells protects tissue and promotes functional recovery in spinal cord-injured rats. Journal of Neuroscience Research. 2014 Mar;**92**(3):307-317. DOI: 10.1002/jnr.23311. PubMed PMID: 24375695

[51] Liu J, Chen J, Liu B, Yang C, Xie D, Zheng X, Xu S, Chen T, Wang L, Zhang Z, Bai X, Jin D. Acellular spinal cord scaffold seeded with mesenchymal stem cells promotes long-distance axon regeneration and functional recovery in spinal cord injured rats. Journal of Neurological Sciences. 2013 Feb 15;**325**(1-2):127-136. DOI: 10.1016/j.jns.2012.11.022. PubMed PMID: 23317924

[52] Wei X, Wen Y, Zhang T, Li H. Effects of bone marrow mesenchymal stem cells with a cellular muscle bioscaffolds on repair of acute hemi-transection spinal cord injury in rats. Zhongguo Xiu Fu Chong Jian Wai Ke Za Zhi. 2012 Nov;**26**(11):1362-1368. Chinese. PubMed PMID: 23230674

[53] Zhang K, Liu Z, Li G, Lai BQ, Qin LN, Ding Y, Ruan JW, Zhang SX, Zeng YS. Electro-acupuncture promotes the survival and differentiation of transplanted bone marrow mesenchymal stem cells pre-induced with neurotrophin-3 and retinoic acid in gelatin sponge scaffold after rat spinal cord transection. Stem Cell Reviews. 2014 Aug;**10**(4):612-625. DOI: 10.1007/s12015-014-9513-4. PubMed PMID: 24789671

[54] Shirian S, Ebrahimi-Barough S, Saberi H, Norouzi-Javidan A, Mousavi SM, Derakhshan MA, Arjmand B, Ai J. Comparison of capability of human bone marrow mesenchymal stem cells and endometrial stem cells to differentiate into motor neurons on electrospun poly(ε-caprolactone) scaffold. Molecular Neurobiology. 2016 Oct;**53**(8):5278-5287. doi: 10.1007/s12035-015-9442-5. PubMed PMID: 26420037

[55] Wang D, Fan Y, Zhang J. Transplantation of Nogo-66 receptor gene-silenced cells in a poly(D,L-lactic-co-glycolic acid) scaffold for the treatment of spinal cord injury. Neural Regeneration Research. 2013 Mar 15;**8**(8):677-685. DOI: 10.3969/j.issn.1673-5374.2013.08.001. PubMed PMID: 25206713; PubMed Central PMCID: PMC4146076

[56] Wang D, Wen Y, Lan X, Li H. Experimental study on bone marrow mesenchymal stem cells seeded in chitosan-alginate scaffolds for repairing spinal cord injury. Zhongguo Xiu Fu Chong Jian Wai Ke Za Zhi. 2010 Feb;**24**(2):190-196. Chinese. PubMed PMID: 20187451

[57] Xue F, Wu EJ, Zhang PX, Li-Ya A, Kou YH, Yin XF, Han N. Biodegradable chitin conduit tubulation combined with bone marrow mesenchymal stem cell transplantation for treatment of spinal cord injury by reducing glial scar and cavity formation. Neural Regeneration Research. 2015 Jan;**10**(1):104-111. DOI: 10.4103/1673-5374.150715. PubMed PMID: 25788929; PubMed Central PMCID:PMC4357092

[58] Kim YC, Kim YH, Kim JW, Ha KY. Transplantation of mesenchymal stem cells for acute spinal cord injury in rats: Comparative study between intralesional injection and scaffold based transplantation. Journal of Korean Medical Science. 2016 Sep;**31**(9):1373-1382. DOI: 10.3346/jkms.2016.31.9.1373. PubMed PMID: 27510379; PubMed Central PMCID: PMC4974177

[59] Tukmachev D, Forostyak S, Koci Z, Zaviskova K, Vackova I, Vyborny K, Sandvig I, Sandvig A, Medberry CJ, Badylak SF, Sykova E, Kubinova S. Injectable extracellular matrix hydrogels as scaffolds for spinal cord injury repair. Tissue Engineering Part A.

2016 Feb;**22**(3-4):306-317. DOI: 10.1089/ten.TEA.2015.0422. PubMed PMID: 26729284; PubMed Central PMCID: PMC4799710

[60] Park J, Lim E, Back S, Na H, Park Y, Sun K. Nerve regeneration following spinal cord injury using matrix metalloproteinase-sensitive, hyaluronic acid-based biomimetic hydrogel scaffold containing brain-derived neurotrophic factor. Journal of Biomedical Materials Research. Part A. 2010 Jun 1;**93**(3):1091-1099. DOI: 10.1002/jbm.a.32519. PubMed PMID: 19768787

[61] Günther MI, Weidner N, Müller R, Blesch A. Cell-seeded alginate hydrogel scaffolds promote directed linear axonal regeneration in the injured rat spinal cord. Acta Biomaterialia. 2015 Nov;**27**:140-150. DOI: 10.1016/j.actbio.2015.09.001. PubMed PMID: 26348141

[62] Madigan NN, Chen BK, Knight AM, Rooney GE, Sweeney E, Kinnavane L, Yaszemski MJ, Dockery P, O'Brien T, McMahon SS, Windebank AJ. Comparison of cellular architecture, axonal growth, and blood vessel formation through cell-loaded polymer scaffolds in the transected rat spinal cord. Tissue Engineering. Part A. 2014 Nov;**20**(21-22):2985-2997. DOI: 10.1089/ten.TEA.2013.0551. PubMed PMID: 24854680;PubMed Central PMCID: PMC4229864

[63] Kubinová Š, Horák D, Hejčl A, Plichta Z, Kotek J, Proks V, Forostyak S, Syková E. SIKVAV-modified highly superporous PHEMA scaffolds with oriented pores for spinal cord injury repair. Journal of Tissue Engineering and Regenerative Medicine. 2015 Nov;**9**(11):1298-1309. DOI: 10.1002/term.1694. PubMed PMID: 23401421

[64] Kubinová S, Horák D, Hejčl A, Plichta Z, Kotek J, Syková E. Highly superporous cholesterol-modified poly(2-hydroxyethyl methacrylate) scaffolds for spinal cord injury repair. Journal of Biomedical Materials Research. Part A. 2011 Dec 15;**99**(4):618-629. DOI: 10.1002/jbm.a.33221. PubMed PMID: 21953978

[65] Han S, Wang B, Li X, Xiao Z, Han J, Zhao Y, Fang Y, Yin Y, Chen B, Dai J. Bone marrow-derived mesenchymal stem cells in three-dimensional culture promote neuronal regeneration by neurotrophic protection and immunomodulation. Journal of Biomedical Materials Research. Part A. 2016 Jul;**104**(7):1759-1769. DOI: 10.1002/jbm.a.35708. PubMed PMID: 26990583

[66] Ribeiro-Samy S, Silva NA, Correlo VM, Fraga JS, Pinto L, Teixeira-Castro A, Leite-Almeida H, Almeida A, Gimble JM, Sousa N, Salgado AJ, Reis RL. Development and characterization of a PHB-HV-based 3D scaffold for a tissue engineering and cell-therapy combinatorial approach for spinal cord injury regeneration. Macromolecular Bioscience. 2013 Nov;**13**(11):1576-1592. DOI: 10.1002/mabi.201300178. PubMed PMID:24038969

[67] Caron I, Rossi F, Papa S, Aloe R, Sculco M, Mauri E, Sacchetti A, Erba E, Panini N, Parazzi V, Barilani M, Forloni G, Perale G, Lazzari L, Veglianese P. A new three dimensional biomimetic hydrogel to deliver factors secreted by human mesenchymal stem cells in spinal cord injury. Biomaterials. 2016 Jan;**75**:135-147. DOI: 10.1016/j.biomaterials.2015.10.024. PubMed PMID: 26497428

[68] Zeng X, Zeng YS, Ma YH, Lu LY, Du BL, Zhang W, Li Y, Chan WY. Bone marrow mesenchymal stem cells in a three-dimensional gelatin sponge scaffold attenuate inflammation, promote angiogenesis, and reduce cavity formation in experimental spinal cord injury. Cell Transplantation. 2011;**20**(11-12):1881-1899. DOI: 10.3727/096368911X566181. PubMed PMID: 21396163

[69] Rodríguez-Vázquez M, Vega-Ruiz B, Ramos-Zúñiga R, Saldaña-Koppel DA, Quiñones-Olvera LF. Chitosan and Its Potential Use as a Scaffold for Tissue Engineering in Regenerative Medicine. BioMed Research International. 2015;**2015**:821279. DOI: 10.1155/2015/821279. Epub 2015 Oct 4. Review. PubMed PMID: 26504833; PubMed Central PMCID: PMC4609393

[70] Langer R and Tirrell DA. Designing materials for biology and medicine. Nature. 2004:**428**(6982):487-492. View at Publisher DOI:10.1038/nature02388. View at Google Scholar. View at Scopus

[71] Zeng W, Rong M, Hu X, Xiao W, Qi F, Huang J, Luo Z. Incorporation of chitosan microspheres into collagen-chitosan scaffolds for the controlled release of nerve growth factor. PLoS One. 2014 Jul 1;**9**(7):e101300. DOI: 10.1371/journal.pone.0101300. eCollection 2014. PubMed PMID: 24983464; PubMed Central PMCID: PMC4077743

[72] Hiroshi N, Tator CH, Shoichet MS. Bioengineered strategies for spinal cord repair. Journal of Neurotrauma. 2006;**23**:496-507. [PubMed: 16629632]

[73] Wong DY, Leveque JC, Brumblay H, Krebsbach PH, Hollister SJ, Lamarca F. Macro-architectures in spinal cord scaffold implants influence regeneration. Journal of Neurotrauma. 2008;**25**:1027-1037. [PubMed: 18721107]

[74] Krych AJ, Rooney GE, Chen B, Schermerhorn TC, Ameenuddin S, Gross LA, Moore MJ, Currier BL, Friedman JA, Spinner RJ, Yaszemski MJ, Windebank AJ. Relationship between scaffold channel diameter and number of regenerating axons in the transected rat spinal cord. Acta Biomaterialia. 2009. (Epub March 27, 2009)

[75] Moore MJ, Friedman JA, Lewellyn EB, Mantila SM, Krych AJ, Ameenuddin S, Knight AM, Lu L, Currier BL, Spinner RJ, Marsh RW, Windebank AJ, Yaszemski MJ. Multiple-channel scaffolds to promote spinal cord axon regeneration. Biomaterials. 2006;**27**:419-429. [PubMed: 16137759]

[76] Ducker TB, and Hayes GJ. Experimental improvements in the use of Silastic cuff for peripheral nerve repair. Journal of Neurosurgery. 1968;**28**:582

[77] Rutkowski GE, and Heath CA. Development of a bioartificial nerve graft. II Nerve regeneration in vitro. Biotechnology Progress. 2002;**18**:373

[78] Kokai LE, Lin YC, Oyster NM, and Marra KG. Diffusion of soluble factors through degradable polymer nerve guides: Controlling manufacturing parameters. Acta Biomaterialia. 2009:**5**:2540

[79] Nectow AR, Marra KG, Kaplan DL. Biomaterials for the development of peripheral nerve guidance conduits. Tissue Engineering. Part B, Reviews. 2012 Feb;**18**(1):40-50.

DOI: 10.1089/ten.TEB.2011.0240. Epub 2011 Sep 23. Review. PubMed PMID: 21812591; PubMed Central PMCID: PMC3262974

[80] Bellamkonda R. Peripheral nerve regeneration: An opinion on channels, scaffolds and anisotropy. Biomaterials. 2006;**27**:3515-3518. [PubMed: 16533522]

[81] Bellamkonda R, Ranieri JP, Bouche N, Aebischer P. Hydrogel-based three-dimensional matrix for neural cells. Journal of Biomedical Materials Research. 1995;**29**:663-671. [PubMed: 7622552]

[82] Khademhosseini A, Langer R. Microengineered hydrogels for tissue engineering. Biomaterials. 2007;**28**:5087-5092. [PubMed: 17707502]

[83] Clark P, Connolly P, Curtis A, Dow J, Wilkinson C. Cell guidance by ultrafine topography in vitro. Journal of Cell Science. 1991;**99**:73-77. [PubMed: 1757503]

[84] Goldner JS, Bruder JM, Li G, Gazzola D. Neurite bridging across micropatterned grooves. Biomaterials. 2006;**27**:460-472. [PubMed: 16115675]

[85] Yao L, Damodaran G, Nikolskaya N, Gorman AM, Windebank AJ, Pandit A. The effect of laminin peptide gradient of enzymatically cross-linked collagen scaffold on neurite growth. Journal of Biomedical Materials Research. Part A. 2009;**11** (Epub February 11, 2009)

[86] Yao L, O'Brien N, Windebank A, Pandit A. Orienting neurite growth in electrospun fibrous neural conduits. Journal of Biomedical Materials Research. Part B, Applied Biomaterials. 2009;**90**(2):483-491. [PubMed: 19130615]

[87] Yao L, Wang S, Cui W, Sherlock R, O'Connell C, Damodaran G, Gorman A, Windebank A, Pandit A. Effect of functionalized micropatterned PLGA on guided neurite growth. Acta Biomaterialia. 2009;**5**:580-588. [PubMed: 18835227]

[88] Liu C, Pyne R, Kim J, Wright NT, Baek S, Chan C. The impact of prestretch induced surface anisotropy on axon regeneration. Tissue Engineering. Part C, Methods. 2016 Jan 8. [Epub ahead of print] PubMed PMID: 26563431; PubMed Central PMCID:PMC4744876

[89] Lundborg G, Dahlin L, Dohi D, Kanje M, Terada N. A new type of "bioartificial" nerve graft for bridging extended defects in nerves. The Journal of Hand Surgery: British & European Volume. 1997;**22**:299-303

[90] Cai J, Peng X, Nelson KD, Eberhart R, Smith GM. Permeable guidance channels containing microfilament scaffolds enhance axon growth and maturation. Journal of Biomedical Materials Research. Part A. 2005;**75**:374-386

[91] Yoshii S, Oka M, Shima M, Taniguchi A, Akagi M. Bridging a 30-mm nerve defect using collagen filaments. Journal of Biomedical Materials Research. 2003;**67A**:467-474

[92] Yoshii S, Oka M, Ikeda N, Akagi M, Matsusue Y, Nakamura T. Bridging a peripheral nerve defect using collagen filaments. The Journal of Hand Surgery. 2001;**26**:52-59

[93] Chew SY, Mi R, Hoke A, Leong KW. Aligned protein–polymer composite fibers enhance nerve regeneration: A potential tissue-engineering platform. Advanced Functional Materials. 2007;**17**:1288-1296

[94] Zhang BG, Quigley AF, Myers DE, Wallace GG, Kapsa RM, Choong PF. Recent advances in nerve tissue engineering. International Journal of Artificial Organs. 2014 Apr;**37**(4):277-291. DOI: 10.5301/ijao.5000317. Epub 2014 Apr 15. Review. PubMed PMID: 24811182

[95] Hudson TW, Evans GR, Schmidt CE. Engineering strategies for peripheral nerve repair. Clinics in Plastic Surgery. 1999;**26**:617-628, ix

[96] Xie F, Li QF, Gu B, Liu K, and Shen GX. In vitro and in vivo evaluation of a biodegradable chitosan-PLA composite composite peripheral nerve guide conduit material. Microsurgery. 2008;**28**:471

[97] Discher DE, Janmey P, and Wang YL. Tissue cells feel and respond to the stiffness of their substrate. Science. 2005;**310**:1139

[98] Engler AJ, Sen S, Sweeney HL, and Discher DE. Matrix elasticity directs stem cell lineage specification. Cell. 2006;**726**:677

[99] Mammoto A, and Ingber DE. Cytoskeletal control of growth and cell fate switching. Current Opinion in Cell Biology. 2009;**21**:864

[100] Yao S, Liu X, Yu S, Wang X, Zhang S, Wu Q, Sun X, Mao H. Co-effects of matrix low elasticity and aligned topography on stem cell neurogenic differentiation and rapid neurite outgrowth. Nanoscale. 2016 May 21;**8**(19):10252-10265. DOI: 10.1039/c6nr01169a. PubMed PMID: 27124547

[101] Frostick SP, Yin Q, Kemp G: Schwann cells, neurotrophic factors, and peripheral nerve regeneration. Microsurgery. 1998;**18**:397-405

[102] Blottner D, Baumgarten HG: Neurotrophy and regeneration in vivo. Acta Anatomica. 1994;**150**:235-245

[103] Li G, Che MT, Zhang K, Qin LN, Zhang YT, Chen RQ, Rong LM, Liu S, Ding Y, Shen HY, Long SM, Wu JL, Ling EA, Zeng YS. Graft of the NT-3 persistent delivery gelatin sponge scaffold promotes axon regeneration, attenuates inflammation, and induces cell migration in rat and canine with spinal cord injury. Biomaterials. 2016 Mar;**83**:233-248. DOI: 10.1016/j.biomaterials.2015.11.059. PubMed PMID:26774562

[104] Liu H, Yao F, Zhou Y, et al. Porous poly (DL-lactic acid) modified chitosan-gelatin scaffolds for tissue engineering. Journal of Biomaterials Applications. 2005;**19**(4):303-322

[105] Yang Y, Zhao W, He J, Zhao Y, Ding F, Gu X. Nerve conduits based on immobilization of nerve growth factor onto modified chitosan by using genipin as a crosslinking agent. European Journal of Pharmaceutics and Biopharmaceutics. 2011;**79**(3):519-525

[106] Westhauser F, Weis C, Hoellig M, Swing T, Schmidmaier G, Weber MA, Stiller W, Kauczor HU, Moghaddam A. Heidelberg-mCT-Analyzer: A novel method for standardized microcomputed-tomography-guided evaluation of scaffold properties in bone and tissue research. Royal Society Open Science. 2015 Nov 11;**2**(11):150496. DOI: 10.1098/rsos.150496. PubMed PMID: 26716008; PubMed Central PMCID: PMC4680623

[107] Pabari A, Yang SY, Mosahebi A, Seifalian AM. Recent advances in artificial nerve conduit design: Strategies for the delivery of luminal fillers. Journal of Controlled Release. 2011 Nov 30;**156**(1):2-10. DOI: 10.1016/j.jconrel.2011.07.001. Epub 2011 Jul 6. Review. PubMed PMID: 21763371

[108] Rosoff WJ, Urbach JS, Esrick MA, McAllister RG, Richards LJ, Goodhill GJ. A new chemotaxis assay shows the extreme sensitivity of axons to molecular gradients. Nature Neuroscience. 2004 Jun;**7**(6):678-682. Epub 2004 May 25. Erratum in: Nat Neurosci. 2004 Jul;7(7):785

[109] Adams DN, Kao EY, Hypolite CL, Distefano MD, Hu WS, Letourneau PC. Growth cones turn and migrate up an immobilized gradient of the laminin IKVAV peptide. Journal of Neurobiology. 2005;**62**:134-147. [PubMed: 15452851]

[110] Dodla MC, Bellamkonda RV. Anisotropic scaffolds facilitate enhanced neurite extension in vitro. Journal of Biomedical Materials Research. Part A. 2006;**78A**:213-221. [PubMed: 16892507]

[111] Tse TH, Chan BP, Chan CM, Lam J. Mathematical modeling of guided neurite extension in an engineered conduit with multiple concentration gradients of nerve growth factor (NGF). Annals of Biomedical Engineering. 2007 Sep;**35**(9):1561-1572. Epub 2007 May 23

[112] Ng TK, Fortino VR, Pelaez D, Cheung HS. Progress of mesenchymal stem cell therapy for neural and retinal diseases. World Journal of Stem Cells. 2014 Apr 26;**6**(2):111-119. DOI: 10.4252/wjsc.v6.i2.111. Review. PubMed PMID: 24772238; PubMedCentral PMCID: PMC3999769

[113] Ichim TE, Solano F, Glenn E, Morales F, Smith L, Zabrecky G, Riordan NH. Stem cell therapy for autism. Journal of Translational Medicine. 2007;**5**:30. DOI: 10.1186/1479-5876-5-30. PMID: 17597540

[114] Ng TK, Lam DS, Cheung HS. Prospects of stem cells for retinal diseases. Asia-Pacific Journal of Ophthalmology. 2013;**2**:57-63 DOI: 10.1097/APO.0b013e31827e3e5d

[115] Zhang R, Chen H, Zheng Z, Liu Q, Xu L. Umbilical cord-derived mesenchymal stem cell therapy for neurological disorders via inhibition of mitogen-activated protein kinase pathway-mediated apoptosis. Molecular Medicine Reports. 2015 Mar;**11**(3):1807-1812. DOI: 10.3892/mmr.2014.2985. PubMed PMID: 25412281

[116] Badner A, Vawda R, Laliberte A, Hong J, Mikhail M, Jose A, Dragas R, Fehlings M. Early intravenous delivery of human brain stromal cells modulates systemic inflammation and leads to vasoprotection in traumatic spinal cord injury. Stem Cells Translational Medicine. 2016 Aug;**5**(8):991-1003. DOI: 10.5966/sctm.2015-0295. PubMed PMID: 27245367; PubMed Central PMCID: PMC4954452

[117] de Almeida FM, Marques SA, Ramalho Bdos S, Massoto TB, Martinez AM. Chronic spinal cord lesions respond positively to transplants of mesenchymal stem cells. Restorative Neurology and Neuroscience. 2015;**33**(1):43-55. DOI: 10.3233/RNN-140431. PubMed PMID:25537259

[118] Zhang LX, Yin YM, Zhang ZQ, Deng LX. Grafted bone marrow stromal cells: A contributor to glial repair after spinal cord injury. The Neuroscientist. 2015 Jun;**21**(3):277-289. DOI: 10.1177/1073858414532171. Review. PubMed PMID: 24777423

[119] Tsumuraya T, Ohtaki H, Song D, Sato A, Watanabe J, Hiraizumi Y, Nakamachi T, Xu Z, Dohi K, Hashimoto H, Atsumi T, Shioda S. Human mesenchymal stem/stromal cells suppress spinal inflammation in mice with contribution of pituitary adenylate cyclase-activating polypeptide (PACAP). Journal of Neuroinflammation. 2015 Feb 22;**12**:35. DOI: 10.1186/s12974-015-0252-5. PubMed PMID: 25889720; PubMed Central PMCID: PMC4346126

[120] Wang Y, Liu H, Ma H. Intrathecally Transplanting Mesenchymal Stem Cells(MSCs) Activates ERK1/2 in spinal cords of ischemia-reperfusion injury rats and improves nerve function. Medical Science Monitor. 2016 May 2;**22**:1472-1479. PubMed PMID: 27135658; PubMed Central PMCID: PMC4917313

[121] Burnouf T, Strunk D, Koh MB, Schallmoser K. Human platelet lysate: Replacing fetal bovine serum as a gold standard for human cell propagation? Biomaterials. 2016 Jan;**76**:371-387. DOI: 10.1016/j.biomaterials.2015.10.065. Review. PubMed PMID: 26561934

[122] Laner-Plamberger S, Lener T, Schmid D, Streif DA, Salzer T, Öller M, Hauser-Kronberger C, Fischer T, Jacobs VR, Schallmoser K, Gimona M, Rohde E. Mechanical fibrinogen-depletion supports heparin-free mesenchymal stem cell propagation in human platelet lysate. Journal of Translational Medicine. 2015 Nov 10;**13**:354. DOI: 10.1186/s12967-015-0717-4. PubMed PMID: 26554451; PubMed Central PMCID: PMC4641400

[123] Ding Y, Zhang RY, He B, Liu Z, Zhang K, Ruan JW, Ling EA, Wu JL, Zeng YS. Combination of electroacupuncture and grafted mesenchymal stem cells overexpressing TrkC improves remyelination and function in demyelinated spinal cord of rats. Scientific Reports. 2015 Mar 16;**5**:9133. DOI: 10.1038/srep09133. PubMed PMID:25779025

[124] Park YM, Han SH, Seo SK, Park KA, Lee WT, Lee JE. Restorative benefits of transplanting human mesenchymal stromal cells overexpressing arginine decarboxylase genes after spinal cord injury. Cytotherapy. 2015 Jan;**17**(1):25-37. DOI: 10.1016/j.jcyt.2014.08.006. PubMed PMID: 25442787

[125] Lee SH, Kim Y, Rhew D, Kim A, Jo KR, Yoon Y, Choi KU, Jung T, Kim WH, Kweon OK. Effect of canine mesenchymal stromal cells overexpressing heme oxygenase-1 in spinal cord injury. The Journal of Veterinary Science. 2016 Sep 1. [Epub ahead of print] PubMed PMID: 27586469

[126] Wang D, Zhang J. Effects of hypothermia combined with neural stem cell transplantation on recovery of neurological function in rats with spinal cord injury. Molecular Medicine Reports. 2015 Mar;**11**(3):1759-1767. DOI: 10.3892/mmr.2014.2905. PubMed PMID: 25385306; PubMed Central PMCID: PMC4270334

[127] Lee JY, Ha KY, Kim JW, Seo JY, Kim YH. Does extracorporeal shock wave introduce alteration of microenvironment in cell therapy for chronic spinal cord injury? Spine

(Phila Pa 1976). 2014 Dec 15;**39**(26):E1553-E1559. DOI: 10.1097/BRS.0000000000000626. PubMed PMID: 25271504

[128] Shi Y, Hu Y, Lv C, Tu G. Effects of reactive oxygen species on differentiation of bone marrow mesenchymal stem cells. Annals of Transplantation. 2016 Nov 14;**21**:695-700

[129] Wang C, Shi D, Song X, Chen Y, Wang L, Zhang X. Calpain inhibitor attenuates ER stress-induced apoptosis in injured spinal cord after bone mesenchymal stem cells transplantation. Neurochemistry International. 2016 Jul;**97**:15-25. DOI: 10.1016/j.neuint.2016.04.015. PubMed PMID: 27137651

[130] Yang W, Yang Y, Yang JY, Liang M, Song J. Treatment with bone marrow mesenchymal stem cells combined with plumbagin alleviates spinal cord injury by affecting oxidative stress, inflammation, apoptosis and the activation of the Nrf2 pathway. International Journal of Molecular Medicine. 2016 Apr;**37**(4):1075-1082. DOI: 10.3892/ijmm.2016.2498. PubMed PMID: 26936518

[131] Chen M, Hou Y, Lin D. Polydatin protects bone marrow stem cells against oxidative injury: Involvement of Nrf 2/ARE pathways. Stem Cells International. 2016;**2016**:9394150. DOI: 10.1155/2016/9394150. PubMed PMID: 27022401; PubMed Central PMCID: PMC4684894

[132] Chen M, Chen S, Lin D. Carvedilol protects bone marrow stem cells against hydrogen peroxide-induced cell death via PI3K-AKT pathway. Biomedicine and Pharmacotherapy. 2016 Mar;**78**:257-263. DOI: 10.1016/j.biopha.2016.01.008. PubMed PMID: 26898450

[133] Wang Z, Fang B, Tan Z, Zhang D, Ma H. Hypoxic preconditioning increases the protective effect of bone marrow mesenchymal stem cells on spinal cord ischemia/reperfusion injury. Molecular Medicine Reports. 2016 Mar;**13**(3):1953-1960. DOI: 10.3892/mmr.2016.4753. PubMed PMID: 26783161; PubMed Central PMCID: PMC4768971

[134] Deng WP, Yang CC, Yang LY, Chen CW, Chen WH, Yang CB, Chen YH, Lai WF, Renshaw PF. Extracellular matrix-regulated neural differentiation of human multipotent marrow progenitor cells enhances functional recovery after spinal cord injury. The Spine Journal. 2014 Oct 1;**14**(10):2488-2499. DOI: 10.1016/j.spinee. 2014.04.024. PubMed PMID: 24792783; PubMed Central PMCID:PMC4692164

[135] Zeng X, Ma YH, Chen YF, Qiu XC, Wu JL, Ling EA, Zeng YS. Autocrine fibronectin from differentiating mesenchymal stem cells induces the neurite elongation in vitro and promotes nerve fiber regeneration in transected spinal cord injury. Journal of Biomedical Materials Research. Part A. 2016 Aug;**104**(8):1902-1911. DOI: 10.1002/jbm.a.35720. PubMed PMID: 26991461; PubMed Central PMCID: PMC5101622

[136] Li J, Guo W, Xiong M, Han H, Chen J, Mao D, Tang B, Yu H, Zeng Y. Effect of SDF-1/CXCR4 axis on the migration of transplanted bone mesenchymal stem cells mobilized by erythropoietin toward lesion sites following spinal cord injury. International Journal of Molecular Medicine. 2015 Nov;**36**(5):1205-1214. DOI: 10.3892/ijmm.2015.2344. PubMed PMID: 26398409; PubMed Central PMCID: PMC4601746

[137] Wang YX, Sun JJ, Zhang M, Hou XH, Hong J, Zhou YJ, Zhang ZY. Propofol injection combined with bone marrow mesenchymal stem cell transplantation better improves electrophysiological function in the hindlimb of rats with spinal cord injury than monotherapy. Neural Regeneration Research. 2015 Apr;**10**(4):636-643. DOI: 10.4103/1673 5374.155440. PubMed PMID: 26170827; PubMed Central PMCID:PMC4424759

[138] Yu DS, Liu LB, Cao Y, Wang YS, Bi YL, Wei ZJ, Tong SM, Lv G, Mei XF. Combining bone marrow stromal cells with green tea polyphenols attenuates the blood-spinal cord barrier permeability in rats with compression spinal cord injury. Journal of Molecular Neuroscience. 2015 Jun;**56**(2):388-396. DOI: 10.1007/s12031-015-0564-z. PubMed PMID: 26007330

[139] Yin YM, Lu Y, Zhang LX, Zhang GP, Zhang ZQ. Bone marrow stromal cells transplantation combined with ultrashortwave therapy promotes functional recovery on spinal cord injury in rats. Synapse. 2015 Mar;**69**(3):139-147. DOI: 10.1002/syn.21802. PubMed PMID: 25600592

[140] Xia P, Pan S, Cheng J, Yang M, Qi Z, Hou T, Yang X. Factors affecting directional migration of bone marrow mesenchymal stem cells to the injured spinal cord. Neural Regeneration Research. 2014 Sep 15;**9**(18):1688-1695. DOI: 10.4103/1673-5374.141804. PubMed PMID: 25374590; PubMed Central PMCID: PMC4211189

[141] Chen L, Cui X, Wu Z, Jia L, Yu Y, Zhou Q, Hu X, Xu W, Luo D, Liu J, Xiao J, Yan Q, Cheng L. Transplantation of bone marrow mesenchymal stem cells pretreated with valproic acid in rats with an acute spinal cord injury. Bioscience Trends. 2014 Apr;**8**(2):111-119. PubMed PMID: 24815388

[142] Hou Y, Ryu CH, Jun JA, Kim SM, Jeong CH, Jeun SS. IL-8 enhances the angiogenic potential of human bone marrow mesenchymal stem cells by increasing vascular endothelial growth factor. Cell Biology International. 2014 Sep;**38**(9):1050-1059. DOI: 10.1002/cbin.10294. PubMed PMID: 24797366

[143] Thakkar UG, Vanikar AV, Trivedi HL, Shah VR, Dave SD, Dixit SB, Tiwari BB, Shah HH. Infusion of autologous adipose tissue derived neuronal differentiated mesenchymal stem cells and hematopoietic stem cells in post-traumatic paraplegia offers a viable therapeutic approach. Advanced Biomedical Research. 2016 Mar 16;**5**:51. DOI: 10.4103/2277-9175.178792. PubMed PMID: 27110548; PubMed Central PMCID: PMC4817398

[144] Menezes K, Nascimento MA, Gonçalves JP, Cruz AS, Lopes DV, Curzio B, Bonamino M, de Menezes JR, Borojevic R, Rossi MI, Coelho-Sampaio T. Human mesenchymal cells from adipose tissue deposit laminin and promote regeneration of injured spinal cord in rats. PLoS One. 2014 May 15;**9**(5):e96020. DOI: 10.1371/journal. pone.0096020. PubMed PMID: 24830794; PubMed Central PMCID:PMC4022508

[145] Aras Y, Sabanci PA, Kabatas S, Duruksu G, Subasi C, Erguven M, Karaoz E. The effects of adipose tissue-derived mesenchymal stem cell transplantation during the acute and subacute phases following spinal cord injury. Turkish Neurosurgery. 2016;**26**(1):127-139. DOI: 10.5137/1019-5149.JTN.15724-15.0. PubMed PMID: 26768879

[146] Wu JH, Li M, Liang Y, Lu T, Duan CY. Migration of adipose-derived mesenchymal stem cells stably expressing chondroitinase ABC in vitro. Chinese Medical Journal (England). 2016 Jul 5;**129**(13):1592-1599. DOI: 10.4103/0366-6999.184464. PubMed PMID: 27364797; PubMed Central PMCID: PMC4931267

[147] Kolar MK, et al. The therapeutic effects of human adipose-derived stem cells in a rat cervical spinal cord injury model. Stem Cells and Development. 2014;**23**(14):1659-1674

[148] Zhou HL, Zhang XJ, Zhang MY, Yan ZJ, Xu ZM, Xu RX. Transplantation of human amniotic mesenchymal stem cells promotes functional recovery in a rat model of traumatic spinal cord injury. Neurochemical Research. 2016 Oct;**41**(10):2708-2718. PubMed PMID: 27351200

[149] Chiang CY, Liu SA, Sheu ML, Chen FC, Chen CJ, Su HL, Pan HC. Feasibility of human amniotic fluid derived stem cells in alleviation of neuropathic pain in chronic constrictive injury nerve model. PLoS One. 2016 Jul 21;**11**(7):e0159482. DOI: 10.1371/journal.pone.0159482. PubMed PMID: 27441756; PubMed Central PMCID:PMC4956194

[150] Wang A, Brown EG, Lankford L, Keller BA, Pivetti CD, Sitkin NA, Beattie MS, Bresnahan JC, Farmer DL. Placental mesenchymal stromal cells rescue ambulation in ovine myelomeningocele. Stem Cells Translational Medicine. 2015 Jun;**4**(6):659-669. DOI: 10.5966/sctm.2014-0296. PubMed PMID: 25911465; PubMed Central PMCID: PMC4449103

[151] Chen C, Chen F, Yao C, Shu S, Feng J, Hu X, Hai Q, Yao S, Chen X. Intrathecal injection of human umbilical cord-derived mesenchymal stem cells ameliorates neuropathic pain in rats. Neurochemical Research. 2016 Sep 21. [Epub ahead of print]PubMed PMID: 27655256

[152] Chung HJ, Chung WH, Lee JH, Chung DJ, Yang WJ, Lee AJ, Choi CB, Chang HS, Kim DH, Suh HJ, Lee DH, Hwang SH, Do SH, Kim HY. Expression of neurotrophic factors in injured spinal cord after transplantation of human-umbilical cord blood stem cells in rats. The Journal of Veterinary Science. 2016 Mar;**17**(1):97-102. DOI: 10.4142/jvs.2016.17.1.97. PubMed PMID: 27051345; PubMed Central PMCID: PMC4808649

[153] Zhilai Z, Biling M, Sujun Q, Chao D, Benchao S, Shuai H, Shun Y, Hui Z. Preconditioning in lowered oxygen enhances the therapeutic potential of human umbilical mesenchymal stem cells in a rat model of spinal cord injury. Brain Research. 2016 Jul 1;**1642**:426-435. DOI: 10.1016/j.brainres.2016.04.025. PubMed PMID:27085204

[154] Zhang Y, Yang J, Zhang P, Liu T, Xu J, Fan Z, Shen Y, Li W, Zhang H. Calcitonin gene-related peptide is a key factor in the homing of transplanted human MSCs to sites of spinal cord injury. Scientific Reports. 2016 Jun 14;**6**:27724. DOI: 10.1038/srep27724. PubMed PMID: 27296555; PubMed Central PMCID: PMC4906351

[155] Yaghoobi K, Kaka G, Mansouri K, Davoodi S, Sadraie SH, Hosseini SR. *Lavandula angustifolia* extract improves the result of human umbilical mesenchymal Wharton's jelly stem cell transplantation after contusive spinal cord injury in Wistar rats. Stem Cells International. 2016;**2016**:5328689. DOI: 10.1155/2016/5328689. PubMed PMID: 27057171; PubMed Central PMCID: PMC4769777

[156] Gao S, Ding J, Xiao HJ, Li ZQ, Chen Y, Zhou XS, Wang JE, Wu J, Shi WZ. Anti-inflammatory and anti-apoptotic effect of combined treatment with methylprednisolone and amniotic membrane mesenchymal stem cells after spinal cord injury in rats. Neurochemical Research. 2014 Aug;**39**(8):1544-1552. DOI: 10.1007/s11064-014-1344-9. PubMed PMID: 24890008

[157] Yeng CH, Chen PJ, Chang HK, Lo WY, Wu CC, Chang CY, Chou CH, Chen SH. Attenuating spinal cord injury by conditioned medium from human umbilical cord blood-derived CD34+ cells in rats. Taiwanese Journal of Obstetrics & Gynecology. 2016 Feb;**55**(1):85-93. DOI: 10.1016/j.tjog.2015.12.009. PubMed PMID: 26927256

[158] Abdullah RH, Yaseen NY, Salih SM, Al-Juboory AA, Hassan A, Al-Shammari AM. Induction of mice adult bone marrow mesenchymal stem cells into functional motor neuron-like cells. Journal of Chemical Neuroanatomy. 2016 Nov;**77**:129-142. DOI: 10.1016/j.jchemneu.2016.07.003. PubMed PMID: 27417692

[159] Besalti O, Aktas Z, Can P, Akpinar E, Elcin AE, Elcin YM. The use of autologous neurogenically-induced bone marrow-derived mesenchymal stem cells for the treatment of paraplegic dogs without nociception due to spinal trauma. Journal of Veterinary Medical Science. 2016 Oct 1;**78**(9):1465-1473. PubMed PMID: 27301583; PubMed Central PMCID: PMC5059374

[160] Mohammad MH, Al-Shammari AM, Al-Juboory AA, Yaseen NY. Characterization of neural stemness status through the neurogenesis process for bone marrow mesenchymal stem cells. Stem Cells and Cloning : Advances and Applications. 2016 Apr 18;**9**:1-15. DOI: 10.2147/SCCAA.S94545. PubMed PMID: 27143939; PubMed Central PMCID: PMC4846075

[161] Ebrahimi-Barough S, Hoveizi E, Yazdankhah M, Ai J, Khakbiz M, Faghihi F, Tajerian R, Bayat N. Inhibitor of PI3K/Akt signaling pathway small molecule promotes motor neuron differentiation of human endometrial stem cells cultured on electrospun biocomposite polycaprolactone/collagen scaffolds. Molecular Neurobiology. 2016 Mar 18. [Epub ahead of print] PubMed PMID: 26993294

[162] Qiu XC, Jin H, Zhang RY, Ding Y, Zeng X, Lai BQ, Ling EA, Wu JL, Zeng YS. Donor mesenchymal stem cell-derived neural-like cells transdifferentiate into myelin-forming cells and promote axon regeneration in rat spinal cord transection. Stem Cell Research & Therapy. 2015 May 27;**6**:105. DOI: 10.1186/s13287-015-0100-7. PubMed PMID: 26012641; PubMed Central PMCID: PMC4482203

[163] Zeng X, Qiu XC, Ma YH, Duan JJ, Chen YF, Gu HY, Wang JM, Ling EA, Wu JL, Wu W, Zeng YS. Integration of donor mesenchymal stem cell-derived neuron-like cells into host neural network after rat spinal cord transection. Biomaterials. 2015 Jun;**53**:184-201. DOI: 10.1016/j.biomaterials.2015.02.073. PubMed PMID: 25890718

[164] Faghihi F, Mirzaei E, Ai J, Lotfi A, Sayahpour FA, Ebrahimi-Barough S, Joghataei MT. Differentiation potential of human chorion-derived mesenchymal stem cells into motor neuron-like cells in two- and three-dimensional culture systems. Molecular

Neurobiology. 2016 Apr;**53**(3):1862-1872. DOI: 10.1007/s12035-015-9129-y. Erratum in: Mol Neurobiol. 2016 Apr;53(3):1873. Barough, Somayeh Ebrahimi [Corrected to Ebrahimi-Barough, Somayeh]. PubMed PMID: 25790953

[165] Wakao S, Matsuse D, Dezawa M. Mesenchymal stem cells as a source of Schwann cells: Their anticipated use in peripheral nerve regeneration. Cells, Tissues, Organs. 2014;**200**(1):31-41. DOI: 10.1159/000368188. Review. PubMed PMID: 25765009

[166] Zhao Y, Jiang H, Liu XW, Xiang LB, Zhou DP, Chen JT. MiR-124 promotes bone marrow mesenchymal stem cells differentiation into neurogenic cells for accelerating recovery in the spinal cord injury. Tissue and Cell. 2015 Apr;**47**(2):140-146. DOI: 10.1016/j.tice.2015.01.007. PubMed PMID: 25697062

[167] Li Z, Zhao W, Liu W, Zhou Y, Jia J, Yang L. Transplantation of placenta-derived mesenchymal stem cell-induced neural stem cells to treat spinal cord injury. Neural Regeneration Research. 2014 Dec 15;**9**(24):2197-2204. DOI: 10.4103/1673-5374.147953. PubMed PMID: 25657742; PubMed Central PMCID: PMC4316454

[168] Mannoji C, Koda M, Kamiya K, Dezawa M, Hashimoto M, Furuya T, Okawa A, Takahashi K, Yamazaki M. Transplantation of human bone marrow stromal cell-derived neuroregenerative cells promotes functional recovery after spinal cord injury in mice. Acta Neurobiologiae Experimentalis (Warsaw). 2014;**74**(4):479-488. PubMed PMID:25576978

[169] Zou D, Chen Y, Han Y, Lv C, Tu G. Overexpression of microRNA-124 promotes the neuronal differentiation of bone marrow-derived mesenchymal stem cells. Neural Regeneration Research. 2014 Jun 15;**9**(12):1241-1248. DOI: 10.4103/1673-5374.135333. PubMed PMID: 25206789; PubMed Central PMCID: PMC4146284

[170] Oh SK, Jeon SR. Current concept of stem cell therapy for spinal cord injury: A review. Korean Journal of Neurotrauma. 2016 Oct;**12**(2):40-46. Review. PubMed PMID: 27857906; PubMed Central PMCID: PMC5110917

[171] Park JH, Kim DY, Sung IY, Choi GH, Jeon MH, Kim KK, et al. Long-term results of spinal cord injury therapy using mesenchymal stem cells derived from bone marrow in humans. Neurosurgery. 2012;**70**:1238-1247

[172] Parr AM, Kulbatski I, Tator CH. Transplantation of adult rat spinal cord stem/progenitor cells for spinal cord injury. Journal of Neurotrauma. 2007;**24**:835-845

[173] White SV, Czisch CE, Han MH, Plant CD, Harvey AR, Plant GW. Intravenous transplantation of mesenchymal progenitors distribute solely to the lungs and improve outcomes in cervical spinal cord injury. Stem Cells. 2016 Jul;**34**(7):1812-1825. DOI: 10.1002/stem.2364. PubMed PMID: 26989838

[174] Kim Y, Jo SH, Kim WH, Kweon OK. Antioxidant and anti-inflammatory effects of intravenously injected adipose derived mesenchymal stem cells in dogs with acute spinal cord injury. Stem Cell Research & Therapy. 2015 Nov 26;**6**:229. DOI: 10.1186/s13287-015-0236-5. PubMed PMID: 26612085; PubMed Central PMCID: PMC4660672

[175] Maia L, da Cruz Landim-Alvarenga F, Taffarel MO, de Moraes CN, Machado GF, Melo GD, Amorim RM. Feasibility and safety of intrathecal transplantation of autologous bone marrow mesenchymal stem cells in horses. BMC Veterinary Research. 2015 Mar 15;**11**:63. DOI: 10.1186/s12917-015-0361-5. PubMed PMID: 25879519; PubMed CentralPMCID: PMC4369105

[176] Ninomiya K, Iwatsuki K, Ohnishi Y, Ohkawa T, Yoshimine T. Intranasal delivery of bone marrow stromal cells to spinal cord lesions. Journal of Neurosurgery: Spine. 2015 Jul;**23**(1):111-119. DOI: 10.3171/2014.10.SPINE14690. PubMed PMID: 25840039

[177] Matsushita T, Lankford KL, Arroyo EJ, Sasaki M, Neyazi M, Radtke C, Kocsis JD. Diffuse and persistent blood-spinal cord barrier disruption after contusive spinal cord injury rapidly recovers following intravenous infusion of bone marrow mesenchymal stem cells. Experimental Neurology. 2015 May;**267**:152-164. DOI: 10.1016/j.expneurol.2015.03.001. PubMed PMID: 25771801

[178] Cruz-Martinez P, Pastor D, Estirado A, Pacheco-Torres J, Martinez S, Jones J. Stem cell injection in the hindlimb skeletal muscle enhances neurorepair in mice with spinal cord injury. Regenerative Medicine. 2014;**9**(5):579-591. DOI: 10.2217/rme.14.38. PubMed PMID: 25372077

[179] Lerner MZ, Matsushita T, Lankford KL, Radtke C, Kocsis JD, Young NO. Intravenous mesenchymal stem cell therapy after recurrent laryngeal nerve injury: a preliminary study. Laryngoscope. 2014 Nov;**124**(11):2555-2560. DOI: 10.1002/lary.24798. PubMed PMID: 25043703

[180] Zhang D, He X. A meta-analysis of the motion function through the therapy of spinal cord injury with intravenous transplantation of bone marrow mesenchymal stem cells in rats. PLoS One. 2014 Apr 1;**9**(4):e93487. DOI: 10.1371/journal.pone.0093487. PubMed PMID: 24690752; PubMed Central PMCID:PMC3972121

[181] Jarocha D, Milczarek O, Wedrychowicz A, Kwiatkowski S, Majka M. Continuous improvement after multiple mesenchymal stem cell transplantations in a patient with complete spinal cord injury. Cell Transplantation. 2015;**24**(4):661-672. DOI: 10.3727/096368915X687796. PubMed PMID: 25807231

[182] Wei L, Zhang J, Xiao XB, Mai HX, Zheng K, Sun WL, Wang L, Liang F, Yang ZL, Liu Y, Wang YQ, Li ZF, Wang JN, Zhang WJ, You H. Multiple injections of human umbilical cord-derived mesenchymal stromal cells through the tail vein improve microcirculation and the microenvironment in a rat model of radiation myelopathy. Journal of Translational Medicine. 2014 Sep 8;**12**:246. DOI: 10.1186/s12967-014-0246-6. PubMed PMID:25196350; PubMed Central PMCID: PMC4174271

[183] Morita T, Sasaki M, Kataoka-Sasaki Y, Nakazaki M, Nagahama H, Oka S, Oshigiri T, Takebayashi T, Yamashita T, Kocsis JD, Honmou O. Intravenous infusion of mesenchymal stem cells promotes functional recovery in a model of chronic spinal cord injury. Neuroscience. 2016 Oct 29;**335**:221-231. DOI: 10.1016/j.neuroscience.2016.08.037. PubMed PMID: 27586052

[184] Schimke MM, Marozin S, Lepperdinger G. Patient-specific age: The other side of the coin in advanced mesenchymal stem cell therapy. Frontiers in Physiology. 2015 Dec 2;**6**:362. DOI: 10.3389/fphys.2015.00362. Review. PubMed PMID: 26696897; PubMed Central PMCID: PMC4667069

[185] Ketterl N, Brachtl G, Schuh C, Bieback K, Schallmoser K, Reinisch A, Strunk D. A robust potency assay highlights significant donor variation of human mesenchymal stem/progenitor cell immune modulatory capacity and extended radio-resistance. Stem Cell Research & Therapy. 2015 Dec 1;**6**:236. DOI: 10.1186/s13287-015-0233-8. PubMed PMID: 26620155; PubMed Central PMCID:PMC4666276

[186] Neuhuber B, et al. Axon growth and recovery of function supported by human bone marrow stromal cells in the injured spinal cord exhibit donor variations. Brain Research. 2005;**1035**(1):73-85

[187] Ribeiro TB, Duarte AS, Longhini AL, Pradella F, Farias AS, Luzo AC, Oliveira AL, Olalla Saad ST. Neuroprotection and immunomodulation by xenografted human mesenchymal stem cells following spinal cord ventral root avulsion. Scientific Reports. 2015 Nov 9;**5**:16167. DOI: 10.1038/srep16167. PubMed PMID: 26548646; PubMed Central PMCID: PMC4637826

[188] Jung DI, Ha J, Kang BT, Kim JW, Quan FS, Lee JH, Woo EJ, Park HM. A comparison of autologous and allogenic bone marrow-derived mesenchymal stem cell transplantation in canine spinal cord injury. Journal of Neurological Sciences. 2009 Oct 15;**285**(1-2):67-77. DOI: 10.1016/j.jns.2009.05.027. PubMed PMID: 19555980

[189] Torres-Espín A, Redondo-Castro E, Hernandez J, Navarro X. Immunosuppression of allogenic mesenchymal stem cells transplantation after spinal cord injury improves graft survival and beneficial outcomes. Journal of Neurotrauma. 2015 Mar 15;**32**(6):367-380. DOI: 10.1089/neu.2014.3562. PubMed PMID: 25203134; PubMed Central PMCID: PMC4361361

[190] Vaněček V, Zablotskii V, Forostyak S, Růžička J, Herynek V, Babič M, Jendelová P, Kubinová S, Dejneka A, Syková E. Highly efficient magnetic targeting of mesenchymal stem cells in spinal cord injury. International Journal of Nanomedicine. 2012;**7**:3719-3730. DOI: 10.2147/IJN.S32824. PubMed PMID: 22888231; PubMed CentralPMCID: PMC3414205

[191] Hu SL, Zhang JQ, Hu X, Hu R, Luo HS, Li F, Xia YZ, Li JT, Lin JK, Zhu G, Feng H. In vitro labeling of human umbilical cord mesenchymal stem cells with superparamagnetic iron oxide nanoparticles. Journal of Cellular Biochemistry. 2009 Oct 1;**108**(2):529-535. DOI: 10.1002/jcb.22283. PubMed PMID: 19623584

[192] Siddiqui AM, Khazaei M, Fehlings MG. Translating mechanisms of neuroprotection, regeneration, and repair to treatment of spinal cord injury. Progress in Brain Research. 2015;**218**:15-54. DOI: 10.1016/bs.pbr.2014.12.007. Review. PubMedPMID: 25890131

[193] Morales II, Toscano-Tejeida D, Ibarra A. Non pharmacological strategies to promote spinal cord regeneration: A view on some individual or combined approaches. Current Pharmaceutical Design. 2016;**22**(6):720-727. Review. PubMed PMID: 26635267

[194] Dong Y, Yang L, Yang L, Zhao H, Zhang C, Wu D. Transplantation of neurotrophin-3-transfected bone marrow mesenchymal stem cells for the repair of spinal cord injury. Neural Regeneration Research. 2014 Aug 15;**9**(16):1520-1524. DOI: 10.4103/1673-5374.139478. PubMed PMID: 25317169; PubMed Central PMCID: PMC4192969

[195] Zhang RP, Wang LJ, He S, Xie J, Li JD. Effects of magnetically guided, SPIO-labeled, and neurotrophin-3 gene-modified bone mesenchymal stem cells in a rat model of spinal cord injury. Stem Cells International. 2016;**2016**:2018474. DOI: 10.1155/2016/2018474. PubMed PMID: 26649047; PubMed Central PMCID: PMC4663356

[196] Gong Y, Wang H, Xia H. Stable transfection into rat bone marrow mesenchymal stem cells by lentivirus-mediated NT-3. Molecular Medicine Reports. 2015 Jan;**11**(1):367-373. DOI: 10.3892/mmr.2014.2727. PubMed PMID: 25333669

[197] Cho H, Choi YK, Lee DH, Park HJ, Seo YK, Jung H, Kim SC, Kim SM, Park JK. Effects of magnetic nanoparticle-incorporated human bone marrow-derived mesenchymal stem cells exposed to pulsed electromagnetic fields on injured rat spinal cord. Biotechnology and Applied Biochemistry. 2013 Nov-Dec;**60**(6):596-602. DOI: 10.1002/bab.1109. PubMed PMID: 24033637

[198] Liu Z, He B, Zhang RY, Zhang K, Ding Y, Ruan JW, Ling EA, Wu JL, Zeng YS. Electroacupuncture promotes the differentiation of transplanted bone marrow mesenchymal stem cells preinduced with neurotrophin-3 and retinoic acid into oligodendrocyte-like cells in demyelinated spinal cord of rats. Cell Transplantation. 2015;**24**(7):1265-1281. DOI: 10.3727/096368914X682099. PubMed PMID: 24856958

[199] Shin DA, Pennant WA, Yoon DH, Ha Y, Kim KN. Co-transplantation of bone marrow-derived mesenchymal stem cells and nanospheres containing FGF-2 improve cell survival and neurological function in the injured rat spinal cord. Acta Neurochirurgica (Wien). 2014 Feb;**156**(2):297-303. DOI: 10.1007/s00701-013-1963-y. PubMed PMID: 24352373225

[200] Abbaszadeh HA, Tiraihi T, Noori-Zadeh A, Delshad AR, Sadeghizade M, Taheri T. Human ciliary neurotrophic factor-overexpressing stable bone marrow stromal cells in the treatment of a rat model of traumatic spinal cord injury. Cytotherapy. 2015 Jul;**17**(7):912-921. DOI: 10.1016/j.jcyt.2015.03.689. PubMed PMID: 25939801

[201] Chen D, Zeng W, Fu Y, Gao M, Lv G. Bone marrow mesenchymal stem cells combined with minocycline improve spinal cord injury in a rat model. International Journal of Clinical and Experimental Pathology. 2015 Oct 1;**8**(10):11957-11969. PubMed PMID: 26722382; PubMed Central PMCID: PMC4680327

[202] Zhou YJ, Liu JM, Wei SM, Zhang YH, Qu ZH, Chen SB. Propofol promotes spinal cord injury repair by bone marrow mesenchymal stem cell transplantation. Neural Regeneration Research. 2015 Aug;**10**(8):1305-1311. DOI: 10.4103/1673-5374.162765. PubMed PMID: 26487860; PubMed Central PMCID: PMC4590245

[203] Neirinckx V, Agirman G, Coste C, Marquet A, Dion V, Rogister B, Franzen R,Wislet S. Adult bone marrow mesenchymal and neural crest stem cells are chemoattractive and accelerate motor recovery in a mouse model of spinal cord injury. Stem Cell Research &

Therapy. 2015 Nov 4;**6**:211. DOI: 10.1186/s13287-015-0202-2. PubMed PMID: 26530515; PubMed Central PMCID: PMC4632651

[204] Ge L, Liu K, Liu Z, Lu M. Co-transplantation of autologous OM-MSCs and OM-OECs: A novel approach for spinal cord injury. Reviews in the Neurosciences. 2016 Apr 1;**27**(3):259-270. DOI: 10.1515/revneuro-2015-0030. PubMed PMID: 26574889

[205] Wu S, Cui G, Shao H, Du Z, Ng JC, Peng C. The Cotransplantation of olfactory ensheathing cells with bone marrow mesenchymal stem cells exerts antiapoptotic effects in adult rats after spinal cord injury. Stem Cells International. 2015;**2015**:516215. DOI: 10.1155/2015/516215. PubMed PMID: 26294918; PubMed Central PMCID: PMC4532957

[206] Torres-Espín A, Redondo-Castro E, Hernández J, Navarro X. Bone marrow mesenchymal stromal cells and olfactory ensheathing cells transplantation after spinal cord injury – A morphological and functional comparison in rats. European Journal of Neuroscience. 2014 May;**39**(10):1704-1717. DOI: 10.1111/ejn.12542. PubMed PMID: 24635194

[207] Oraee-Yazdani S, Hafizi M, Atashi A, Ashrafi F, Seddighi AS, Hashemi SM, Seddighi A, Soleimani M, Zali A. Co-transplantation of autologous bone marrow mesenchymal stem cells and Schwann cells through cerebral spinal fluid for the treatment of patients with chronic spinal cord injury: Safety and possible outcome. Spinal Cord. 2016 Feb;**54**(2):102-109. DOI: 10.1038/sc.2015.142. PubMed PMID: 26526896

[208] Kanno H, Pearse DD, Ozawa H, Itoi E, Bunge MB. Schwann cell transplantation for spinal cord injury repair: Its significant therapeutic potential and prospectus. Reviews in the Neurosciences. 2015;**26**(2):121-128. DOI: 10.1515/revneuro-2014-0068. Review. PubMed PMID: 25581750

[209] Amr S, Amr H. Bridging defects in chronic spinal cord injury using peripheral nerve grafts; results of indwelling catheter implantation; a comparative clinical study. In: Research on Complication – Current Issues and Technology. Brisbane, Australia: iConcept Press Limited; 2016. ISBN 978-1-922227-37-9

[210] Georgiou M, Bunting SCJ, Davies HA, Loughlin AJ, Golding JP, Phillips JB. Engineered neural tissue for peripheral nerve repair. Biomaterials. 2013;**34**:7335-7343

[211] Goraltchouk A, Scanga V, Morshead CM, Shoichet MS. Incorporation of protein eluting microspheres into biodegradable nerve guidance channels for controlled release. Journal of Controlled Release. 2006;**110**(2):400-407

[212] Yang Y, De LL, Rives CB, Jang JH, Lin WC, Shull KR, Shea LD. Neurotrophin releasing single and multiple lumen nerve conduits. Journal of Controlled Release. 2005;**104**(3):433-446

[213] Xu X, Yee WC, Hwang PY, Yu H, Wan AC, Gao S, Boon KL, Mao HQ, Leong KW, Wang S. Peripheral nerve regeneration with sustained release of poly (phosphoester) microencapsulated nerve growth factor within nerve guide conduits. Biomaterials. 2003:**24**(13):2405-2412

[214] Benoit JP, Faisant N, Venier-Julienne MC, Menei P. Development of micro-spheres for neurological disorders: From basics to clinical applications. Journal of Controlled Release. 2000;**65**:285-296. [PubMed: 10699288]

[215] Chan JM, Zhang L, Yuet KP, Liao G, Rhee JW, Langer R, Farokhzad OC. PLGA-lecithin-PEG coreshell nanoparticles for controlled drug delivery. Biomaterials. 2009;**30**:1627-1634. [PubMed: 19111339]

Chapter 12

Mesenchymal Stem Cell in the Intervertebral Disc

Dan Long, Shaoyu Liang, Haiyan Liu, Xiuping Wu,
Zhongyang Li, Haozhong Wang, Shishu Huang and
Jiancheng Zeng

Additional information is available at the end of the chapter

http://dx.doi.org/10.5772/intechopen.69296

Abstract

Degeneration of the intervertebral disc (IVD) is a major spinal disorder that causes back pain. Nucleus pulposus (NP) in the central of IVD dehydrates and become more fibrous in the IVD degeneration. NP cells undergo apoptosis with the degeneration of extracellular matrix (ECM) components. To replenish the NP cells and core ECM, bone marrow mesenchymal stromal cells (BMSCs) have been highlighted in the regeneration of IVD degeneration. BMSCs differentiate into NP-like cells with the secretion of ECM components, which may not only replenish the number of NP cells but also stimulate NP reconstruction. This further maintains tissue homeostasis. Up to date, the disc progenitor cells (DPCs) have been identified with the characteristics of multidifferentiation and stem cell phenotype. These cells are involved in the IVD diseases and show regenerative potentials. However, the differences between the BMSCs and DPCs remain elusive, in particular, the cellular connection *in vivo*. As such, this chapter will discuss the findings of the two cell types and propose a novel concept in the understanding of the biology of IVD.

Keywords: low back pain, intervertebral disc, nucleus pulposus, progenitor cells, extracellular matrix

1. Introduction

Low back pain (LBP) is the second most common symptom in the United States. Of the US population, 85% people experience an episode of LBP at some point in their lifetime. For individuals under 45 years, LBP remains the most common cause of disability and is generally associated with a work-related injury. In 2005, an estimate of 85.9 billion dollars was spent in the related treatment of back and neck pain· The relevant statistics indicated that the healthcare expenditures increased 65% between 1997 and 2005 without evidence of improvement in health status.

2. The shielded structure and rigid environment of intervertebral disc

An intervertebral disc (IVD) is a cylindrical structure, comprising a well-hydrated central nucleus pulposus (NP), an annulus fibrosus (AF) consisting of firm and flexible collagenous lamellae which surrounds the NP, and cartilaginous endplates forming an interface between the disc and adjacent vertebrae (**Figure 1**).

During the development of mammals, the vertebral column derives from the aggregation of mesenchymal cells around the notochord [1]. Following segmentation, motion segments emerge with large number of cells accumulating in the developing AF but fewer cells in the rapidly growing vertebral bodies. The cells in the AF become highly orientated, laying down the disc matrix in a similar orientation to form the concentric annular lamellar structure [1, 2]. Notochordal cells are named by their typical morphology of the notochord (physaliferous), a population of large cells with small and densely packed nuclei and cytoplasmic matrix vacuoles in human nucleus pulposus, are presumed remnants of the embryonic notochord that guided formation of the spine and the nuclei polposi [1]. The abundance of notochordal cells within NP declines with age at a rapid rate which varies among different species; where, by early adulthood in the human and species including that of chondrodystrophoid dog, nucleus

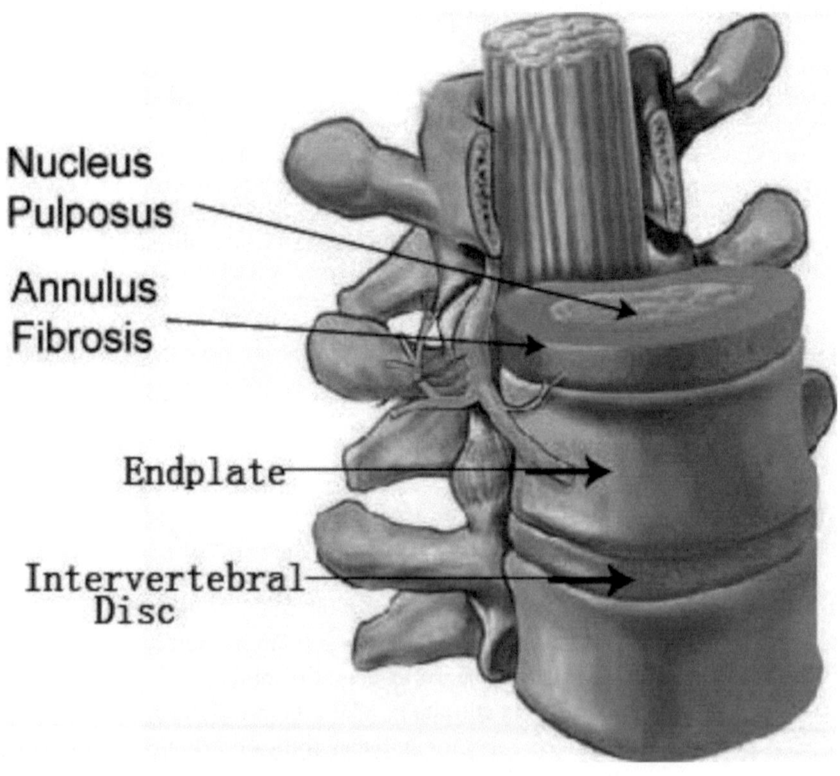

Figure 1. Schematic cross-section of an intervertebral disc.

becomes repopulated by chondrocyte-like cells that are thought to be originated from the adjacent endplate or inner AF regions [3]. All the previous results are solely based on the morphological detection and the existence of notochordal cells is believed to be significantly associated with aging. However, a recent study shows notochordal cells exist in human young and middle age by immunohistochemistry of notochordal cell markers. The occurrence of notochordal cells with immunohistochemical phenotype significantly correlates with granular matrix changes and cleft formation in the nucleus pulposus [4].

A network of microscopic blood vessels penetrates the endplates to principally provide nutrition for the disc and normally disappears around the time of skeletal maturity [3]. With a sparse vascular supply in the outer lamellae of the annulus, mature discs are totally reliant on diffusion of essential solutes across the endplates for nutrition and metabolic exchange [5]. The inner part of the IVD, particularly the NP, is completely avascular and aneural in the largest of the mature human lumbar IVD, where some cells can be 20 mm away from the nearest direct blood supply thereby making the NP severely hypoxic [5]. Mature IVD is composed of heterogeneous cell populations. A majority of the AF cells originate from the mesenchyme and exhibit many characteristics of fibroblasts and chondrocytes, such as the ability to synthesize the type I and II collagen and aggregating proteoglycans [3]. The morphology of AF cells may reflect their adaptation within the special biochemical and structural environment, as these cells appear ellipsoidal and align with the oriented collagen fibers within the lamellas [6]. Cells in the outer AF region display thin cytoplasmic projections that stain positive for both actin microfilaments and vimentin intermediate filaments, which have been associated with tissue regions subjected to compression [7]. Cells within the inner AF regions are often rounded, sparsely distributed, and surrounded by a pericellular matrix region rich in types III and VI collagen [7]. The NP is a gelatinous structure comprised primarily of aggrecan and type II collagen together with the small amounts of collagen type VI, IX, and XI. Cells are sparsely distributed in the NP and may also extend small cytoplasmic processes and, similar to chondrocytes, these cells highly express vimentin intermediate filaments, F-actin, and cytokeratins [7].

The most prominent feature of the IVD is its high content of extracellular matrix (ECM), which is substantially maintained by the cells within IVD, of which, the disc matrix is an elaborate structure of macromolecules that attract and hold water. The major structural components of the macromolecule are collagens and proteoglycans [8]. It is estimated that the ratio of type II collagen and the proteoglycan aggrecan in the AF is 1:20 [9]. Collagens provide firm and tensile strength whereas proteoglycans, through interactions with water, give the tissues stiffness, viscoelasticity and resistance to compression [8, 9]. Collagenous proteins comprise 70% of the outer annulus dry weight, but only account for 20% of NP [8, 9]. On the contrary, NP has a higher proteoglycan concentration, with up to 50% of the nucleus dry weight in adolescence. Given the co-existence of multiple matrix components and their high contents in IVD, the integrity of the IVD partially relies on the proper balance between the matrix synthesis and degradation, and the failure of which is suggested being a cause of the disc degeneration [9].

IVD degeneration is associated with the LBP. The IVD, especially the inner fibrosus (IF) and nucleus pulposus (NP), is virtually avascular and therefore highly hypoxic. At the cranial and caudal ends of each disc are the cartilaginous endplates that separate the vertebral bone from the disc itself and are believed to be the major channel of nutrient diffusion in IVD. Recent studies have reported changes in tissue structure, various cellular parameters and composition of matrix macromolecules in degenerated discs. Disc degeneration is characterized by decreased water and proteoglycan content and loss of the gel-like appearance of NP. Disc degeneration is thought to be contributed by increased cell senescence and dysregulated cellular activities. The IVD has limited nutrient, oxygen supply, and constant high mechanical stress. These may lead to difficulty for IVD to regenerate itself in IVD degeneration and injuries.

3. The finding of disc progenitor cells

Adult tissue-specific stem cells are a rare heterogeneous population of multipotent cells that can be isolated from many different adult and fetal tissues, including bone marrow, muscle, fat, hair follicles, tooth root, placenta, dermis, perichondrium, articular cartilage, umbilical cord, lung, and liver [10]. These cells show extensive proliferation, produce differentiated progeny, and functionally repair damaged tissues [11]. Adult stem cells normally reside in a specific cellular microenvironment (niche) that constitutes a privileged setting for the support of self-renewal [12]. There are three general properties unique for all the stem cells, regardless of their source. Clonogenicity, the ability of a single cell to proliferate independently to form a colony, is a property commonly ascribed to stem cells, although many clonogenic cells are limited in their capacity for expansion *ex vivo* [13]. Secondly, stem cells can give rise to specialized cells. When unspecialized stem cells give rise to specialized cells, the process is called differentiation [13]. Differentiation is triggered by the signals inside and outside cells. The internal signals from genes are interspersed across long strands of DNA and carry coded instructions for all cellular structures and functions [13]. The external signals comprise physical contact with neighboring cells, chemicals secreted by other cells, and certain molecules in the microenvironment [14]. The interaction of signals during differentiation causes the cell's DNA to acquire epigenetic marks that restrict DNA expression in the cell and can be passed on through cell division [14]. Most adult stem cells are multipotent, capable of differentiating into at least three lineages (osteogenic, chondrogenic, and adipogenic) when cultured under defined *in vitro* conditions [14]. Thirdly, adult stem cells can go through numerous cycles of cell division while maintaining the undifferentiated state [15]. Stem cells are capable of dividing and renewing themselves for long periods. Unlike terminal stage cells, which do not normally replicate themselves, stem cells may replicate many times or proliferate. A starting population of stem cells that proliferates for many months in the laboratory can yield millions of cells [16]. If the resulting cells continue to be undifferentiated, like the original stem cells, the cells are said to be capable of self-renewal.

Bone marrow mesenchymal stromal cells (BMSCs) are composed of heterogeneous population of undifferentiated and committed cells [17]. The regenerative properties ascribed to BMSCs are characterized into three aspects: the plasticity to differentiate toward target cell types, the activation of the proliferation of resident cells, and the improvement of nutrient supply via paracrine effects. One of the remarkable phenomena of IVD degeneration is a reduction of proteoglycan content, partially caused by the apoptosis/necrosis of the nucleus pulposus (NP) cells in the IVD. The biotherapeutic treatment, therefore, aims to replenish the local resident cells and structural extracellular matrix (ECM) within IVD [18, 19]. Three-dimensional (3D) cultures have been used to induce BMSCs to differentiate into chondrocyte-like cells. These include cell pellet, alginate bead, hydrogel, and engineered 3D scaffold [20–24]. Chondrocyte-like phenotype can also be obtained via a single monolayer co-culture of BMSCs, either with NPC or annulus fibrosis (AF) cells with cell-cell contact [25, 26]. Importantly, the chondrocyte-like cells have been shown to possess NPC phenotype [27, 28]. Via intradiscal injection into degenerative IVD, BMSCs are able to survive and commence proliferation under severe hypoxic environment [29–32]. The production of ECM elevates in the NP post-transplantation, including aggrecan, collagen type, and glycosaminoglycans [33]. Animal studies have validated the effect of BMSCs. BMSCs are capable of replenishing NPCs and evoking their production of ECM components. This arrests the progressive decrease of disc height, as well as to partially maintain or even restore minimal disc height in mildly degenerative IVD [34]. Therefore, intradiscal transplantation of BMSCs shed some light on the maintenance of IVD homeostasis. However, the utility of BMSCs is still a subject of debate due to many unanswered questions. The method of transplantation, the choice of carrier, and the fate of BMSCs after delivery need further investigation. Notably, the intradiscal-delivered BMSCs have been found to leak from IVD and generate osteophytes [35]. Although embedding BMSCs in tissue-engineered scaffold before transplantation can alleviate the leakage issue, safety issues remain a concern [36].

IVD cannot self-repair and no cure is currently available for IVD degeneration. Various animal models have suggested the promising potential of mesenchymal stem cell (MSC) implantation to arrest IVD degeneration or even partially regenerate the disc [21, 37]. However, there are two major issues in MSC therapies: first, most studies are focused on the exogenous stem cells but the limitation is their potential immunogenicity. MSCs have indeed been shown to halt degeneration processes but are rarely able to completely regenerate the degenerative disc as the disc degeneration often continues after a certain period [37]. Besides, the therapy is invasive and therefore may potentially lead to complications such as infection and discitis. Second, all studies are carried out in quadruped animals and these models do not more closely resemble humans in terms of biomechanical loading in the spine, diffusion distances for nutrients and metabolites to the NP, age-related declination of notochordal cells, and the occurrence of age-related disc degeneration. In addition, most studies have been monitored only for relatively short time, in the range of weeks after treatments and their efficacy in long term remains elusive.

Recently, several studies have reported that cells derived from IVD tissue have multi-differentiation potential and possess mesenchymal stem cell-like features *in vitro*. NPCs express many MSC surface markers and are potent in differentiating into chondrogenic, osteogenic, and to some extent adipogenetic lineages [38, 39]. Similar multipotency of annulus fibrosus cells (AFCs) from scoliotic IVD was confirmed [40]. In degenerated and nondegenerated (scoliosis) IVD tissues, cells express stem/progenitor markers such as *OCT3/4, CD105, CD90, STRO-1* and *NOTCH1* [41]. A population of NPCs from nonchondrodystrophic canine IVD possesses neurogenic differentiation potential *in vivo* and expresses stemness genes, including *Sox2, Oct3/4, Nanog, CD133, Nestin* and *NCAM* [42]. Interestingly, this subset of NPCs expresses higher level of *Nanog gene* compared to BMSCs and is negative in the expression of protein 0 and *Brachyury gene*, which are positive in unsorted NPCs [42]. However, these data were drawn from models of mice, rats, dogs, rabbits, and even Chinese hamsters [43]. Differences between the IVDs of human and these animals, however, are large and present at multiple levels. These include anatomical structure, cellular and biochemical components, mechanical loading, and age-related changes. Nonhuman primates are closely related to humans and have been shown to be excellent model organisms for many health and disease conditions in human. The structure of the spine of the primates, including baboon and the higher species Rhesus monkeys, is similar to that of human, except with some deviations in the number of the vertebra and the spine curvature. The monkeys spend much of their time in semi-erect and erect positions, possibly indicating the loading conducted through the vertebral column closely parallel to those encountered in humans [44, 45]. Histology and microscopic features of monkey IVD also suggest its high similarity to human IVD [45]. Furthermore, microarchitecture of glycosaminoglycans and collagens in the IVD of Rhesus monkey has also been shown to be similar to human IVD at the ultrastructural level [45]. More importantly, recent MRI studies have demonstrated that IVD degeneration develops in healthy monkeys at 5 years of age, the human age equivalent of 17.5 years [45]. With increasing disc degeneration, changes in disc height, MRI signals within NP and hyperostotic spondylotic can all be detected [46]. Such changes are also reported to correlate with radiographic and histopathologic changes [46]. Nonhuman primates, particularly Rhesus monkey, are also considered an advanced model to study IVD degeneration. Therefore, study on cells derived from normal IVD of Rhesus monkey further confirmed the existence of IVD disc progenitor cells (DPCs), which possess clonogenicity, multipotency, and differentiation after serial expansion *in vitro* and *in vivo* [47].

Thus, endogenous DPCs have become an enticing subject in the IVD study. However, whether IVD aging/degeneration is associated with or resulted from the diminishing of endogenous DPCs remains unknown. A study has identified a population of NPCs from mice and humans expressing tyrosine kinase receptor Tie 2, a novel surface marker of BMSCs, and disialoganglioside 2 (GD2), a hematopoietic stem cells (HSCs) surface protein [48]. $Tie2^{+}GD2^{+}$ NPCs are clonally multipotent and generate NP-like tissue in *in vivo* serial transplantation [48]. Importantly, $Tie2^{+}GD2^{-}$ NPCs are the precursor of $Tie2^{+}GD2^{+}$ NPCs and the frequency of these progenitor cells decreases with aging and the severity of degeneration of the IVD [48]. Interestingly, DPCs from healthy IVD possess higher differentiation capacity toward chondrogenic lineage and NP-like cells compared with DPCs from degenerative IVD [49]. Therefore, the number and functionality of DPCs are associated with the degenerative process. Further validation of this theory may promote understanding of the etiology of IVD degeneration and contribute to the development of novel biotherapies.

Taking together, DPCs, as an endogenous cell population, may be more suitable in the biotherapeutic treatment of IVD diseases and become a new target for IVD regeneration. However, before any therapeutic application or pre-clinical/clinical trial, several research gaps need to be addressed. First, the mechanism of hypoxia-induced $Tie2^{+}$ expression on $Tie^{+}GD2^{+}$ NPCs awaits further elucidation. NPCs were sensitive to oxygen tension and hypoxia-inducible factors (HIF) reduce the susceptibility of hypoxic apoptosis of NPCs. Whether Tie2 couples with HIFs to resist hypoxic stress in NPCs is worthy of being studied. Second, the progenitor niche components of DPCs need to be identified. The fate of DPCs emerges with the pathological change of IVD. This suggests the existence of regulatory components within IVD, modulating the survival and self-renewal of these cells.

4. Conclusions

In conclusion, the studies of DPCs extend the current knowledge regarding the biology of endogenous IVD cells. Combined with tissue engineering and cell therapy, the application of DPCs would pave the way for the manipulation of IVD diseases and provide new hope that may contribute to IVD regeneration.

Acknowledgements

This work was supported by the Natural Science Foundation of China (81472078).

Declaration of interest

All authors state that they have no conflicts of interest.

Author details

Dan Long[1†], Shaoyu Liang[2†], Haiyan Liu[3], Xiuping Wu[3], Zhongyang Li[4], Haozhong Wang[4], Shishu Huang[4*] and Jiancheng Zeng[4]

*Address all correspondence to: ss.huang@siat.ac.cn

1 Key Laboratory of Transplant Engineering and Immunology of The Ministry of Health, Regenerative Medical Research Centre, West China Hospital, Sichuan University, Chengdu, China

2 Department of Science and Technology, West China Hospital, Sichuan University, Chengdu, China

3 The Affiliated Dental Hospital of Shanxi University, Taiyuan, China

4 Department of Orthopaedic Surgery, West China Hospital, Sichuan University, Chengdu, China

† These authors have contributed equally to this work.

References

[1] Humzah MD, Soames RW. Human intervertebral disc: Structure and function. The Anatomical Record. 1988;**220**:337-356

[2] Goto S, Uhthoff HK. Notochord action on spinal development. A histologic and morphometric investigation. Acta Orthopaedica Scandinavica. 1986;**57**:85-90

[3] Kim KW, Lim TH, Kim JG, et al. The origin of chondrocytes in the nucleus pulposus and histologic findings associated with the transition of a notochordal nucleus pulposus to a fibrocartilaginous nucleus pulposus in intact rabbit intervertebral discs. Spine. 2003;**28**:982-990

[4] Christoph W, Nerlich AG, Schaaf R. Immunohistochemical identification of notochordal markers in cells in the aging human lumbar intervertebral disc. European Spine Journal. 2010;**19**:1761-1770

[5] Gries NC, Berlemann U, Moore RJ, Vernon-Roberts B. Early histologic changes in lower lumbar discs and facet joints and their correlation. European Spine Journal. 2000;**9**:23-29

[6] Mercer S, Bogduk N. The ligaments and annulus fibrosus of human adult cervical intervertebral discs. Spine. 1999;**24**:619-626

[7] Hayes AJ, Benjamin M, Ralphs JR. Role of actin stress fibers in the development of the intervertebral disc: Cytoskeletal control of extracellular matrix assembly. Developmental Dynamics. 1999;**215**:179-189

[8] Adams MA, Dolan P, McNally DS. The internal mechanical functioning of intervertebral discs and articular cartilage, and its relevance to matrix biology. Matrix Biology. 2009;**28**:384-389

[9] Mwale F, Roughley P, Antoniou J. Distinction between the extracellular matrix of the nucleus pulposus and hyaline cartilage: A requisite for tissue engineering of intervertebral disc. European Cells & Materials. 2004;**8**:58-63

[10] Alsalameh S, Amin R, Gemba T, et al. Identification of mesenchymal progenitor cells in normal and osteoarthritic human articular cartilage. Arthritis & Rheumatology. 2004;**50**:1522-1532

[11] Phinney DG, Kopen G, Isaacson RL, et al. Plastic adherent stromal cells from the bone marrow of commonly used strains of inbred mice: Variations in yield, growth, and differentiation. Journal of Cellular Biochemistry. 1999;**72**:570-585

[12] Fuchs E, Tumbar T, Guasch G. Socializing with the neighbors: Stem cells and their niche. Cell. 2004;**116**:769-778

[13] Friedenstein AJ, Chailakhjan RK, Lalykina KS. The development of fibroblast colonies in monolayer cultures of guinea-pig bone marrow and spleen cells. Cell and Tissue Kinetics. 1970;**3**:393-403

[14] Taga T, Fukuda S. Role of IL-6 in the neural stem cell differentiation. Clinical Reviews in Allergy & Immunology. 2005;**28**:249-256

[15] Zouboulis CC, Adjaye J, Akamatsu H, Moe-Behrens G, Niemann C. Human skin stem cells and the ageing process. Experimental Gerontology. 2008;**43**:986-997

[16] Bodnar MS, Meneses JJ, Rodriguez RT, et al. Propagation and maintenance of undifferentiated human embryonic stem cells. Stem Cells and Development. 2004;**13**:243-253

[17] Russell KC, Phinney DG, Lacey MR, et al. In vitro high-capacity assay to quantify the clonal heterogeneity in trilineage potential of mesenchymal stem cells reveals a complex hierarchy of lineage commitment. Stem cells. 2010;**28**:788-798

[18] Park JY, Kuh SU, Park HS, et al. Comparative expression of matrix-associated genes and inflammatory cytokines-associated genes according to disc degeneration: Analysis of living human nucleus pulposus. Journal of Spinal Disorders & Techniques. 2011;**24**:352-357

[19] Huang S, Tam V, Cheung KM, et al. Stem cell-based approaches for intervertebral disc regeneration. Current Stem Cell Research & Therapy. 2011;**6**:317-326

[20] Steck E, Bertram H, Abel R, et al. Induction of intervertebral disc-like cells from adult mesenchymal stem cells. Stem Cells. 2005;**23**:403-411

[21] Sobajima S, Vadala G, Shimer A, et al. Feasibility of a stem cell therapy for intervertebral disc degeneration. The Spine Journal. 2008;**8**:888-896

[22] McCanless JD, Cole JA, Slack SM, et al. Modeling nucleus pulposus regeneration in vitro: Mesenchymal stem cells, alginate beads, hypoxia, bone morphogenetic protein-2, and synthetic peptide B2A. Spine. 2011;**36**:2275-2285

[23] Gupta MS, Cooper ES, Nicoll SB. Transforming growth factor-beta 3 stimulates cartilage matrix elaboration by human marrow-derived stromal cells encapsulated in photocross-linked carboxymethylcellulose hydrogels: Potential for nucleus pulposus replacement. Tissue Engineering Part A. 2011;**17**:2903-2910

[24] Richardson SM, Hughes N, Hunt JA, et al. Human mesenchymal stem cell differentiation to NP-like cells in chitosan-glycerophosphate hydrogels. Biomaterials. 2008;**29**:85-93

[25] Yang SH, Wu CC, Shih TT, et al. In vitro study on interaction between human nucleus pulposus cells and mesenchymal stem cells through paracrine stimulation. Spine. 2008;**33**:1951-1957

[26] Umeda M, Kushida T, Sasai K, et al. Activation of rat nucleus pulposus cells by coculture with whole bone marrow cells collected by the perfusion method. Journal of Orthopaedic Research. 2009;**27**:222-228

[27] Mwale F, Girard-Lauriault PL, Wang HT, et al. Suppression of genes related to hypertrophy and osteogenesis in committed human mesenchymal stem cells cultured on novel nitrogen-rich plasma polymer coatings. Tissue Engineering. 2006;**12**:2639-2647

[28] Nelea V, Luo L, Demers CN, Antoniou J, et al. Selective inhibition of type X collagen expression in human mesenchymal stem cell differentiation on polymer substrates surface-modified by glow discharge plasma. Journal of Biomedical Materials Research Part A. 2005;**75**:216-223

[29] Yang F, Leung VY, Luk KD, et al. Mesenchymal stem cells arrest intervertebral disc degeneration through chondrocytic differentiation and stimulation of endogenous cells. Molecular Therapy. 2009;**17**:1959-1966

[30] Hee HT, Ismail HD, Lim CT, et al. Effects of implantation of bone marrow mesenchymal stem cells, disc distraction and combined therapy on reversing degeneration of the intervertebral disc. Journal of Bone and Joint Surgery. 2010;**92**:726-736

[31] Serigano K, Sakai D, Hiyama A, et al. Effect of cell number on mesenchymal stem cell transplantation in a canine disc degeneration model. Journal of Orthopaedic Research. 2010;**28**:1267-1275

[32] Hiyama A, Mochida J, Iwashina T, et al. Transplantation of mesenchymal stem cells in a canine disc degeneration model. Journal of Orthopaedic Research. 2008;**26**:589-600

[33] Sakai D, Mochida J, Iwashina T, et al. Regenerative effects of transplanting mesenchymal stem cells embedded in atelocollagen to the degenerated intervertebral disc. Biomaterials. 2006;**27**:335-345

[34] Sakai D, Mochida J, Yamamoto Y, et al. Transplantation of mesenchymal stem cells embedded in Atelocollagen gel to the intervertebral disc: A potential therapeutic model for disc degeneration. Biomaterials. 2003;**24**:3531-3541

[35] Vadalà G, Sowa G, Hubert M, et al. Mesenchymal stem cells injection in degenerated intervertebral disc: Cell leakage may induce osteophyte formation. Journal of Tissue Engineering and Regenerative Medicine. 2012;**6**:348-355

[36] Chik TK, Ma XY, Choy TH, Li YY, et al. Photochemically crosslinked collagen annulus plug: A potential solution solving the leakage problem of cell-based therapies for disc degeneration. Acta Biomater. 2013;**9**:8128-8139

[37] Hiyama A, Skubutyte R, Markova D, et al. Hypoxia activates the notch signaling pathway in cells of the intervertebral disc: Implications in degenerative disc disease. Arthritis & Rheumatology. 2011;**63**:1355-1364

[38] Risbud MV, Guttapalli A, Tsai TT, et al. Evidence for skeletal progenitor cells in the degenerate human intervertebral disc. Spine. 2007;**32**:2537-2544

[39] Blanco JF, Graciani IF, Sanchez-Guijo FM, et al. Isolation and characterization of mesenchymal stromal cells from human degenerated nucleus pulposus: Comparison with bone marrow mesenchymal stromal cells from the same subjects. Spine. 2010;**35**:2259-2265

[40] Feng G, Yang X, Shang H, et al. Multipotential differentiation of human anulus fibrosus cells: An in vitro study. The Journal of Bone and Joint Surgery American. 2010;**92**:675-685

[41] Brisby H, Papadimitriou N, Brantsing C, et al. The presence of local mesenchymal progenitor cells in human degenerated intervertebral discs and possibilities to influence these in vitro: A descriptive study in humans. Stem Cells and Development. 2013;**22**:804-814

[42] Erwin WM, Islam D, Eftekarpour E, et al. Intervertebral disc-derived stem cells: Implications for regenerative medicine and neural repair. Spine. 2013;**38**:211-216

[43] Alini M, Eisenstein SM, Ito K, et al. Are animal models useful for studying human disc disorders/degeneration? European Spine Journal. 2008;**17**:2-19

[44] Longo UG, Ripalda P, Denaro V, et al. Morphologic comparison of cervical, thoracic, lumbar intervertebral discs of cynomolgus monkey (*Macaca fascicularis*). European Spine Journal. 2006;**15**:1845-1851

[45] Kramer PA, Newell-Morris LL, Simkin PA. Spinal degenerative disk disease (DDD) in female macaque monkeys: Epidemiology and comparison with women. Journal of Orthopaedic Research. 2002;**20**:399-408

[46] Noury F, Mispelter J, Szeremeta F, et al. MRI methodological development of intervertebral disc degeneration: A rabbit in vivo study at 9.4 T. Magnetic Resonance Imaging. 2008;**26**(10):1421-1432

[47] Huang S, Leung VY, Long D, et al. Coupling of small leucine-rich proteoglycans to hypoxic survival of a progenitor cell-like subpopulation in Rhesus Macaque intervertebral disc. Biomaterials. 2013;**34**:6548-6558

[48] Sakai D, Nakamura Y, Nakai T, et al. Exhaustion of nucleus pulposus progenitor cells with ageing and degeneration of the intervertebral disc. Nature Communications. 2012;**3**:1264

[49] Mizrahi O, Sheyn D, Tawackoli W, et al. Nucleus pulposus degeneration alters properties of resident progenitor cells. The Spine Journal. 2013;**13**;803-814

Chapter 13

Transportation of Mesenchymal Stem Cells for Clinical Applications

Tomoki Aoyama

Additional information is available at the end of the chapter

http://dx.doi.org/10.5772/67716

Abstract

Cell-processing procedures are conducted in accordance with Good Manufacturing Practices, and clinical procedures are performed by highly optimised methods. A high-quality transportation system is essential for safe and effective handling of mesenchymal stem cells (MSCs) between cell-processing and transplantation stages. For MSC transportation, either frozen cell or non-frozen cell transportation is performed. There are many requirements for transporting a package by either type of transportation. In frozen cell transportation, some issues have yet to be resolved: the primary receptacle and cryoprotectant reagents. In non-frozen cell transportation, control of cell metabolism and protection from environmental changes are more serious problems. Stabilisation of temperature, shock resistance, gas control, and an ultraviolet radiation (UVR) shielding technology should be considered. The transportation system should be established in compliance with the guidelines. Both development of a high-quality transportation package and establishment of a high-quality transportation system are important for the effective use of MSCs in clinical applications.

Keywords: transport, mesenchymal stem cell, regenerative medicine, clinical use

1. Introduction

Cell-processing procedures are conducted in accordance with Good Manufacturing Practices, and clinical procedures such as transplantation and infusion are performed using highly optimised techniques [1]. Thus, a high-quality transportation system is necessary for safe and effective handling of materials and mesenchymal stem cell (MSC) products between cell-processing and transplantation stages. Frozen cell transportation is conventionally used in laboratories. Nonetheless, there may be some disadvantages of this method for clinical applications. For example, the tubes usually used for frozen transportation are not completely sealed receptacles, and the cryoprotection reagents have some effects on the human body. In non-frozen cell transportation, control of cell metabolism and protection from environmental changes are serious problems. A transportation system should be established to prevent man-made incidents. In this study, the elements required for transportation of MSCs are discussed.

2. Conditions required for MSC transportation

There are many possible scenarios of MSC transportation. When the MSCs are isolated immediately from materials in a hospital, establishment of a transportation system inside the hospital is needed (**Figure 1A**). If further manipulation of MSCs or mass culture is performed, MSC materials such as bone marrow, cord blood, and adipose tissue are transported from the

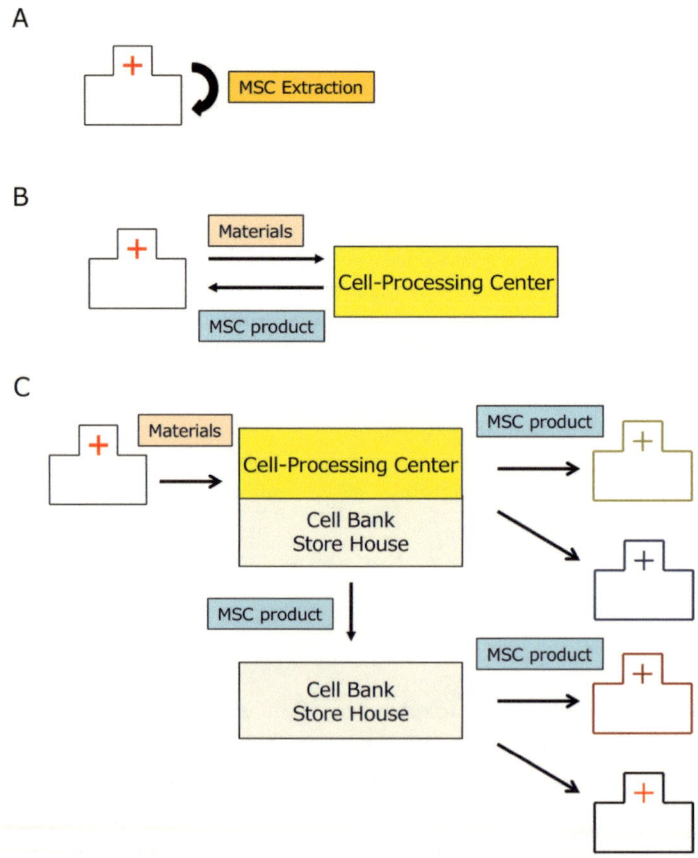

Figure 1. The scenarios of MSC transportation. (A) Extraction of materials in a hospital. (B) Extracted materials are transported to a cell-processing centre, and MSC products are transported to the hospital. (C) In some cases, MSCs are stored in a cell bank.

hospital to a cell-processing centre (CPC). At a CPC, the materials are cultured and manipulated and then transported to the hospital (**Figure 1B**). These transportation scenarios (**Figure 1A**, **B**) work effectively with autogenic transplantation. When allogeneic transplantation is performed, some temporary storage is used. In such cases, other facilities such as cell banks or stock houses are added [2] (**Figure 1C**).

Basically, when the transportation is performed within a hospital, there are not many requirements for the transportation package and transport system if a good environment is maintained in the hospital. Some packaging technology is needed when outside transportation is performed among a hospital, CPC, and cell bank (**Figure 1B**, **C**). In the outside environment, the transported materials and MSC products should be protected from some external environmental factors such as micro-organisms, temperature, shocks, humidity, ultraviolet radiation (UVR), and atmospheric pressure. The transportation package should withstand such external environmental stimuli.

3. Requirements for transportation and three-level packaging

There are some requirements for transportation of MSCs for clinical use:

- Leakproofness
- Sterility
- Temperature stabilisation
- Shock resistance
- Gas stability
- UV shielding
- Monitoring

Each requirement is important, but the degree of importance is different in each transportation scenario (**Figure 1**, **Table 1**). A single type of package cannot fulfil all the requirements. A three-level packaging system is recommended by the World Health Organisation (WHO) [3] (**Figure 2**). Each package satisfies each requirement. Tests required for compliance with the criteria include a 9-m drop test, a puncture test, and a stacking test for the packages [3].

3.1. The primary receptacle

The materials and the cell products are inserted into the first package. The primary receptacle should be completely sealed to prevent invasion of micro-organisms and leakage of contents. The materials constituting the primary receptacle should be chosen carefully to prevent elution of chemical materials. Packaging the contents into the primary receptacle is performed in the CPC area, and the primary receptacle should be sterilised.

	Sterility	Shock resistance	Temperature stabilisation	Monitoring
Inside hospital	✓✓	✓	✓	✓
From hospital to CPC (materials)	✓✓	✓✓	✓✓	✓✓
From CPC to hospital (products)				
Non-frozen	✓✓✓	✓✓✓	✓✓✓	✓✓✓
Frozen	✓✓	✓	✓✓	✓
From cell bank to hospital (products)				
Non-frozen	✓✓✓	✓✓✓	✓✓✓	✓✓✓
Frozen	✓✓	✓	✓✓	✓
From stockhouse to hospital (products)				
Non-frozen	✓✓✓	✓✓✓	✓✓✓	✓✓✓
Frozen	✓✓	✓	✓✓	✓

MSC, mesenchymal stromal cell; CPC, cell-processing centre.

Table 1. Technologies needed for each processing scenario of MSC materials and products.

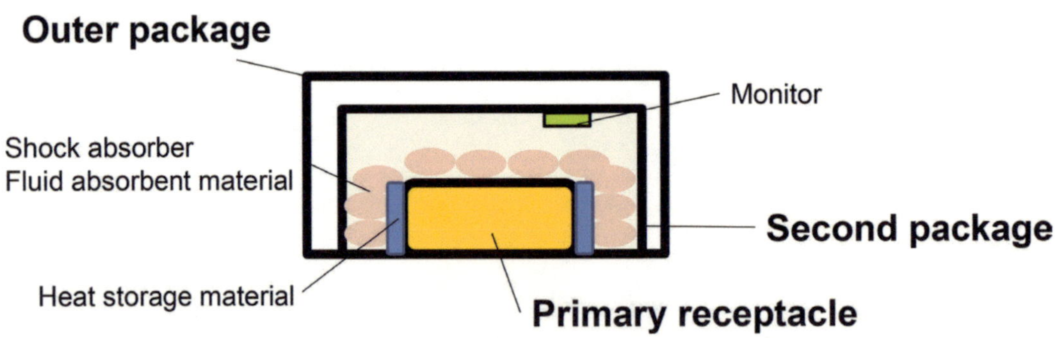

Figure 2. A three-level packaging system of a transportation box in regenerative medicine.

3.2. The second package

The primary receptacle is inserted into the second package (**Figure 2**). Sterilisation treatment of the second package is not necessary if the complete sterilisation and sealing are achieved for the primary receptacle. Some materials for absorbing liquids are needed if the first receptacle is broken and the contents leak out. A shock absorber and temperature control system such as materials for hot storage are included in the second package. For monitoring the temperature and shocks, a sensing device is also included.

3.3. The outer package

The second package is placed into the outer package (**Figure 2**). The outer level is the packaging most exposed to external environmental stimuli. The characteristics of heat insulation, toughness, waterproofness, and UV shielding are necessary. For the handling, slings and belts are attached to the outer package. A notice, security lock, and invoice are attached as acquired.

4. Requirements for the packages

4.1. Leakproofness

Tight sealing is required from the standpoint of leakproofness and sterility. According to the 'Guidance on Regulations for the Transport of Infectious Substances', leakproofness is required in the primary receptacle and second package [3]. The guidance itself is used for transportation of 'infectious substances', but it defines the 'patient specimens' and 'biological products' as 'infectious substances'. Thus, this guidance is to be used as a reference. If the primary receptacle loses leakproofness and springs a leak, some absorbent materials in the second package should absorb all fluid [3]. Whatever the intended temperature of the consignment, the primary receptacle and secondary package should withstand (without leakage) the internal pressure no less than 95 kPa in the range of temperatures from -40 to +55°C [3].

4.2. Sterility

The importance of sterility does not need to be explained. To guarantee sterility, the primary receptacle should be completely sealed. Therefore, the semi-closed cap-type receptacle typically used for frozen transportation is not recommended as a primary receptacle. Besides, the primary receptacle itself should be sterilised.

4.3. Temperature stabilisation

Stabilisation of temperature is employed to control cell metabolism and to prevent degradation of cells and products. There are two categories for temperature stabilisation: frozen cell or non-frozen cell transportation.

In the field of frozen cell transportation, research and development on the freezing procedure are conducted for the purpose of storage. The success of frozen cell transportation and storage depends on the temperature control and cryoprotective agents [4]. According to Hubel, understanding the mobility of water is paramount for good insight into biochemical reactions during freezing [4]. The temperature dependence of protein activity obeys the Arrhenius equation [4, 5]:

$$K = A \cdot e^{-Ea/(RT)} \tag{1}$$

where k is the rate constant, T is the absolute temperature (Kelvins), A is the pre-exponential factor, E_a is the activity energy for the reaction (Joules), and R is the universal gas constant. The optimal temperature for transportation and storage should be selected below the threshold temperature to manage activity of a protein [4]. Distinct changes are observed near -53°C [6, 7]. A substrate fails to bind to ribonuclease A at -58°C [8]. Some authors reported that activity of β-glucosidase is observed even at -70°C [9, 10]. These results suggest that a trace protein activity may persist even at a low temperature, below -80°C [4]. To stabilise proteins and minimise cell dehydration, pharmaceutical agents are used for cryoprotection. Cryoprotectants are classified into permeating and non-permeating agents depending on their ability to cross the cell membrane [11, 12]. The permeating cryoprotectants, such as dimethyl sulphoxide (DMSO), are used for clinical preservation and banking of MSCs for the treatment of graft-versus-host disease (GVHD) [13]. The International Stem Cell Initiative (ISCI) recommends the use of dry shippers involving liquid nitrogen (-196°C) for frozen stem cell transportation [2]. Nonetheless, some issues should be resolved regarding cryopreservation and frozen cell transportation. A significant loss of living cells after recovery occurs after cryopreservation. The proportion of apoptotic and senescent cells increases during cryopreservation in comparison with fresh live culture [14–16]. Not only cell viability but also alteration of MSC characteristics is to be considered. Immunosuppressive properties of MSCs are altered after cryopreservation although these alterations may be favourable for the treatment of GVHD [17, 18]. DMSO is the most widely used cryoprotectant in the world, but it has some adverse effects on patients such as nausea, headache, hypotension, hypertension, and diarrhoea [19]. Further research and development on cryoprotectants, freezing and thawing technologies, cell containers, the primary receptacle, and administration method are needed for frozen cell transportation.

In non-frozen cell transportation, control of cell metabolism is compromised. There is little information about the non-frozen cell transportation of MSCs. Information on the storage of platelets serves as a useful reference. Platelets are stored between 20 and 24°C with continuous agitation, suspended in plasma with a citrate-based anticoagulant [20]. Platelets at 4°C undergo a shape change, α-granule release, aggregation, apoptosis, and a significant reduction of the lifespan. When MSCs that are detached from a culture dish are incubated at 4, 20, or 37°C in a culture medium for 24 h, the living cell numbers after 37°C incubation are higher. Nevertheless, dead cell numbers increase after 37°C incubation (**Figure 3**). This result shows that cell metabolism is not sufficiently controlled during 37°C incubation. The development of a preservation solution and verification of suitable temperature are necessary for non-frozen-MSC transportation.

4.4. Shock and vibration resistance

There is little or no information about the bioresponse of MSCs to shocks and vibration during transportation. One study evaluated the shocks and vibration associated with truck and railway transportation [21]. Peak shock acceleration for the longitudinal axis was 2.8g, for the transverse axis 2.3g, and for vertical axis 7.0g on a truck, and all axes were 4.7g on a train when a 22-ton cargo was transported [22]. Peak vibration acceleration for the longitudinal axis was 0.27g; for the transverse axis, it was 0.27g; and for the vertical axis, 0.52g on a truck.

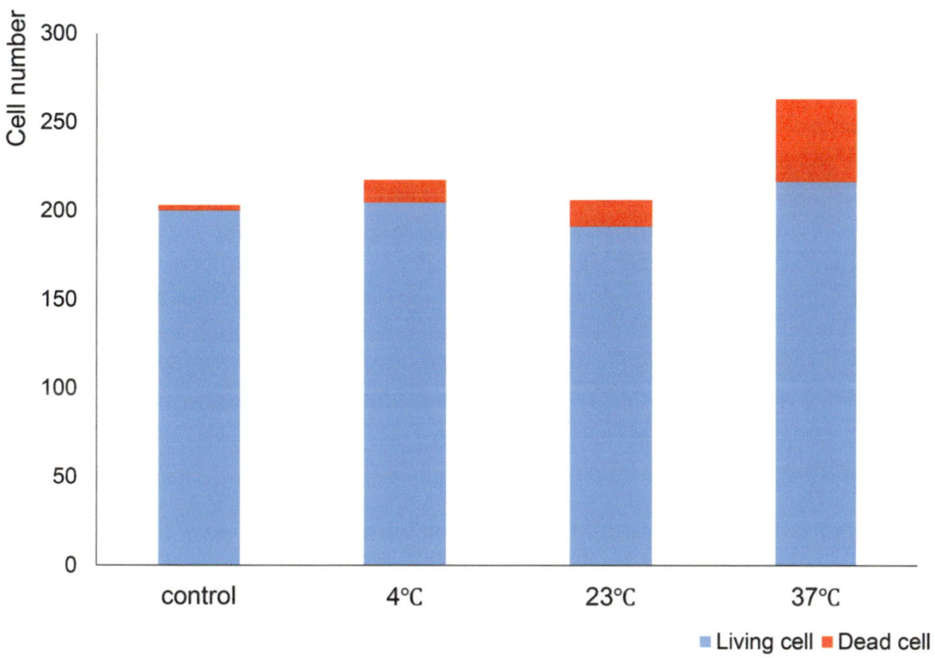

Figure 3. The cell number of MSCs peeled from a culture dish and incubated at 4, 20, or 37°C in a culture medium for 24 h. Living cell and dead cell numbers were determined by means of the Live/Dead Staining Kit II (Takara Bio Inc., Shiga, Japan).

The frequency range was 0–1900 Hz [22]. Peak vibration acceleration for the longitudinal axis was 0.10*g*, transverse axis 0.19*g*, and vertical axis 0.52*g* on a train. The frequency range was 0–750 Hz [23]. Strong shocks should be prevented from destroying the extracellular matrix of products. Microvibration should be considered for prevention of changes in the characteristics of MSCs. The lipid membrane, primary cilium, extracellular matrix, and intracellular cytoskeleton respond to mechanical stimuli. They transduce mechanical stimuli via ion channels, integrins, and focal adhesion kinase and thus alter gene expression and protein synthesis. As a consequence, survival, apoptosis, proliferation, and differentiation properties are affected [24]. For example, MSCs were stimulated by vibration from 1 to 1000 Hz, and myogenic transcription factors were found to be upregulated by low frequencies (1 and 50 Hz), but osteogenic transcription factors were upregulated by high frequencies (500 and 1000 Hz) [25]. These data suggest that some shock absorber is needed to prevent the shocks and vibration.

4.5. Gas stability

In frozen-MSC transportation, gas control is not needed, but in non-frozen-MSC transportation, some gas control technology is necessary. When we consider the gas control, the knowledge about cellular senescence is crucial. Cellular senescence was discovered in the 1960s by Hayflick [26]. He showed that human diploid fibroblasts have a limited ability to proliferate; it is called senescence in ex vivo culture [26]. Some stressors, such as dysfunctional telomeres, genotoxic stressors, DNA damage, UVR, and oxidative stress, induce cellular senescence [27]. This phenomenon is induced by the upregulation of $p16^{INK4A}$ in MSCs [28]. The expression of $p16^{INK4A}$ is associated with oxidative stress. When MSCs are cultured under hypoxic conditions (1% pO_2), their proliferation is increased by 8–20 population doublings as compared with normoxic conditions (20% pO_2) [29]. The $p16^{INK4A}$ gene is downregulated in hypoxic culture [29]. An antioxidant agent that induces 5% pO_2 conditions [30] and 3% hydrogen gas treatment [31] prolongs the replicative lifespan. Not only cellular proliferation but also the characteristics of MSCs are influenced by gas conditions. The osteogenic differentiation potential is not influenced, but adipogenic and chondrogenic differentiation abilities are enhanced under hypoxic culture conditions [29]. Cell surface antigens are downregulated, and the ability to suppress T-cell proliferation is diminished by oxidative stress [32]. These results suggest that some gas control system is required for non-frozen-MSC transportation.

4.6. UVR shielding

UVR is a natural component of sunlight and is invisible to the human eye. There are three types of UVR by wavelength: UV-A (315–400 nm), UV-B (280–315 nm), and UV-C (100–280 nm). UV-A and UV-B can reach the Earth's surface, but the ozone layer filters out UV-C [33]. Ageing of skin associated with UVR exposure is referred to as photoageing. Sun-exposed areas of the skin, such as the face and neck, undergo premature ageing [33]. UV-A induces production of reactive oxygen species (ROS) and activates cell cycle checkpoint proteins such as p53 and p21 [34]. UV-B exposure causes structural rearrangement of nucleotides that leads to defects in DNA strands. The UVR-induced ROS production and DNA damage to epidermal stem cells lead to photoageing [33]. Although there are many reports about the harmful effects of UVR on epidermal stem cells, there is little information about UVR effects on MSCs. The report on the gene expression profile, when MSCs were exposed to low-dose UV-B, showed that only a minimal change in gene expression was induced [35]. Further research is needed and some UVR shielding technology may be required, especially in case of air transportation.

4.7. Monitoring

Temperature data loggers are required for proof that transportation was completed without a problem. In the absence of temperature data loggers, consideration should be given to the use of chemical or other indicators to provide information on temperature during transportation [2]. If the shock resistance, gas control, and UV shielding are crucial for the transportation of MSCs, then acceleration sensors, gas monitoring, and radiation monitoring should be considered.

5. Transportation systems

Establishment of a transportation system is required in compliance with the government regulations in each country and international guidelines. At present, there is no specific guidance on transportation for regenerative medicine. Nonetheless, some guidelines issued by the ISCI [2] and WHO [3] and specific legislation in Europe [36] can be used for reference.

http://dx.doi.org/10.5772/67716

Education of delivery personnel to prevent man-made incidents, preparation of the documentation for clearance, a traceability system, and security control should be provided. Further research and consensus guidelines are needed for the transportation of materials related to MSCs and cell products for clinical use.

Acknowledgements

This work was supported by JSPS KAKENHI grant number 26282154. The author acknowledges members of the BioL Foundation.

Abbreviations

MSC	Mesenchymal stem cell
GMP	Good Manufacturing Practices
CPC	Cell-processing centre
GVHD	Graft-versus-host disease
DMSO	Dimethyl sulfoxide
$p16^{INK4A}$	Cyclin-dependent kinase 4 inhibitor p16
DNA	Deoxyribonucleic acid
UVR	Ultraviolet radiation
ROS	Reactive oxygen species

Author details

Tomoki Aoyama

Address all correspondence to: blue@hs.med.kyoto-u.ac.jp

Human Health Sciences, Graduate School of Medicine, Kyoto University, Kyoto, Japan

References

[1] Codinach M, Blanco M, Ortega I, Lloret M, Reales L, Coca MI, Torrents S, Doral M, Oliver-Vila I, Requena-Montero M, Vives J, Garcia-López J. Design and validation of a consistent and reproducible manufacture process for the production of clinical-grade bone marrow-derived multipotent mesenchymal stromal cells. Cytotherapy. 2016;**18**:1197-1208. doi:10.1016/j.jcyt.2016.05.012

[2] Andrews PW, Baker D, Benvinisty N, Miranda B, Bruce K, Brüstle O, Choi M, Choi YM, Crook JM, de Sousa PA, Dvorak P, Freund C, Firpo M, Furue MK, Gokhale P, Ha HY, Han E, Haupt S, Healy L, Hei DJ, Hovatta O, Hunt C, Hwang SM, Inamdar MS, Isasi RM, Jaconi M, Jekerle V, Kamthorn P, Kibbey MC, Knezevic I, Knowles BB, Koo SK, Laabi Y, Leopoldo L, Liu P, Lomax GP, Loring JF, Ludwig TE, Montgomery K, Mummery C, Nagy A, Nakamura Y, Nakatsuji N, Oh S, Oh SK, Otonkoski T, Pera M, Peschanski M, Pranke P, Rajala KM, Rao M, Ruttachuk R, Reubinoff B, Ricco L, Rooke H, Sipp D, Stacey GN, Suemori H, Takahashi TA, Takada K, Talib S, Tannenbaum S, Yuan BZ, Zeng F, Zhou Q. Points to consider in the development of seed stocks of pluripotent stem cells for clinical applications: International Stem Cell Banking Initiative (ISCBI). Regenerative Medicine. 2015;**10**(2 Suppl):1-44. doi:10.2217/rme.14.93

[3] World Health Organization. Guidance on regulations for the transport of infectious substances 2013-2014: applicable as from 1 January 2013.

[4] Hubel A, Spindler R, Skubitz AP. Storage of human biospecimens: selection of the optimal storage temperature. Biopreservation and Biobanking. 2014;**12**:165-175. doi:10.1089/bio.2013.0084

[5] Arrhenius S. On the reaction rate of the inversion of non-refined sugar upon souring. Zeitschrift für Physikalische Chemie. 1889;**4**:226-248.

[6] Doster W, Cusack S, Petry W. Dynamical transition of myoglobin revealed by inelastic neutron scattering. Nature. 1989;**337**:754-756.

[7] Hartmann H, Parak F, Steigemann W, Petsko GA, Ponzi DR, Frauenfelder H. Conformational substates in a protein: structure and dynamics of metmyoglobin at 80 K. Proceedings of the National Academy of Sciences of the United States of America. 1982;**79**:4967-4971.

[8] Rasmussen BF, Stock AM, Ringe D, Petsko GA. Crystalline ribonuclease A loses function below the dynamical transition at 220 K. Nature. 1992;**357**:423-424.

[9] Bragger JM, Dunn RV, Daniel RM. Enzyme activity down to -100 degrees C. Biochimica et Biophysica Acta. 2000;**1480**:278-282.

[10] More N, Daniel RM, Petach HH. The effect of low temperatures on enzyme activity. Biochemistry Journal. 1995;**305**:17-20.

[11] Meryman HT. Cryoprotective agents. Cryobiology. 1971;**8**:173-183.

[12] Meryman HT. Cryopreservation of living cells: principles and practice. Transfusion. 2007;**47**:935-945.

[13] Marquez-Curtis LA, Janowska-Wieczorek A, McGann LE, Elliott JA. Mesenchymal stromal cells derived from various tissues: biological, clinical and cryopreservation aspects. Cryobiology. 2015;**71**:181-197. doi:10.1016/j.cryobiol.2015.07.003

[14] Chinnadurai R, Garcia MA, Sakurai Y, Lam WA, Kirk AD, Galipeau J, Copland IB. Actin cytoskeletal disruption following cryopreservation alters the biodistribution of human

mesenchymal stromal cells in vivo. Stem Cell Reports. 2014;**3**:60-72. doi:10.1016/j.stemcr.2014.05.003

[15] Chinnadurai R, Copland IB, Garcia MA, Petersen CT, Lewis CN, Waller EK, Kirk AD, Galipeau J. Cryopreserved mesenchymal stromal cells are susceptible to T-cell mediated apoptosis which is partly rescued by IFNγ licensing. Stem Cells. 2016;**34**:2429-2442. doi:10.1002/stem.2415

[16] Pollock K, Sumstad D, Kadidlo D, McKenna DH, Hubel A. Clinical mesenchymal stromal cell products undergo functional changes in response to freezing. Cytotherapy. 2015;**17**:38-45. doi:10.1016/j.jcyt.2014.06.008

[17] Holubova M, Lysak D, Vlas T, Vannucci L, Jindra P. Expanded cryopreserved mesenchymal stromal cells as an optimal source for graft-versus-host disease treatment. Biologicals. 2014;**42**:139-144. doi:10.1016/j.biologicals.2014.01.003

[18] François M, Copland IB, Yuan S, Romieu-Mourez R, Waller EK, Galipeau J. Cryopreserved mesenchymal stromal cells display impaired immunosuppressive properties as a result of heat-shock response and impaired interferon-γ licensing. Cytotherapy. 2012;**14**:147-152. doi:10.3109/14653249.2011.623691

[19] Alessandrino P, Bernasconi P, Caldera D, Colombo A, Bonfichi M, Malcovati L, Klersy C, Martinelli G, Maiocchi M, Pagnucco G, Varettoni M, Perotti C, Bernasconi C. Adverse events occurring during bone marrow or peripheral blood progenitor cell infusion: analysis of 126 cases. Bone Marrow Transplantation. 1999;**23**:533-537.

[20] Capocelli KE, Dumont LJ. Novel platelet storage conditions: additive solutions, gas, and cold. Current Opinion in Hematology. 2014;**21**:491-496. doi:10.1097/MOH.0000000000000081

[21] Maheres SJ, Lahti EA, Ross SB. Used fuel disposition campaign storage and transportation. Transportation shock and vibration literature review. FCRD-UFD-2013-000169. PNNL-22514. 2013.

[22] Magnuson CF. Shock and vibration environments for large shipping container during track transport (Part I). SAND77-1110. Sandia National Laboratories, Albuquerque, New Mexico. 1977.

[23] Magnuson CF. Shock and vibration environments during normal rail transportation of heavy cargo. SAND82-0819. Sandia National Laboratories, Albuquerque, New Mexico. 1982.

[24] Liu YS, Lee OK. In search of the pivot point of mechanotransduction: mechanosensing of stem cells. Cell Transplantation. 2014;**23**:1-11. doi:10.3727/096368912X659925

[25] Childs PG, Boyle CA, Pemberton GD, Nikukar H, Curtis AS, Henriquez FL, Dalby MJ, Reid S. Use of nanoscale mechanical stimulation for control and manipulation of cell behaviour. Acta Biomaterialia. 2016;**34**:159-168. doi:10.1016/j.actbio.2015.11.045

[26] Hayflick L. The limited in vitro lifetime of human diploid cell strains. Experimental Cell Research. 1965;**37**:614-636.

[27] Turinetto V, Vitale E, Giachino C. Senescence in human mesenchymal stem cells: functional changes and implications in stem cell-based therapy. International Journal of Molecular Sciences. 2016;**17**:pii: E1164. doi:10.3390/ijms17071164

[28] Shibata KR, Aoyama T, Shima Y, Fukiage K, Otsuka S, Furu M, Kohno Y, Ito K, Fujibayashi S, Neo M, Nakayama T, Nakamura T, Toguchida J. Expression of the p16INK4A gene is associated closely with senescence of human mesenchymal stem cells and is potentially silenced by DNA methylation during in vitro expansion. Stem Cells. 2007;**25**:2371-2382.

[29] Jin Y, Kato T, Furu M, Nasu A, Kajita Y, Mitsui H, Ueda M, Aoyama T, Nakayama T, Nakamura T, Toguchida J. Mesenchymal stem cells cultured under hypoxia escape from senescence via down-regulation of p16 and extracellular signal regulated kinase. Biochemical and Biophysical Research Communications. 2010;**391**:1471-1476. doi:10.1016/j.bbrc.2009.12.096

[30] Ito A, Aoyama T, Yoshizawa M, Nagai M, Tajino J, Yamaguchi S, Iijima H, Zhang X, Kuroki H. The effects of short-term hypoxia on human mesenchymal stem cell proliferation, viability and p16(INK4A) mRNA expression: investigation using a simple hypoxic culture system with a deoxidizing agent. Journal of Stem Cells and Regenerative Medicine. 2015;**11**:25-31.

[31] Kawasaki H, Guan J, Tamama K. Hydrogen gas treatment prolongs replicative lifespan of bone marrow multipotential stromal cells in vitro while preserving differentiation and paracrine potentials. Biochemical and Biophysical Research Communications. 2010;**397**:608-613. doi:10.1016/j.bbrc.2010.06.009

[32] Denu RA, Hematti P. Effects of oxidative stress on mesenchymal stem cell biology. Oxidative Medicine and Cellular Longevity. 2016;**2016**:2989076. doi:10.1155/2016/2989076

[33] Panich U, Sittithumcharee G, Rathviboon N, Jirawatnotai S. Ultraviolet radiation-induced skin aging: the role of DNA damage and oxidative stress in epidermal stem cell damage mediated skin aging. Stem Cells International. 2016;**2016**:7370642. doi:10.1155/2016/7370642

[34] Peng Y, Xuan M, Leung VYL, Cheng B. Stem cells and aberrant signalling of molecular systems in skin aging. Ageing Research Reviews. 2015;**19**:8-21. doi:10.1016/j.arr.2014.10.006

[35] Wong DY, Ranganath T, Kasko AM. Low-dose, long-wave UV light does not affect gene expression of human mesenchymal stem cells. PLoS One. 2015;**10**:e0139307. doi:10.1371/journal.pone.0139307

[36] Beck G, Schenerman M, Dougherrty J, Cordoba-Rodriguez R, Joneckis C, Mire-Sluis A, McLeod LD. Raw material control strategies for bioprocesses. Bioprocess International. 2009;**9**:18-33.